Designing Babies

ALSO BY ROBERT L. KLITZMAN, MD

A Year-long Night: Tales of A Medical Internship

In a House of Dreams and Glass: Becoming a Psychiatrist

Being Positive: The Lives of Men and Women with HIV

The Trembling Mountain: A Personal Account of Kuru, Cannibals, and Mad Cow Disease

Mortal Secrets: Truth and Lies in the Age of AIDS

When Doctors Become Patients

Am I My Genes?: Confronting Fate and Family Secrets in the Age of Genetic Testing

The Ethics Police?: The Struggle to Make Human Research Safe

Designing Babies

HOW TECHNOLOGY IS CHANGING
THE WAYS WE CREATE CHILDREN

ROBERT L. KLITZMAN, MD

To Eileen

with best regards

OXFORD
UNIVERSITY PRESS

OXFORD
UNIVERSITY PRESS

Oxford University Press is a department of the University of Oxford. It furthers
the University's objective of excellence in research, scholarship, and education
by publishing worldwide. Oxford is a registered trade mark of Oxford University
Press in the UK and certain other countries.

Published in the United States of America by Oxford University Press
198 Madison Avenue, New York, NY 10016, United States of America.

CIP data is on file at the Library of Congress
ISBN 978-0-19-005447-2

9 8 7 6 5 4 3 2 1

Printed by Sheridan Books, Inc., United States of America

In memory of my mother

"God created man in His own image . . . male and female created He them. And God blessed them; and God said unto them: 'Be fruitful, and multiply, and replenish the earth.'"

Genesis 1:22–23

"Buying is much more American than thinking."

Andy Warhol

CONTENTS

PART I | Introduction

1 "A Take-Home Baby": Reproductive Dreams and Journeys 3

2 "Not Part of the Parent Club": Problems Conceiving a Healthy Child 22

PART II | Choosing Reproductive Technologies

3 Becoming an Infertility Patient: Choosing Treatment 31

4 Choosing Eggs 41

5 Choosing Sperm 70

6 Choosing Embryos to Avoid Disease 75

7 "Family Balancing": Choosing Sex and Other Traits 108

8 "Two Kids for the Price of One?": Choosing Twins 123

9 Choosing Wombs 140

10 Choosing Children: Whether to Adopt 146

PART III | Choosing Adults

11 Choosing Doctors 155

12 "Will They Be Good Enough Parents?":
    Choosing Patients 170

PART IV | Confronting Stresses

13 "How Much Is a Child Worth?":
    Choosing Budgets 195

14 Emotional Roller Coasters: Countering Other
    Stresses 209

15 Choosing Supports 217

16 "Meant to Be?": Choosing Spirituality and
    Religion 232

PART V | Choosing Our Future

17 Choosing Education 243

18 Choosing Policies 252

19 Designing Our Future World 272

    Appendices 277
        Appendix A   Glossary of Terms and Abbreviations 279
        Appendix B   Methods 283
        Appendix C   Sample Questions 287
    Acknowledgments 293
    About the Author 295
    Notes 297
    Index 329

Designing Babies

I | Introduction

1 | "A Take-Home Baby"

REPRODUCTIVE DREAMS AND JOURNEYS

"Do you want to be the father of my child?" a friend, Abby, asked me a few years ago.

"What do you mean?" I asked, surprised.

"I'm looking for a sperm donor." She was now 37 years old and wanted to have a child. Four years earlier, she had broken up with her long-term boyfriend and since then had been single. Two years before she asked me this question, she had applied to an agency to adopt a child. I had written a reference letter for her, stating that she would be an excellent parent—warm, loving, and resourceful. She was a painter, with a large group of artistic friends who would be helpful and supportive. Unfortunately, the agency rejected her.

Her fatherhood question perplexed me. We were fond of each other but recently had seen each other only a few times a year and were less close than in the past.

"I'm serious!" she said, sensing my hesitation. "I've thought of going to a sperm bank, but I'd prefer to use someone I know. You'd be perfect!"

The proposition was tempting. I do not have children. Here, perhaps, was my chance, my only opportunity to have one. "Let me think about it," I stammered, uncertain and partly to buy time.

"You wouldn't have to do anything other than donate the sperm," she added, matter-of-factly.

Over the next several days, I began to ponder her question. Giving my sperm to someone felt odd, unfamiliar. But, I thought, that alone should not determine my decision. Presumably, she would take the lead in raising the child. Yet I wondered how involved I would or should be in helping her raise our offspring. Creating a child who contained roughly half my genes and would, presumably, look and act like me in certain ways but who would belong to someone else felt weird. I would be the biological father, but would or should I also take on the social role of being one? What if she wanted me to have more, or less, of a role? A child of mine, my own flesh and blood, would also presumably be raised mostly by her. She was a bit disorganized, which could be endearing in a friend but not necessarily ideal in a parent. What if I disagreed with her about how to raise our child?

I wondered what to do and how other people made these decisions, and I began to investigate, asking others and searching the medical literature. Increasingly, I heard of other friends going to infertility clinics. Female friends had asked their sisters for eggs. Gay and lesbian friends were having children, buying other people's eggs or sperm. Walter and John, a gay couple I knew in Los Angeles, deliberated for almost a year over whether to select an ambitious 6-foot, blond, blue-eyed Princeton pre-med tennis player or a laid-back UCLA folk-rock singer who worked as a volunteer, helping refugees. Individuals selling their eggs face potential medical risks from the medications and surgical procedures involved and generally proceed because they need or want the money.

We are entering a new world of mechanical reproduction and face myriad questions.

Little was known about the specifics of reproduction until 1674 when Antonie Van Leeuwenhoek, a cloth merchant in Delft, examined his own semen under a microscope and discovered sperm. For decades afterwards, sperm were each thought to contain a *homunculus*—a little person. He and others assumed that sperm contained the future human being, and that the egg merely nourished it.[1]

In 1780, Lazzaro Spallanzani, a remarkable Italian Catholic priest and biologist, discovered that reproduction among animals requires both sperm and eggs. He performed the first in vitro fertilization (IVF), with frogs, and the first artificial fertilization, in a dog.[2] He put "breeches of waxed taffeta" on male frogs which "notwithstanding the incumbrance, seek the females with equal eagerness, and perform as well as they can, the act of generation."[3] But in those situations, he found that "the eggs are never prolific, for want of having been bedewed with semen, which sometimes may be seen in the breeches in the form of drops. That these drops are real seed, appeared clearly from the artificial fecundation that was obtained by means of them."[4,5]

He also discovered that microbes can move through air and were killed by boiling water and that bats fly at night due not to eyesight but to some other sense (as we now know: their hearing).

In 1884, Dr. William Pancoast, a professor at Philadelphia's Jefferson Medical College, saw a married couple who had been unable to get pregnant. The husband had had gonorrhea and now lacked sperm. Pancoast anesthetized the woman and, unbeknown to the couple, inserted sperm into her from his "best-looking" male student. Only later did Pancoast reluctantly tell the husband, who suggested not informing the woman. Pancoast died in 1898, keeping this intervention secret.

Not until 1909, 25 years after the procedure and 11 years after his death, did one of his five students, Addison Davis Hard, report these events. Hard revealed that he had in fact traveled from Minnesota to New York to meet the couple's offspring, now a 25-year-old businessman—raising questions of whether Hard had in fact been the sperm donor).[6,7,8] Dr. Hard argued that this procedure could improve the genes of the human species. But other physicians condemned it as immoral and "against the laws of God."

Only slowly did doctors pursue, improve, and offer these procedures.

In 1978, the first child was born using IVF—Louise Brown in England. Media critics decried the advent of "test tube babies" as "unnatural" and therefore dangerous and wrong. Yet since then, assisted reproductive technologies (ARTs) have mushroomed, enabling countless adults who were previously not able to have children to now do so. The first baby was born in the United States through IVF in 1981. At least 10% of all couples are infertile, and increasingly thousands of fertile individuals, like my friend Abby, seek infertility treatment, including fertile women who have delayed childbirth or are at risk of transmitting a serious genetic disease to their offspring. Several advances have further fueled this technology. Doctors discovered ways of stimulating ovaries through follicle-stimulating and later gonadotrophin-releasing agonists and freezing and thawing of embryos, first leading to a birth in 1984.[9] In 1992, researchers learned to inject single sperm directly into an egg with a pipette—in a process called intracytoplasmic sperm injection (ICSI). (Note: A glossary in the Appendix explains all technical terms and abbreviations.) In the US, the number of babies born using ARTs more than doubled between 1996 and 2006 and almost quadrupled between 1996 and 2016.[10] These advancing technologies have produced over 1 million babies in the US.[11] ARTs now account for about 7% of all births in Denmark, almost 6% in Greece and the Czech Republic, 5% in Belgium,[12] and 2% of all births in the US, with about 4% in certain states.[13] These proportions are rapidly rising and will soon no doubt at least reach 12% in the US—in the US, 17% of all women aged 25–44 (6.9 million women) or their male partners and around 20% of women 35–55 (4 million women) or their male partners have already used infertility services.[14]

Sociodemographic and attitudinal changes have both followed and fueled this rapid growth. ARTs have also allowed thousands of gay and lesbian couples and single fertile individuals, like Abby, to have children. In industrialized countries, many women intentionally delay childbirth in order to embark on careers and are now becoming pregnant through these new procedures. Since 1970, the percentage of women over 35 who are giving birth for the first time has increased 50%, and the age of first-time mothers has increased by almost four years.[15] More and more, many of us would not exist if our parents had not used these technologies.

An ancient problem, infertility has new, effective solutions. Some of the earliest and most common sculptures of early civilizations—from Cycladic art of the ancient Minoans to the Mayans—depict pregnant women. Countless artifacts from prehistoric South America, the South Pacific, sub-Saharan Africa, and Renaissance Europe portray fertility and pregnancy. Ancient Incas, Papua New Guineans, and Nigerians all crafted powerful wooden and clay statuettes of pregnant fertility goddesses. In Genesis, Sarah asks God whether she will bear a child and then gives birth at age 90. Renaissance artists repeatedly painted childbirth in religious contexts— golden showers in the Annunciation, Jesus' birth in the manger, and the Madonna breast-feeding baby Jesus. Religious and folk remedies, rites, and rituals to produce children are as old as humanity itself—we doubtlessly wouldn't have survived as a species otherwise. The first story in the Bible is creation, culminating in the creation of human beings—the last thing God did before resting.

We can also now affect the DNA of our offspring, eliminating certain genes and ensuring that others are present. Potential parents at risk of transmitting lethal mutations to their offspring can now avoid doing so. Scientific advances and media headlines suggest new uses of these technologies to design babies.

A few years ago, I wrote a book, *Am I My Genes? Confronting Fate and Family Secrets in the Age of Genetic Testing*,[16] exploring how patients confront genetic mutations and decide whether to undergo genetic testing and how to proceed. "The biggest issue I face," one woman at risk for Huntington's disease, an untreatable, fatal neurological and psychiatric disorder caused by a mutation, told me, "is whether to have children on my own, or adopt or screen embryos or abort fetuses, or adopt a child, or not have kids at all. If I abort a fetus with my mutation, am I saying that my life has not been worth living?"

Her question has haunted me ever since.

I began to wonder about these rapid, extraordinary advances—what was happening and whether it was good or bad.

I asked a friend from medical school, Jim, who worked in this field, about his experiences. "Why don't you come by my office one day, and see what we're doing?" he responded.

I took Amtrak to a city in a nearby state, where his office occupied a sleek, modern office building. In the waiting room sat young men and women who differed from those in every other doctor's waiting room I had ever visited. Eager, healthy, and energetic, they looked like they were ready to try out for a sports team. None had bandages, splints, walkers, crutches, or canes or seemed frail, depressed, old, disabled, distraught, confused, or in pain. They had infertility.

A nurse ushered me into my friend's pristine, wood-paneled office. "It's amazing what we do here," Jim said. Tall, with wavy blond hair and piercing bright blue eyes, he had trained in obstetrics and gynecology (OB/GYN) and then subspecialized in reproductive endocrinology and infertility (or REI). Endocrinologists study hormones within the body, including those involved in reproduction. Doctors in this field are called "REIs."

Jim escorted me into a long, narrow, brightly lit corridor. We donned fresh white disposable paper gowns, light blue hats, and bright yellow shoe coverings. He then led me into a large, cool, dimly lit, windowless room with light gray walls—his operating room. As we shuffled in, a hush swallowed us. We looked like priests entering a sacred, inner temple chamber. Cold air and the sharp smell of iodine antiseptic stung my nostrils.

He introduced me to Jane, a 37-year-old, fair-skinned, blond-haired patient. She lay on her back in a thin white cotton patient gown, with her legs spread and her knees up, looking as if she were about to give birth. He sat down on a low, round, gray swivel stool between her legs.

For several years, she and her husband had tried to get pregnant without success. Her OB/GYN had prescribed medications to increase her ovulation—the release of eggs by her ovaries. But Jane still did not get pregnant. The OB/GYN had then referred her to Jim, who tried intrauterine insemination (or IUI), collecting sperm from her husband and inserting it directly into her uterus to facilitate fertilization of her eggs.

To form a human, an egg and sperm must meet and fuse to create an embryo—a single large cell that soon splits into two identical cells. These two cells then each divide to make four cells, which in turn split to form eight, until a large mass appears that begins to differentiate, instructed by its genes, and eventually becomes a person. Usually, a sperm must hit and fertilize an egg in the fallopian tubes, which lead from the ovaries to the uterus, or in the uterus itself. But, for several reasons, any of these various steps can fail, hampering conception of a child. ARTs artificially replicate and facilitate these processes.

Jane underwent three IUIs. They all failed. This procedure works only 15%–20% of the time. Jim was now going to try a more invasive approach—IVF. He carefully prepared a thin, long, flexible tube and threaded it between her legs, into her vagina, through her uterus, and then into one of her ovaries, to suck out eggs. He would attempt to remove eggs manually from her ovaries and fertilize them with sperm in a Petri dish, hoping to create embryos to implant back into her uterus.

On a small, round, black-and-white television screen, I watched an ultrasound of the thin tube as he pushed it up against the edge of a hollow, pale gray, oval sac—a follicle. Women are born with all the follicles and oocytes (or eggs) they will ever have—about 400,000. Each follicle contains an immature egg that eventually grows and matures.

On the TV screen, the mouth of the snake-like tube thrust up against the pudgy follicle. Inside lay a small, round, black ball—a human egg. The suction tube tried to catch the plump egg, which wobbled and squirmed, resisting. Finally, the mouth of the tube captured the edge of the dark ball. Most of the egg still extruded, ballooning out, but the long tube slowly began to swallow it. Eventually, the egg collapsed, squeezed into the long, narrow cavity, and was quickly slurped down. Jim then maneuvered the tube to the edge of another bloated follicle and repeated the procedure. He carefully placed each egg in a Petri dish, retrieving around 30 in all.

"We also collect and analyze the sperm," he told me, as he stood up. Under a microscope, he inserted a clear glass dish, from another patient's husband, and showed me squiggly sperm, swimming like tiny tadpoles, each seemingly with a mind of its own. Porn videos and magazines filled a small windowless room in Jim's office suite, where this husband had masturbated to produce a fresh sample. Jim then squirted these sperm onto a Petri dish of eggs he had retrieved from the man's wife. If men have few or defective sperm, Jim sometimes also injects a single sperm directly into an egg, a technique known as ICSI.

"Now, we wait three to five days until embryos form," Jim said, sliding the disk onto the shelf of a 2-foot high stainless-steel incubator.

He pulled out another disk he had stored several days earlier. On it sat six or seven tiny black specks, each smaller than a sesame seed. Under the microscope, each speck—an embryo—consisted of a small clump of tiny round cells. A few clumps formed symmetrical spheres, looking like mini, translucent soccer balls. Two or three of these bunches, though, seemed lopsided, misshapen.

"I choose the embryos that look best," he told me, "most symmetrical. Jane wants twins—one boy and one girl. So, I'll probably put two or three embryos back in." He would pull off a cell from each embryo to test its genes, to determine its sex. As women age, the quality of their eggs diminishes and the risks of genetic abnormalities increase. The future child could have aneuploidy—an abnormal number of chromosomes (bundles of thousands of genes)—resulting in Down syndrome or other diseases. The risks of aneuploidy increase with women's age, particularly for prospective mothers in their late 30s. Jim would thus test the embryos' genes to screen out these disorders. He could also perform preimplantation genetic diagnosis (PGD) to test the embryos for scores of genes associated with other diseases as well, such as breast cancer. For several years, many physicians had conducted preimplantation genetic screening (PGS) to assess the overall number of chromosomes, and PGD to test for particular individual single-gene (monogenic) mutations within a chromosome. The field is now seeking to change these terms, and refer instead to three types of Preimplantation Genetic Testing (PGT)—PGT-A (for aneuploidy), PGT-M (for monogenic single-gene mutations) and PGT-SR (for chromosomes with structural rearrangements). But the terms PGD and PGS remain far more commonly used. Each year, physicians use PGD (PGT-M and PGT-SR) to screen embryos for more diseases. Prospective parents who might pass on certain disorders, including couples who are both carriers for recessive conditions such as cystic fibrosis, can thus avoid transmitting these problems. As scientists discover genes associated with not only diseases but various characteristics such as deafness, dwarfism, homosexuality, or intelligence, doctors will be capable of screening embryos for these traits as well.

In the end, Jim would choose two or three embryos, based on their appearance and DNA. If Jane had any viable leftover embryos, he would freeze them for possible future use. Nine months later, if all proceeded right and Jane didn't miscarry, she would give birth.

IVF costs more than IUI and works more often, though success decreases with the woman's age. The odds that this cycle of IVF would give Jane, at age 37, a healthy child—a "take-home baby"—were around 20%. If she were in her mid-20s, the odds would be around 40%. If she were 40, the likelihood would be only around 1%–2%. Moreover, about one in five of these pregnancies end in miscarriage.[7]

Unfortunately, Jane's insurance does not cover IVF, so she is paying out-of-pocket $24,000. If she wants to start IVF again, she would have to pay another $24,000. She is an elementary school teacher. Her husband is an electrician. They do not have much money. If this second IVF fails, they will have to decide what to do. As she gets closer to 40 and the quantity of her eggs decreases even more, they could instead try using another woman's egg, adopt a child, or remain childless.

Jim would have to wait several days to know how Jane's embryos would look or whether she would have any viable extras. IVF cycles can fail because the embryos turn out to have problems.

"We store embryos in the back," he told me. He led me to a windowless refrigerated room filled with large white metal drums, each about 2 feet wide and 3 feet high. He unscrewed the top of one container. Frigid white vapor wafted out. Inside lay row upon row of tiny clear vials in white plastic partitions. "We have tens of thousands here. Many patients have their children, but have good extra embryos that we retain. Unfortunately, couples never get back to us about what to do with the extras. So, we keep these— probably indefinitely. It would look bad to throw them out. Just imagine the headlines and right-wing backlash—'Doctor throws out future human beings.'"

Until now, human birth required sexual intercourse. But no more. The mechanics involved in sex have now been broken, technologically, into their constituent parts.

At each of these steps, doctors and patients face quandaries, including how long to try medications before IUI and IUI before IVF; whether to keep trying IVF if it fails; which embryos to select or reject and how to decide; whether to choose embryos based on sex, predispositions to various diseases, or other characteristics; how many embryos to transfer; and what to do with leftovers. These decisions have profound implications, redefining who reproduces, and when, how, and to what degree.

Countries vary widely in whether and how they address these issues. Many Islamic countries ban use of donor sperm. Other nations prohibit it for single women.

European governments tightly regulate this burgeoning industry but in return cover certain costs. By contrast, in the US, few regulations affect these decisions. Federal guidelines cover only a few general procedural issues. The Food and Drug Administration (FDA) mandates that doctors screen donated eggs and sperm, which will be inserted into a woman's body, for several infectious diseases.[18] The Clinical Laboratory Improvement Act establishes basic quality standards for laboratory accuracy and reliability for test results.[19] The Centers for Disease Control collects several sets of IVF statistics from clinics through the Society for Assisted Reproductive Technology,[20] an affiliate of the American Society for Reproductive Medicine (ASRM),[21] and maintains a publicly accessible database.

But, unlike many other countries, US guidelines do *not* address various critical decisions, such as when someone else's eggs may be used and how much such egg donors (or sellers) or gestational surrogates are paid. Moreover, physicians are now free to select embryos based on various mutations and traits; and may implant into wombs only embryos that are, for instance, male. ASRM guidelines address several areas, including certain aspects of egg, sperm, and embryo donation as well as surrogacy. Providers who violate these guidelines may have ASRM membership revoked or face other "name and shame" tactics but presumably still practice. Yet clinics frequently fail to follow ASRM guidelines that seek, for instance, to prevent young women 18–21 years old from selling their own eggs, since these individuals may not fully understand the risks to their own body.[22,23] The effectiveness of ASRM guidelines is often limited.

In the US, the ART industry has successfully opposed regulations that exist in western European countries. In 2005, for instance, the United Kingdom stopped anonymous sperm donation, in the interests of children born through that method, requiring that all such children be allowed at age 18 to know and contact their biological fathers. Yet in the US, anonymous sperm donation remains the norm.

The US essentially lets individual physicians make their own decisions about several crucial ethical issues, including much of what individuals understand when buying or selling human eggs, how much eggs cost, whether offspring will ever know who created them, and how much doctors profit from these lucrative procedures. Many observers vie w the relatively unregulated US industry as "the Wild West."[24]

Research is finding that IVF and ICSI can harm offspring. IVF singleton pregnancies, compared to spontaneous conceptions, cause about twice as much perinatal morality, preterm delivery, lower birth weight, preeclampsia, and other problems.[25] Children born through IVF, especially ICSI, have a 30–40% increased risk of birth defects[26] and higher rates of intellectual disability (ID).[27] Even when adjusted for maternal age and other factors, children born through ART are almost 60% more likely to develop ID, and two and a half times more likely to develop severe ID. Almost 9% sets of twins born less than 32 weeks using IVF, and 1 out of 30 children born using ICSI develop ID.

These issues extend globally, far beyond the US and other national borders. Global reproductive tourism has grown into a multibillion-dollar industry. The European Union, for example, bans buying and selling body parts, including human eggs; and most European countries bar having a woman carry a fetus for someone else ("renting wombs" in industry parlance). Consequently, each year thousands of adults crisscross the world, buying and selling human eggs, sperm, and embryos, "renting wombs," and "designing babies"—choosing what particular genes their offspring will or will not possess.

Only three countries—India, Russia, and the US—legally allow egg-buying. All EU nations ban such compensation (except for reimbursement of minimal basic expenses such as transportation).[28]

In the US, hundreds of egg donor agencies have consequently sprouted but remain essentially unregulated, raising concerns.[29,30,31,32] These companies advertise widely in elite college newspapers, seeking bright young women willing to sell their eggs. Prospective parents can now purchase eggs online, clicking from drop-down menus to choose eye color, hair color, ethnicity, height, education, religion, and other traits. The prices of eggs increase with the college's average SAT scores, triggering concerns about eugenics.[33]

For several years, the ASRM had stated that paying donors more than $10,000 was "not appropriate."[34,35,36] In 2013, however, Lindsey Kamakahi, a young woman who had sold her eggs and received no more than $10,000, sued the ASRM, alleging that this payment guideline constituted price fixing. In 2016, the ASRM settled the lawsuit[37,38,39] and eliminated recommended ceilings on egg compensation. Donors, agencies, and doctors can now charge as much as they want for human eggs. Prices will surely rise and vary widely based on the perceived social desirability of donors' characteristics. In fact, the term "egg donor," though still commonly used in the field, is generally a misnomer. In the US, most women who use other people's eggs buy them, and especially since the end of price caps, egg "seller" and "supplier" are more precise terms.

Recently, US companies have also begun creating their own embryos, buying eggs and sperm and then owning and selling the resultant embryos.[40] Companies may create these embryos "on demand"—for instance, if a couple wants a child with blond hair and blue eyes—simultaneously manufacturing additional embryos to freeze and sell. But in the US, no laws or guidelines exist concerning this market. We prohibit the sale and purchase of people (i.e., slavery) but not of embryos. These companies are not required to monitor how many such embryos for full siblings they sell, either overall or in a particular region. The offspring who result could grow up and, unknowingly, marry and reproduce.

In the US, insurance coverage is increasing slightly but remains very limited. Jane and most other patients have to pay out-of-pocket. Infertility treatment costs an average of over $24,000 in the US,[41] constituting 50% of patients' annual disposable income. In 2011, in California, the total cost for a successful delivery was $112,799 for IVF.[42] Yet it costs only around $4,000 in the UK, Australia, and Germany[43] and 6% of annual income in Australia.[44] In Europe, ART, though more regulated, is also more generously covered by health insurance. Among the 24 EU countries in 2013, six offered complete coverage; 15 provided partial coverage, based on factors such as the woman's age (e.g., covering only women under 40), marital status (e.g., including only married heterosexual couples), or overall percent of costs (e.g., 40%); and three countries had no coverage.[45] India provides no coverage; Japan, some. Yet in countries that cover costs, patients may have to wait a year or still pay out-of-pocket for medications.

In the US, several states' laws address insurance coverage for infertility but vary considerably regarding which patients, when, and how much. Fourteen states dictate that certain kinds of insurance plans must cover certain infertility procedures but differ about whether patients must be married and heterosexual, must be younger than a certain age, must use only their own eggs and sperm, must not be on Medicaid, and must have first tried to conceive for a certain number of years, as well as how much insurance companies must offer and whether religious institutions must comply.[46] Only seven states mandate that employers cover IVF, yet the amounts of coverage range widely, and are often limited to only a single cycle or a fixed, relatively small amount of money. Fifteen states require group insurers to offer some kind of infertility coverage to employers, but employers can opt out. States that mandate some coverage, even if limited, have increased ART use.

Scientists can also now alter the genes of an embryo, forever changing that organism—and its descendants. In the early 2000s, scientists began to discover how bacteria in yogurt and elsewhere possess unique abilities to snip out the genes of viruses that attack it. Researchers quickly adapted this technology, known as "clustered regularly interspaced short palindromic repeats" or CRISPR to other species. This procedure can remove or add genes associated with various traits. This technique can also add genes to bacteria, viruses, and mosquitoes to make superbugs, and in 2016, the Pentagon listed it as a weapon of mass destruction.[47]

CRISPR has advanced rapidly. In 2015, Chinese scientists shocked the world by announcing that they had begun using this technology to alter the genes of human embryos.[48,49,50] They said that they had not yet transferred these engineered embryos into wombs, but a worldwide outcry led to a summit in Washington, DC, later that year to set guidelines. Research organizations from China, the US, the UK, and elsewhere agreed to certain parameters, most importantly, not yet letting the embryos progress to the point of producing human beings.[51] This research rushes swiftly forward.[52]

Every day I receive emails, advertising do-it-yourself CRISPR kits. In July 2017, CRISPR was successfully used to fix deleterious mutations in human embryos.[53]

In July 2018, the Nuffield Council on Bioethics, the major organization that develops bioethics policy guidelines for the UK, released a report that encourages the development of gene editing in human embryos, providing that such efforts follow certain ethical guidelines. This report even mentioned possible "clinical licensing" of this technology.

In November 2019, He Jiankui, a Chinese researcher, stunned the world, stating that he had edited human embryos and transferred them into a womb, producing twin girls named Lulu and Nana. The global scientific community swiftly condemned his actions.

He said that he had worked with HIV-infected fathers to prevent their offspring from becoming infected, by disabling the CCR5 gene that creates receptors that enable HIV to enter cells. But blocking this gene also increases risks of other viruses entering cells. Moreover, doctors can already prevent fathers with HIV from infecting their offspring–through sperm-washing, Gene editing also remains risky due to so-called 'off-target' effects—accidentally cutting out other, needed bits of DNA. In addition, genes may have multiple, unknown functions, such that impeding a gene could cause other, unanticipated harms. He failed to seek approval from a research ethics committee, and did not conduct sufficient preliminary research in animals to demonstrate that such editing would work. The deletion that he says resulted may not even be the one he intended.[54]

The Chinese government subsequently placed him under house arrest. In March 2019, a World Health Organization meeting in Geneva recommended the establishment of a moratorium on transfer of gene-edited human embryos into wombs, and a registry of all gene-edited trials.[55] Nonetheless, in June 2019, a Russian scientist announced plans to edit the same gene in embryos made by HIV-infected women.[56]

Clearly, questions arise of for how long such a moratorium could last, when it should end, who should decide and based on what criteria, and what should be done if not all countries or researchers in fact comply.

Dr. He's actions highlight how readily such gene editing may be used in ways that governments cannot wholly regulate or oversee. Regulatory hurdles vary significantly between countries and may be much less in China and several other countries. Many observers fear that not all scientists worldwide will obey the moratorium. Moreover, patients and doctors regularly cross borders. In 2016, for example, a child was born in Mexico using mitochondrial replacement therapy. From the egg of a prospective mother who had a mutation in her mitochondria, the doctor had removed the nucleus and inserted it into another woman's egg, from which the nucleus had been removed. He then fertilized this newly created egg with the father's sperm, and eventually thereby produced a "three-parent-baby." The parents created the initial embryos in Massachusetts; the doctor was based in New York. However, since the FDA bans this technology in the US, the physician and patients all traveled to Mexico to perform the procedure.[57]

In 1888, the English scientist Francis Galton drew on the work of his half-cousin, Charles Darwin, to argue that society should encourage the selection of "the fittest" genes. "Eugenics" means "good genes" in Greek but is now popularly defined as the purposeful selection of certain genes over others. Many observers worry that patients' choices of egg donors today, based on traits such as height, blond hair and blue eyes, and Ivy League education begin to constitute eugenics.[58]

Given how rapidly CRISPR is advancing, it will probably be integrated into IVF in the near future, raising additional conundrums. Hank Greely, a Stanford law professor, has argued that humans will soon use sex just for recreation and use IVF—with its pipettes and syringes—for reproduction, choosing which genes we do or do not want to give our offspring.[59] We need to be prepared to address these swift advances.

A 2017 National Academy of Science report concluded that clinical trials using CRISPR to create humans should be permitted only to prevent "a serious disease or condition" in the "absence of reasonable alternatives,"[60] but such an alternative already exists—testing embryos through PGD. This current, widespread technology for genetically selecting human embryos to transfer into wombs raises the closest set of ethical and social concerns to those posed by CRISPR. PGD, providing the most related situations from which we now need to learn, reveals the kinds of problems for which we need to prepare.

Countries vary widely in their approaches toward PGD (and other ARTs), and hence will likely differ in their ultimate stances toward CRISPR. Doubtlessly, gene editing will continue to present risks, including off-target effects and longer term risks to the offspring, posing dilemmas about when the procedure will be "safe enough," particularly at two pivotal points in its future use—when children are initially created, first as part of initial research studies, and later as part of broader clinical "roll out." Clearly, sufficient prior animal models will be imperative, yet to prove full safely, the first human offspring will also need to be followed through adulthood and the birth of their own children, whose health will need to be assessed as well. Even then, potential dangers may lurk, as they do in most approved invasive medical procedures.

Overall, both CRISPR and PGD pose similar underlying ethical tensions of how physicians and others should balance parents' rights to the kind of child they desire against possible risks to the future child, rights of the future child to have an open future[61] and concerns about possible eugenics and threats to social justice. PGD today provides crucial examples and insights regarding how these tensions now play out and get addressed.

Most human traits seen as socially desirable, such as intelligence, appear to result from combinations of multiple genes, along with social and environmental factors. Studies suggest that intelligence is 50% genetic, but scientists haven't figured out which genes. Hundreds of bits of DNA have been examined, but the one found thus far to have the most predictive power increases IQ test scores by only one point. Undoubtedly, complexes of many genes and other factors interact with each other.

Importantly, Jim and other REIs are already confronting these questions of how much and in what ways to alter our descendants' DNA.

Still, most people know little about fertility, infertility, or these new technologies for affecting the genes of our descendants. Most college-aged men and women overestimate when a woman's fertility declines and overestimate the odds of getting pregnant after one sexual encounter and the success rates of IVF.[62] Many heterosexual couples rarely disclose that they used someone else's eggs, sperm, or embryos or possess mutations or are eliminating these genes from their descendants. These topics involve taboos.

Periodically, scandals and other news concerning IVF and "designer babies" appear on television, in magazines, and in newspaper headlines, often sensationalized and "hyped," with images to elicit disgust or suggest miracles and offer hope. In January 2009, Nadya Denise Doud-Suleman gave birth to octuplets through IVF while on welfare, unleashing a torrent of both outrage and awe. The media quickly dubbed her "Octomom" and her physician, Dr. Michael Kamrava, "Octodoc."

In 2015, media headlines reported on a lawsuit against Sophia Vergara, then the world's highest-paid actress, by her ex-fiancé, who sought custody of embryos they had together created. He argued that they belonged to him and that he should thus be allowed to implant them into a woman and raise the kids despite Ms. Vergara's objections. The court said no.

Science fiction novels and films have also voyaged into these realms, depicting possible future misuses of these technologies. *Brave New World* describes a future society that manufactures people in artificial wombs. In *The Boys from Brazil*, former Nazis, hiding in Brazil, plot to produce and raise an army of Aryan clones. *Gattaca* depicts a future world that segregates individuals at birth into social classes, based on their genes. In *The Handmaid's Tale,* a totalitarian theocracy forces fertile women to reproduce for "morally fit" mothers. Nobel Prize winner Kazuo Ishiguro's *Never Let Me Go* portrays a future institution that creates and raises human clones solely to donate their organs to others.

But technologies for selecting and creating human beings are now used daily, presenting countless doctors and patients with major dilemmas. Prospective parents who cannot have children on their own or are at risk of transmitting mutations to their descendants wrestle with whether to use these mechanical interventions. No other human choices are as life-altering as selecting genes to create entirely new descendants and members of our species. More than ever before, we are directly shaping our own genetic evolution.

"Designing babies" may conjure up science fiction scenarios of altering genes to create brilliant, 6-foot 2-inch, blond, blue-eyed superathletes; but doctors and patients are already using new technologies to affect the genetic makeup of future generations, forcing us to confront quandaries about our responsibilities toward our own and other species, as creations of nature. Adults extend themselves into the future, symbolically and genetically, through children, providing vital meaning, but are now doing so ever more technologically and commercially.

Therefore, we now face crucial moral, social, cultural, psychological, and existential conundrums about how to employ these technologies; whether to monitor or control them and, if so, how; and more broadly, where as a species we are or should be heading; and what responsibilities, if any, we have in these realms.

To answer these questions and know how to respond, it is vital to understand how these technologies are *now* already being used, which are being adopted, how, by whom, when, how much, who decides and how.

The medical literature has suggested a few findings but left many key questions unanswered. Several have articles examined a few aspects of these issues—how many patients have sought treatment, at what age, with what rates of success, and how they dealt with stresses involved—but have not explored this emerging complex world and the multiple uses of these technologies as a whole. Many of these prior studies were also quantitative rather than qualitative and thus did not always convey the lived experiences of these prospective patients.

Numerous questions therefore remain about how prospective parents and medical providers currently view and make decisions about these challenges and what the broader implications are. Larger issues emerge, too. New scientific advances commonly glitter with initial promise but soon generate not only benefits but also risks. Nuclear fusion gives us cheap nuclear energy as well as nuclear bombs. Email and the Internet enable immediate, wide communication but also identity theft and hacking. Major questions surface about what current uses of these technologies suggest about our future.

I decided to explore this new and rapidly evolving world. One of my professors, the anthropologist Clifford Geertz, argued that to comprehend any social situation, we should try to obtain a "thick description" of the lives and decisions of the individuals involved—not by imposing external preconceptions or theoretical structures on them but by trying to grasp their own views in their own words.[63] This approach can elucidate not only what these individuals are doing but what they think they are doing and how they understand their activities and the challenges they confront.

I thus interviewed and surveyed hundreds of infertility treatment providers and patients—to fathom how they respond to these perplexities—to map this new terrain. I polled hundreds of doctors about their knowledge, attitudes, and behaviors[64,65] and examined hundreds of ART websites that screen embryos and buy and sell human eggs.[66,67,68,69]

But, strikingly, no book had explored the full breadth of these issues. Thus, I decided to write this volume, describing what I learned.

To reveal this new land most fully, I draw here on the stories of patients and clinicians (physicians, psychotherapists, and nurses)—to paint at once both a group portrait and a landscape. Overall, as described more fully in Appendix B, I focus here on the experiences of 37 individuals whom I interviewed in significant depth—17 physicians, 10 other infertility care providers, and 10 patients. One of these physicians and three of the other infertility providers are also, themselves, infertility patients. I systematically analyzed these interview data to examine what themes emerged. This book also draws on extensive conversations I have had more informally over several years with dozens of additional fertility doctors, other healthcare providers, and patients, exploring their experiences and challenges, to comprehend these issues as fully as possible and further enhance my understanding of these realms.

These clinicians and patients revealed their hopes, dreams, obstacles, concerns, and fears—"what keeps me up at night." They all recognized that they are engaged in a radical, novel, and rapidly evolving enterprise, creating children who would not otherwise exist, navigating an extraordinary, unprecedented terra incognita.

What I saw astonished me. These men and women shed light on the myriad, unforeseen facets and ramifications of these new technologies and dramatically altered my views.

I began by investigating genetic testing of embryos but quickly realized that potential parents wrestle with not just this one choice but many others as well. Selecting the genes of embryos constitutes just one small part of much bigger phenomena of using technology to create children. Would-be parents encounter quandaries about not only *whether* to screen embryos but for *which genes* to do so; how much to try; at what expense; whether instead to use other people's eggs, sperm, embryos or wombs and, if so, whose; and how to choose doctors, and talk about these issues with each other and with clinicians, family members, current and future offspring, friends, bosses, and fellow patients. Healthcare providers grapple with additional conundrums of whether particular patients will be "good enough parents."

Prospective parents all share a common goal, a "take-home baby," but range widely in age, education, finances, marital status, sexual orientation, personality, geography, and prior medical experiences and thus start at different places. Most would-be parents are fertile, have healthy children, and lack genes that cause dangerous diseases; but countless other potential parents do not fit this description. While 10% of all adults are biologically infertile, others are fertile but possess lethal genes that can kill their offspring. Single, lesbian or gay potential parents face structural barriers. To have a healthy child, these varied groups of individuals seek three ingredients—a womb and healthy eggs and sperm to form an embryo—but face choices about whether to use medications, IUIs, IVF, PGD, other people's sperm, eggs, or womb, adopt a child, or remain childless. Each of these options poses challenges. Since most IUIs and IVF cycles fail, prospective parents usually have to decide whether to keep trying or to change strategies. Various factors influence these choices, including abilities to find and pay for each of these options, and previous infant deaths or abortions due to mutations. Mutations themselves range in severity, predictability, treatability and age at onset. Relationships with physicians, nurses, other healthcare providers, clergy, spouses and other family members also mold these decisions. Many infertile adults feel awkward discussing sex and reproduction, even with their doctors.

These individuals face not only choices about each of these technologies but broader human uncertainties, challenges, stresses, and questions about how much to pay, what financial sacrifices to make, and how to cope. These quandaries raise broader social concerns as well about how we see and define "reproduction," "sex," "birth," "children," "parents," "siblings," "families," "identity," "life," and the rights of future generations.

The chapters of this book explore how individuals confront these dilemmas. Part I provides an overview. Chapter 2 probes patients' experiences of infertility. Part II examines these reproductive choices—whether to pursue these technologies and, if so, which; whether to select IVF, other people's eggs, sperm, embryos, wombs, or children (through adoption); whether to select embryos either to avoid disease or to seek certain non-medical traits (such as sex); how many children to create; and what to do with leftover embryos. Part III investigates how patients choose doctors and vice versa. Part IV explores how patients and their providers decide how much to spend, how to cope and find support, and how to respond to larger religious and metaphysical issues. Part V delves into the implications for future education, research, policy, and society.

To protect confidentiality, I have used pseudonyms and disguised a few identifying details. This "de-identification" may make it harder for readers to get to know some of these men and women, but, by the end, these individuals emerge, I think, as full and unique people. For emphasis, I have italicized a few of their comments.

These patients and providers differentiated between these technologies and the "natural, old-fashioned way" of having children. They are creating humans who could not have been made "naturally"—that is, without the use of sophisticated technologies. In making these varied choices, these clinicians and prospective parents are artificially making and designing life itself.

While some observers may use the term "designer babies" narrowly, as referring only to the selection of genes associated with socially desirable characteristics, I soon saw how this notion extends far more broadly in this new, rapidly expanding universe. I thus use this term, in part metaphorically, to suggest the ever-widening arrays of technologies that doctors and prospective parents are now using to affect the genes of their offspring—testing embryos for increasing numbers of purposes; buying particular eggs, sperm, and embryos, and "renting" wombs. This wide range of medical interventions and machinery are swiftly evolving.

I began wary, as many people are, of the idea of designing babies but came to see the perspectives of the individuals involved and the many human sides and complexities involved. To my surprise, I ended up supporting several, but by no means all, of the uses of these technologies. The elimination of genes for terrible diseases is at times legitimate, but questions of whether and how we should employ these technologies in other ways pose larger quandaries.

I present here several providers and patients charting and illuminating paths that countless others will soon take. The insights in this book, revealing how these men and women view, make, and experience these decisions and what they recommend to others, can help prospective parents and their families considering these interventions. These rapidly evolving technologies are crucial for the rest of us to grasp.

The individuals here exemplify broad themes that emerge—underlying issues that consistently arose in the research I conducted. These narratives thus reflect not isolated cases but rather cross-cutting phenomena, offering insights that can help us respond to these double-edged scientific advances, now altering humankind. These technologies offer both boons and potential perils. Unfortunately, we are prepared for neither set of outcomes. Whether as family members, clinicians, policymakers, citizens, or members of a species, nation, and world, we need to understand this new world, which will increasingly be affecting us all.

2 | "Not Part of the Parent Club"

PROBLEMS CONCEIVING A HEALTHY CHILD

"It was hard being the 'child-free friend' for a long time," Yvonne, a Philadelphia social worker told me. For several years, she has been struggling to get pregnant without success. "People looked at us and assumed we didn't want kids; that we valued our careers more. They judged us. I was very afraid that my husband and I would never become parents. Ever since preschool, I grew up, taking care of my baby dolls, saying, 'When I'm a mom, I'm going to blah blah blah.' I see these assumptions about little kids: that boys will be daddies and girls will be mommies. A friend said, 'My husband and I are starting to talk babies.' That's such a happy feeling as a couple starts to feel ready.

"When having a child becomes hard, it's a huge fall—the death of that fantasy. Every month, when I get my period, I re-experience that loss. Each period is its own little miscarriage, because I hope I might be pregnant. From the day I ovulate, to when my cycle should end, I think, 'I know not to get my hopes up.' Then, I'm seven days past ovulation: 'OK, just one more week!' In those last few days, this whole mental monster takes over. The fall is horrible: it becomes shorthand between women who are dealing with in-fertility: 'I'm waiting to get my period,' 'I'm 10 days past ovulation.' I know how stressful that is. Outsiders don't know this shorthand. People don't think about infertility as a *grief process*, but it's similar to the death of a spouse or loved one—the death of your imagined family.

"There are raw days. And they can overlap with special occasions. If Christmas is the day after I got my period, it might be a miserable holiday. Birthdays are among the worst—a reminder that time is passing, you're getting older."

Eventually, through in vitro fertilization (IVF), Yvonne had a baby girl. She then wanted to have a second child as a sibling for her daughter. "My young daughter is five, and more and more says, 'Why don't we have a baby? I want a brother or sister.' I say, 'We would have liked it, too. But it hasn't happened.' I'm always honest, especially when my daughter can hear. I'm not going to pretend we were planning just for one child. She says about a couple: 'Oh, they're going to get married and have a baby!' I say, 'Well, *if* they choose to.'

"It's hard having an only child. My mother-in-law says, 'Don't you want another baby?' That hurts! Of course I do!"

Yvonne was raised Jewish, but her husband is from a very conservative Catholic family that is wary of these technologies, exacerbating these tensions. She tried several intrauterine inseminations (IUIs) and had two miscarriages.

She then learned that she had a chromosomal abnormality, Emanuel syndrome (a translocation, in which parts of two chromosomes switch), for which she could genetically test embryos to prevent transmission. But she is now "in limbo," unsure, debating whether to screen embryos, "so that we can look back and know we did everything we could."

"I walk through stores, envying pregnant women I see, thinking they're gorgeous and glorious. I want to be like them. Other times, I see a happy couple—the woman with a big belly—and just hate them! That's totally irrational. But it's like that. I stand in drug stores, deciding between tampons and a pregnancy test!"

Infertile individuals who want to have children and get pregnant but then miscarry can feel devastated. Seeing peers having children can prompt pangs of failure.

Having children can give meaning and purpose to individuals and couples. Every species has biological urges to have sex and reproduce; otherwise we would not exist. Mammals also possess strong instincts to raise offspring. In humans, not only biological but deep psychological and social drives propel these goals as well. Women may see motherhood as a social norm, a "natural instinct," or a developmental stage in a relationship. Feminist scholars argue that our society too strongly ties women's value to procreation,[1] which leads prospective mothers to see failures to achieve this aim as tremendous loss.

To understand how infertility patients view reproductive technologies, it is important to grasp first where they are coming from—what difficulties led them to seek treatment.

As Yvonne suggested, extended family members can push prospective parents, either implicitly or explicitly, to have children, unaware of the stress that impediments to these desires can exacerbate. Small everyday interactions can take on far greater significance.

"Why Don't You Have Kids Yet?": Social Pressures

Particularly for heterosexual women such as Yvonne, the fact that friends of the same age are having children can heighten stresses. Gay men and lesbians generally feel much less pressure, though norms are shifting in these communities, too.

"All their peers are getting pregnant, or having their third or second pregnancies," Suzanne, a mental health provider at a northern California IVF clinic, said. "Some infertility patients feel a lot of family pressure to have a child. But there's more subtle *peer* pressure: everybody else is getting ahead, while they feel they're not. Some people feel ashamed they can't do what everybody else seems to do readily—that something is inherently wrong with them. They feel thwarted in achieving their life goals—not part of 'the parent club.'"

"Unknowingly, family members, primarily patients' parents, put pressures on them," Joe, a physician who specializes in male infertility, told me. "'When are you going to give me grandchildren?'—as though it were a gift for *them*. Families don't understand what the couple is going through; and the couple often doesn't want to talk about it."

Extended family members can become intensely invested in the birth of a new child, without recognizing these stresses. Reproduction involves two prospective parents transmitting to future generations *their* genes, along with those of *their* parents, siblings, and others. Present offspring will become brothers or sisters or cousins. A would-be parent's parents would become grandparents, and siblings would become aunts and uncles. Portions of all of these individuals' genes will continue on through others. Infertility and its treatment thus become profoundly *social* phenomena.

"Look at the magazines at the drugstore checkout: movie stars having babies," Helen, a Wisconsin mental health provider who specializes in infertility issues, observed. "It's a status thing. Bosses make childless employees work Fridays and holidays because parents are prioritized above them."

A sense of failure can especially trouble women who have heretofore succeeded in their careers. These new frustrations can be hard to handle. "Often, this is the first time in these women's lives that things have gone really wrong," Helen continued. "There's lost innocence. They just don't get it. They feel victimized—for the first time really experiencing life as unfair. If they haven't learned to cope with that, it can be even worse. Other women understand that life is unfair, which can sometimes be protective: 'My whole life has been like this! Here we go again.' But a lot of women think, 'it's so unfair.' If I say, 'Life is unfair,' they don't really get it. Gradually, some move on and get involved in gardening or get dogs. We don't know."

Due to feelings of shame and failure, some patients avoid communicating their distress. While gay men frequently discuss with many people efforts to use these technologies, heterosexual women often tell few. But secrecy can heighten stress.[2] Shame and social isolation can become mutually self-perpetuating. "Initially, women are more likely to tell others," Suzanne, the IVF clinic psychotherapist, elaborated, "because the assumption is: 'We're having a little difficulty, but everything will be fine very soon.' Then, when not reaching their goal, shame develops. Other people now know it, too, which becomes hard. Patients tend to withdraw and isolate themselves. Other people are moving on with their lives, getting pregnant, raising children. Patients no longer feel they belong with their peer group."

As Yvonne suggests, miscarriages can feel devastating, dashing women's hopes. "When I have a miscarriage, nobody wants to talk to me or be around," Francine, a 33-year-old Miami legal assistant, said. She has undergone three IUIs, four IVFs, one with preimplantation genetic diagnosis, and consulted six different clinics, all without success. She and her husband, who owns his own construction company, now plan to go to Mexico for an experimental immunological treatment. "When my sister had cancer, people visited, watched her kids, cleaned her house, bought her dinner," Francine said. "Every day, she got a care package or note: 'I'm thinking of you.' But when you're going through infertility, nobody does that. Friends should call, and ask if they can do something. You still have to go to work every day. You can't call your boss and say, 'I just had a miscarriage'—at least most people don't. You've got to work and suffer through the day. When you get home, you don't feel like cooking. You're sad, upset, and often in physical pain because you're having to pass tissue. I've sent flowers to other women and called: 'I know you're going through a very difficult time. Can I bring you a meal?'" Unlike other disorders, infertility does not impede one's daily abilities to function. Hence, while prospective parents may see miscarriage as a death, outsiders may fail to understand and even be wary.

"I feel a lot of frustration at my body not cooperating with me, not doing what it's supposed to," Francine continued. This unpredictability can challenge multiple aspects of life. Women struggle with how to view this problem and themselves—with when, if ever, they may have a medical diagnosis or identity as "infertile." They may perceive themselves as needing assisted reproductive technologies but as healthy, not as having a medical diagnosis or entering a potentially tainted category. "There's a stigma being labeled as infertile," Francine added. "It's a *bad thing*. Women don't see themselves as infertile: 'My doctor said I need IVF. I'm not *an infertile person*. An infertile person is someone who's been doing this a long time!'"

The resentment that Yvonne and others now experience can bewilder them. "It brought out a lot of emotions in me that I am not proud of," Amanda, a 36-year-old married Delaware high school administrator, said. "I used to sit in the mall at the food court and watch pregnant women go by. I was happy for them, but miserable—angry that some people could get pregnant so easily. I'm not a 'poor-me' type person but felt *how come everybody else can get pregnant, and I can't?* Seeing 16-year-old kids pregnant made me angry and sad. I hated that it brought these feelings out in me."

Even Karen, a California infertility doctor who herself underwent infertility treatment, faced difficulties. "I live near an elementary school, and would now always go out of my way to avoid the building. I couldn't stand seeing those kids. It was the first time in my life that I really wanted something, and couldn't make it happen. It was out of my personal control. Brutal."

As a doctor, she found it tough: seeing her patients who gave birth but did not appreciate their own ease in doing so. "I resented patients of mine who took their pregnancies for granted—especially those who had five or six children, and didn't want to quit smoking or drinking. Infertility goes on for so long, with so much disappointment, and these heavy-duty drugs make you totally crazy. I was a basket case."

"Less Macho?" The Stresses of Infertility for Men

"My most difficult decisions involve men who want to have children, but are not making sperm," Joe, a physician who specializes in male infertility, told me. "We go into the testes and search for sperm—a large, expensive procedure—and come up with nothing."

Males and females have roughly equal rates of infertility but experience it very differently. Women race against a "biological clock"—the fact that problems having a healthy baby skyrocket after age 40. Since they give birth, women tend to get blamed for infertility more and face painful expectations from others who assume that getting pregnant is easy. Yet, approximately 10% of men are biologically infertile because low sperm counts,[3] and for unclear reasons, this proportion has been increasing.[3,4]

Men with little or no sperm often feel guilty, ashamed, and less "macho." "Some men don't follow-up for a long time," Joe continued. "Others adopt, or go for donor insemination. Many feel terribly guilty—that it's their fault, even though it isn't. Often, it was just a birth anomaly. But it's difficult to explain that to them—for them emotionally to believe that. They never thought they would have this problem. Some come back to talk. We refer them for counseling. They don't distinguish between their sexual prowess and fertility—between the ability to have sex and procreation. If they do one, they think they should be able to do both. Otherwise, they feel 'less a man.'"

In certain societies, infertile men can face even more shame. "In most Middle Eastern cultures, both Muslim and Jewish, and a lot of Mediterranean ones, including Greeks and Italians, men take reproduction as a birthright, and never think about it," Joe added. "If they are in a relationship, they're going to be fathers. Other men think they did something to cause their infertility. Bodybuilders think their anabolic steroids caused their sterility. Occasionally, they stop the steroids, but remain sterile—they may have been sterile to begin with. Some men took steroids only briefly to get a boost in the gym. When they learn they are sterile, they automatically assume it was the drug. It probably wasn't, but they still feel terribly guilty."

Many are uncomfortable discussing infertility even with their doctors, hampering communication and referrals. "Some patients are embarrassed to discuss infertility," Joe elaborated. "They may be more comfortable talking about their sexual ability, but Viagra and infertility are not the same! Fertility often doesn't come up."

Men may therefore face particular barriers. "For women, it's bad enough—and for men, probably even worse," Joe observed. "Men don't want to go through these exams: It's humiliating. Some remain silent and fail to get help. One young man was told there was 'no possibility' he would ever get his wife pregnant. They waited a couple of years because he was just depressed! As the wife got older, the pressure was on. We were able to help them. They now have three kids."

Marital Tensions

"We don't blame our husbands," said Francine, the Miami legal assistant planning to travel to Mexico for treatment after four unsuccessful IVF cycles. "Regardless of whether it's female- or male-factor, we're undergoing treatments because we love them and want to have a baby with them. If it were as simple as just having a baby with somebody else, we could. It's a lot easier to come home to somebody who is understanding, kind to you, and doesn't use your diagnosis against you, rather than saying, 'It's your fault.' Sperm donors are easy to come by, husbands are not."

Spouses' differing views of their infertility can strain not just each individual but their relationship, intensifying pre-existing conflicts. When the man, rather than the woman, is infertile, relationships are even more unstable.[5]

But the problem of infertility, even if biologically caused by only one member of a couple, surely belongs to both of them. The couple faces the impasse together, though they may not equally share this view.

Many men fail to fully grasp the issues and instead blame infertility on their wife or girlfriend. "We're the ones at the doctor being probed, talking to our girlfriends who've been through similar experiences," Francine elaborated. "Husbands sometimes say, 'I thought it was *you*.' Sometimes they end up being accusatory. Not intentionally, but it comes out as: 'It's *your* problem; not *mine*.' I try to explain better why I'm upset or anxious: 'This is our *collective* problem,' I told my husband. 'If you want to have a baby with me, you need to try to help me through this process. I don't want you to drink during this cycle and end up in a collection room with no sperm, while I'm on the table having my eggs collected and there's nothing to inseminate them with.'"

Marital tensions easily mount. "I called my husband, almost hysterical because I didn't think I could take the stress of waiting another day for the lab results," Francine continued. "His response was very laid back: 'It's just another day!' I thought, '*You're just not emotionally invested!*' He always responds, 'I'm emotionally invested; but nothing can be done!' There's a lack of communication." Yet enhanced discussions about these complexities can help.

In sum, for both men and women, infertility causes psychological and social stresses. Quests for children spring from deep sources—biological urges, social expectations, norms, pressures, dreams, and senses of purpose. Failures to achieve this goal can generate stress, feelings of failure, and jealousy of current parents, aggravated by painful daily reminders. Miscarriages can represent the loss of a vital dream. Yet within a couple, spouses can also experience infertility differently, shaped by personal and cultural views and expectations.

Particularly if they have not had problems getting pregnant, outsiders, including patients' parents, siblings, extended family members, friends, acquaintances, employers, and co-workers, may not appreciate these travails and instead be critical or skeptical. Stigma and misunderstandings persist. Patients respond to these challenges differently, based on several factors—whether they are men or women, single or coupled, straight or gay.

These psychological and social pains of infertility lead many individuals to pursue these technologies; but as we will see, they soon confront additional challenges.

II | Choosing Reproductive Technologies

3 | Becoming an Infertility Patient

CHOOSING TREATMENT

"When I thought about having a child, I figured the birth control goes away and you have a child conceived out of a loving act," Amanda, a Delaware high school administrator, said. She underwent six intrauterine inseminations (IUIs) and in vitro fertilization (IVF) before conceiving twin boys. "I did not anticipate it being in a doctor's office with a doctor helping. It was against everything I had imagined conceiving a child would be. I knew that once you hit 40, your odds of getting pregnant go down, but *that* was the extent of my knowledge. Beyond that, I thought that people who want to get pregnant get pregnant, and have a baby and live happily ever after. *I never thought about this gap*—the idealism of how you conceive a child versus the reality."

Once she and her husband, a clothing salesperson, entered the world of assisted reproductive technologies (ARTs), they had to decide how to proceed. She "had one chemical pregnancy and one miscarriage, and then took a three-month break. I was leaning toward adoption; but my husband said we should try IVF, so that we would never look back and say, 'What if we had done it?' We would have to pay for it out-of-pocket, so we'd only do it once. If it didn't work, we'd adopt. I was waiting for a big light bulb 'ah-hah' moment, where I would know exactly what I wanted to do. Eventually, we tried IVF. I was not hopeful. But it worked! I conceived my twins. They're now six months old."

As she suggests, infertility itself is usually unanticipated, and patient and public understandings of reproduction are limited, especially regarding the harsh realities of the biological clock—that women's fertility plummets after age 40. Prospective parents are often surprised to learn they are unable to have a healthy baby "naturally" and must instead decide about technological interventions.

Providers and patients face questions of whether and when exactly to try IVF and other procedures and do not always agree. Most patients dislike IVF in and of itself but see it as their best option since medications themselves often fail and cause side effects.

Prospective parents commonly want and expect these procedures to be easy and predictable. But frequently, the reality soon proves otherwise. Rates of success are improving but still less than desired.

"Where you start, and where you think you're headed are not always in tune," said Francine, the Miami legal assistant now planning to go to Mexico. "You think you're headed one way: 'this IVF cycle is going to work.' But there are different endings. It would be nice to have a crystal ball that says, 'This is how it ends.' There are so many choices available, it's a *choose-your-own-ending story*. Some endings have a baby that's genetically yours. Others have an adopted child. Or no child. Everyone finds their way along this path differently. You could go through three IVF cycles and say, 'We're not trying anymore.' Or you could do one cycle, switch doctors, do another cycle, go to Mexico, or use donor eggs or a gestational carrier. It evolves as it progresses. You become open to other ideas. You might be willing to use donor eggs only after your second or third failed cycle, not after your first. It's good for me to see this as a journey with a baby at the end, but I don't know how I'm going to get there."

She and countless others remain hopeful but unsure. Whether heterosexual, gay or lesbian, single or partnered, confronting a biological or social impediment to having a healthy child, whether infertile or at risk of transmitting a mutation, prospective parents start from different places but embark on journeys with the same clear goal: a healthy child.

Getting Referred for Reproductive Technologies

"I wish I'd known about IVF before, but I didn't," John, a Texas car mechanic pursuing IVF, told me. He had three children with his first wife, then underwent a vasectomy, divorce, and second marriage. With his new wife, he now wants to have children: "She wants a child of her own."

He reversed the vasectomy but remained unable to conceive a child. Tests then revealed he had a low sperm count. "I only found out about IVF through Internet research. The more I dug, the more I found. My wife's gynecologist had mentioned it, but didn't have any office literature. All they said was: 'There's a reproductive place in another city. You can go and talk to them.'"

For many patients, the first step in using these technologies is getting a referral to an infertility specialist, yet even that step can pose challenges. Not all patients who want children and might benefit from infertility treatment receive such a referral. In most countries, only about half of infertile women who desire children, even if covered by insurance, ever seek professional help.[1] Although 10% of couples are infertile, only 2% of infant births in the United States and several other countries use ARTs.[2] In one study of US women trying to get pregnant, only 11% saw a doctor, got advice, or had diagnostic tests and only 4% sought treatment.[3] Among infertile women in France, only 45% saw a doctor—of whom 45% ended up getting treatment.[4] In an Internet study of women trying to conceive, only 44% had ever consulted a physician.[5]

Cost is a major barrier.[6] Even among infertile women with household incomes three times over the poverty level, almost half have never used infertility services. Among infertile women with private health insurance, over half never use these treatments.[7] Efforts to increase insurance coverage have succeeded slightly and should continue but usually remain limited.[8]

Patients face other social, cultural, institutional, logistical, and attitudinal impediments as well. In the US, black and Hispanic women are less likely to receive infertility services than Asian or white women, partly due to less income, education, and encouragement from friends and family, and more perceived stigma, and moral concerns.[9]

Referrals to infertility specialists and clinics come mostly from other doctors and often don't occur.[10] In one British clinic, about a third of patients were referred by their obstetricians and gynecologists (OB/GYNs) and a third by general or primary care providers.[11]

Yet many questions emerge about when and how these referrals occur—how patients even come to learn about their needs for IVF, which providers in fact refer patients, when, why, and how. In other medical fields, characteristics of the individual referring physician, the patient, and the particular type of specialist affect whether patients get referred to various specialists.[12]

Added barriers can hamper infertility referrals. Many OB/GYNs would discourage use of these technologies if the woman plans to be a single parent (17%) or is unmarried (14%) or lesbian (14%). Male and religious doctors were about twice and four times more likely, respectively, to discourage treatment in each of these situations.[13]

Providers' lack of knowledge can also stymie referrals. Prospective parents who can transmit dangerous mutations to their offspring or carry recessive genes may not know about the possibility of screening embryos. Most non-ART physicians also know little about these technologies. When I surveyed

220 internists, only 7.1% felt qualified to answer patients' questions about preimplantation genetic diagnosis (PGD). Their views were associated with the number of years since their training and other factors.[14,15] Among the 163 neurologists I polled, 24% had discussed with patients the possibility of prenatal genetic testing and 6% had discussed PGD, but only 5% felt qualified to answer questions about PGD.[16]

Little research has examined how patients select IVF providers. In other areas of medicine, patients choose hospitals for major procedures based on demographics, resources, and types of medical conditions.[17] Pregnant women choose hospitals for delivery based on sociodemographics, health status, and perceptions of the hospital's quality.[18] But how infertile women find their ways to treatment raises separate questions since they are not yet pregnant and may not even know that such treatment might help them.

Who Refers Patients for Treatment?

For many patients, the first step in receiving infertility treatment is getting a referral, without which patients may not even know about their options. At least seven types of other healthcare providers, I found, can potentially refer patients to a fertility specialist—most commonly OB/GYNs but also general practitioners, internists, oncologists, and genetic counselors. Yet these clinicians do not always refer patients or do so when and how they optimally should. In these other fields, various factors facilitate or impede ARTs consultations. In fields outside of infertility, many doctors fail to appreciate fully how much a woman's fertility decreases with age, and they fail to refer patients until too late.

Many patients start to address infertility with their ongoing OB/GYN, who can thus play a key role as "gatekeeper." Yet, patients may end up discussing infertility with several other clinicians as well, shuttling back and forth from their long-term OB/GYN to one or more reproductive endocrinology and infertility specialists (REIs) and back to their OB/GYN. Over many years, patients have frequently come to trust their OB/GYN. Even after receiving a referral, patients may reconsult their long-term OB/GYN about specialists' treatment plans and recommendations. Checking back periodically, infertility referrals are not necessarily one-time events but parts of ongoing processes with feedback loops.

Doctors vary widely in not only whether and when but *how* they refer patients for infertility treatment—whether they supply specific names and contact details of REIs or merely vague information about the existence of various ARTs as options and whether they follow up on such referrals and/

or initiate diagnostic tests and/or treatment themselves. Physicians who are not REIs range from clearly and readily referring patients to merely mentioning the possibility in passing, offering little, if any, encouragement. Over time, awareness of infertility issues has been increasing, but providers in other fields may still end up giving inadequate or inaccurate information. Failure to refer 42-year-olds "happens a lot less lately," Steve, a Virginia infertility doctor in a private group practice, observed. "Whether because patients have figured it out on their own, or older doctors have now retired. But . . . some OB/GYNs say, 'Oh, you're 42, you're going to be fine!'"

At times, providers delay referrals for ART due to limited knowledge regarding the availability or effectiveness of ever-evolving treatments and/or other barriers or do not early enough inform or encourage infertile patients about these options. "We've gotten a lot better, but you still hear it: 'Just go home and relax. Give it some time. Everything is going to work out for you,'" Peter, a West Coast REI, said. "The biggest tragedy is women in their early 40s. . . . Women wait and show up in our office, and have been doing nothing for two years, and should've been having fertility treatments. There's just a general ignorance about the biological clock. Even among OB/GYNs, although we are much better at it. . . . The population doesn't get it, and when they ask their own doctor, they get misinformation. Women say, 'Oh, I was getting concerned, and my doctor said, 'Oh don't worry about it. You're going to be OK.'"

Impediments to Referrals

In several other fields, physicians often have gaps in general or specific knowledge about infertility, treatment, and genetics. "Generalists do not commonly go to infertility courses, because they don't see it every day," Joe, a physician, said.

Providers in other fields can either over- or underestimate IVF success rates, affecting whether, how, and when they discuss treatment with patients. "I recently asked OB/GYNs," Edward, an Ohio infertility doctor, said, "if you refer a woman to an IVF clinic, what's the chance [you] would deliver a baby?' They said, 'About a 70 to 90% chance per cycle.' . . . Unfortunately, that perception—that IVF is always going to work—is out there."

In other medical specialties, doctors tend to focus on existing offspring, rather than possible future ones. "For pediatricians or internists, future pregnancies might be an afterthought," Henry, a Midwest medical center infertility doctor, said. "Most of their focus is on evaluating and treating the diagnosed individual. It may slip their memory to refer someone" to avoid problems in "future pregnancies."

Even genetic counselors may know little about PGD, though it could help their patients avoid transmitting serious mutations to offspring. "We tend to use genetic counselors that have familiarity with these techniques," Edward continued. "But a lot of docs out there don't."

For patients at risk of transmitting lethal mutations, providers may see PGD as a guaranteed answer, without grasping the potential limitations. "The knee-jerk reaction is always awe—it's amazing we have this technology," Jennifer, a Northwest medical center physician, continued. "The assumption is that it's a panacea. It's going to solve all of our problems for families with genetic disorders. . . . Patients get misled by thinking it's a trivial procedure, the perfect solution."

Clinicians may have particular knowledge gaps concerning genetics, especially regarding rare mutations that can impair reproduction. Patients may recognize these deficiencies. "Even my new doctor has never really heard of it," said Cathy, the Alabama call center employee about trisomy-22, a relatively rare chromosomal abnormality from which her infant daughter died. "He didn't seem clueless" but didn't "seem like he'd heard of it."

Financial concerns can also impede providers in other fields from referring patients earlier. OB/GYNs and REIs may "overlap" with each other, blurring certain boundaries. "There are politics about gynecologists not wanting to pass patients on to REIs, and keeping these patients longer than they should," Diane, an IVF clinic nurse, observed. "There are different interests—doctors making money, doing different things." Jill, an Illinois REI, described this tension even more pointedly: "OB/GYNs probably shouldn't be taking care of these patients. But there's conflict: the OB/GYN might *want* to take care of them—it's *revenue*. Yet the patients would benefit from earlier referral." REIs and their staff may be biased, perceiving competition from others. Nonetheless, their views are important to note.

Patient-Related Obstacles

"Only about 25% of infertile couples actually ever make it to the office," Steve, the Virginia infertility doctor, said, "and fewer than half of those will ever pursue a cycle". Even after getting a referral to an infertility specialist, patients may fail to proceed or may drop out, feeling uncomfortable. These gaps are hard to fathom since only patients who have learned about and decided to consult an REI end up seeing one. "I don't know how many *don't* get here because they don't get the opportunity to know about it," Calvin, a southern California REI, said. "It's hard to know how much we're missing. Some people see it as science fiction."

Patients may also delay treatment because they fail to appreciate how much age reduces fertility. Most of the public do not grasp this precipitous drop,[19,20] partly since the popular media have reported celebrities and other women in their mid- to late 40s giving birth.[21,22] While some subfertile women are aware that age impedes infertility, 85% mistakenly think that IVF can reverse these effects.[23] Women in their early and mid-30s, in particular, may insufficiently grasp that this clock is ticking.

The media can mislead both patients and doctors, highlighting needs for wide public education. "Women stand in line in the grocery store and read magazines where such-and-such a celebrity has just had twins—and she's 52," Diane, an IVF nurse, observed. "Nobody says it's donor eggs. So, plenty of women come here and think: 'Well, *she* had a baby at 52 or 48. Why can't I? I feel healthy. I'm not overweight, I eat right, take care of myself and feel good. If she can do it, I can.' I tell her the fine print: 'it's donor eggs.' She looks at me like I'm lying."

Implications

"Education is greatly needed, but it's hard to know exactly *when* and *how* to educate these folks," Jennifer, a physician, said. Greater awareness about infertility and its treatment can help but faces challenges regarding whom, what, and how best to teach.

Doctors outside of infertility may be busy and/or not readily open to additional input. "If doctors are *asking* for a talk on reproduction, then the speaker can bring up these issues," Bill, a New England REI, explained. "But you can't go knocking on doors saying, 'Can I talk to your staff for a few minutes about PGD?'"

Questions surface regarding how much education is realistic for clinicians in other fields, given that these technologies are advancing swiftly. "It's an awful lot of information to know," Roger, a New York infertility specialist, confessed. "If I had to know as much as an internist, it would be asking too much. Doctors are pretty smart. They don't like talking about things they aren't very comfortable with. They recognize they can cause more harm than good."

Referrals can thus involve complex systems. Patients may first have to mention their infertility problem to a provider or learn about the presence of mutations for which embryos can be screened. Over time, patients may follow that referral to a specific provider or choose a different one.[24]

Frequently, healthcare systems use primary care doctors as gatekeepers to reduce access to specialists and thereby costs.[25,26] But primary care physicians have been found to misdiagnose patients and order unnecessary tests and procedures prior to making referrals to specialists.[27] Mutual communication between primary care physicians and specialists can help but may not occur.[28]

Both doctors and patients can feel awkward discussing infertility. In areas other than infertility, clinicians often feel "uncomfortable" discussing "taboos" of sex and reproduction. Among family practice physicians, for instance, only 48% of females and 58% of males are comfortable obtaining sexual histories from patients of the opposite gender.[29] Enhanced education of providers in other fields concerning the nature, benefits, and risks of infertility treatment can help.

Which Procedures to Pursue?

"What treatments did I want? Everything?" Francine, the Mexico-bound Miami legal assistant, asked. "In the beginning, it wasn't clear for me. I just did whatever they told me. Then, I started to do my own research. What they were telling me wasn't inaccurate; but I didn't exactly fall inside this perfect little box.

"It's an evolution," she continued. "IVF was beyond my comprehension or anything I thought I was willing to do. I never thought I'd be able to give myself a shot. Now I give myself hormone shots all the time. I never thought I'd consider using a gestational carrier. But now maybe a gestational carrier is the answer. I find myself considering things I never would have." Francine and others commonly shift their expectations and beliefs.

Typically, women undergoing IVF must self-administer shots of follicle-stimulating hormone or luteinizing hormone every 24 hours for 7–12 days (to stimulate egg production) and injections of progesterone intramuscularly for 8–10 weeks after the embryo is transferred into the womb.[30,31]

Even after following through on a referral, prospective parents face multiple decisions. Like Jane, in my colleague Jim's office, patients may start simply with medications and, if those fail, progress to IUI, IVF, PGD; purchasing eggs, sperm, or embryos; or renting wombs. Among women aged 25–44 with current fertility problems who received medical treatment, about two-thirds used medication, a fifth used IUIs, and about

10% each underwent surgery (e.g., for blocked fallopian tubes), or ARTs, including IVF.[32]

These technologies vary in risks, benefits and costs. Usually, infertile patients first undergo tests to determine whether infertility stems from the man, the woman, or both. Before proceeding to IVF, women ordinarily attempt medications alone. OB/GYNs, rather than REIs, often prescribe these drugs, which help many women but frequently fail and cause side effects. "Some patients take a few years, trying things other than IVF," Karen, the California physician who has herself undergone infertility treatment, said. "I went for about a year and a half. Clomid makes you get more eggs, and Pergonal increases your egg production. I did those many times, and did IUI. I didn't want to do that longer. It was tortuous. Every month, I get a pimple before I get my period. Now, I would get a pimple and say, 'Oh shit!'" It meant that her efforts at pregnancies had failed.

Critics have argued that physicians overuse certain procedures. Doctors often perform intracytoplasmic sperm injection (ICSI), for instance, when it may not be needed, trying to increase the odds of fertilization even in the absence of male infertility problems, for which it is designed. ICSI is used in 93.3% of IVF cycles with male factor infertility, but in 66.9% of cycles *without* male factor infertility, though these injections decrease rates of implantation, pregnancy and live births,[33] and increase costs and risks of birth defects and intellectual disabilities.[34,35]

At multiple stages, clinicians must weigh these varied factors. "How much risk do I want a particular patient to undertake?" Max, a private practice REI, asked, "when a patient is responding aggressively, and I want to prevent ovarian hyperstimulation syndrome [OHSS], which can be severe and require hospitalization?" Hormonal injections in IVF cause OHSS in around 3% of patients,[36] enlarging ovaries and precipitating other symptoms. Though usually minor, OHSS can be severe and fatal. "Do I cancel the treatment? Do I coax her to get fewer good eggs?" Max continued. "Do I retrieve and freeze fertilized eggs, and then transfer embryos later? Those decisions are difficult. I try to make recommendations based on standard studies. But at times, no matter what I do, a patient gets sick."

In choosing technologies, doctors and patients both face challenges. "I didn't want to do IVF," Tammy, a 25-year-old graduate student told me. "But if that's the only choice, it's fine with me." She had polycystic ovarian syndrome and had been trying to get pregnant for five years with her husband, who had a very low sperm count.

IVF commonly presents complications and dilemmas. "Straight, traditional IVF is an oxymoron," Brenda, a Vermont infertility psychotherapist, said.

Many prospective parents are thus apprehensive. "The fear of it initially—the unknown—prevented me from doing IVF earlier," said Roxanne, a 41-year-old Jewish marketer from Michigan. For several years she had attempted, without success, to get pregnant naturally. "I knew a little about the actual IVF process, but didn't realize it wasn't so terrible." She tried several IUIs and IVF and eventually had her son. Now, she was again undergoing treatment to have a second child. "I would like to have a sibling for my son."

Over time, patients may waver in how much to pursue various procedures, not knowing a priori. "People should probably now just skip all of the other junk, and just go straight to IVF to have better success rates," Karen, the California physician, said. "You're more likely to get pregnant and get a baby faster, and don't have to go through these other treatments—which are agonizing."

But concerned about the ever-ticking clock, many women proceed rapidly to IVF, which they later regret. "I started IVF too quickly, and maybe wouldn't have needed it, but I'm impatient," Karen confessed. "When I want to do something, I want to get it done."

For various reasons, many prospective parents investigate, but ultimately eschew, IVF. They may want a child but feel overwhelmed by the medical, financial, and psychological obstacles. "Most of the patients I see are very motivated," Ginger, a mental health professional at a small IVF clinic, said. She herself underwent infertility treatment for her first child and adopted her second. "But other patients come once, and that's the end of it. That's as far as they're pursuing IVF, or donor gametes," that is, eggs or sperm. "Sometimes it's their personality: they realize they're not equipped to deal with this. Other people realize they can't do this financially."

About half of infertile adults who want a child and would likely benefit from these technologies do not use them. These individuals may be too embarrassed to discuss infertility, do not receive referrals from their primary care or other providers, or are wary due to financial, moral, or other conundrums.

Once they receive a diagnostic workup to determine whether the infertility results from the male or female member of the couple or both, treatment can begin. But these patients and their providers must then decide which interventions to try. Medications alone and IUIs often fail, leading patients to consider other technologies, including IVF and PGD, which then present additional dilemmas.

4 | Choosing Eggs

Just as my friend, Abby, asked me to help her have a child, countless prospective parents seek other people's gametes—that is, sperm or eggs. Heterosexual couples generally want to use their own gametes but may be unable to do so. Only around 40% of in vitro fertilization (IVF) cycles with a patient's own gametes produce a "take-home baby."

Yet as Francine and others suggested, women who have failed to get pregnant using their own eggs can try using another woman's but then face challenges. According to Helen, a Wisconsin mental health provider, every woman who's thought of doing egg donation "asks: What about when the children say, you're not my real mother'? Patients say, 'I don't want somebody else's body part inside me, *it won't be my real child* inside me. It's like being a science experiment. How will I care about it? How will it care about *me*?' These days, doctors tell patients to inform the child about the biological mother. But patients usually say, 'I don't really want to tell the kid, because it's going to reject me!' Donor eggs are generally very private. The parents don't want the information to get out. It feels 'other'; too far away; weird. Adoption feels much more open. Yet, people can adapt."

Since the US is one of the only countries that legally permit individuals to buy and sell human eggs, most prospective parents here who use other people's eggs buy them from strangers, leading to a proliferation of egg donor agencies.

Nonetheless, many women resist using donor eggs, only wanting a child who is, genetically, 100% linked to them. "The hardest thing we struggled with was when to throw in the towel and pursue having a child who is not biologically related to us," sighed Roxanne, a 41-year-old Michigan Jewish woman who initially feared IVF, through which she finally had a son, and who is now seeking a sibling for him.

Patients wishing to buy eggs can pick from ever-growing assortments. Websites proffer drop-down menus of women selling eggs, letting buyers click on hair and eye color, height, weight, race, ethnicity, education, and other characteristics. But these sellers themselves receive large injections of hormones, facing potential risks of ovarian hyperstimulation syndrome (OHSS).[1] Critics argue that these markets are too capitalistic, risk exploiting young egg sellers, and constitute modern forms of eugenics.

Doctors and patients thus grapple with quandaries including *whether* to use another person's eggs; *when to* use them; *how* to decide (how much to consider the age of each member of the couple and the rights and well-being of the parent vs. the unborn child); *whether to buy strangers' eggs* (or seek to use a friend's or family member's) and, if so, whose; *whether to tell others* that an outsider's egg was used; and whether such sales should always be *anonymous*. To avoid needing others' eggs in the future, young women also face dilemmas of *whether to freeze* their own eggs.

Whether Other Women's Eggs Are Needed

In 2016, tabloid headlines around the world announced that a 72-year-old Indian woman had given birth, using IVF and another woman's egg. She and her elderly husband had long prayed for a child. She named the child Arman, meaning "wish" in Hindi.[2] But should infertility doctors provide these technologies to such older patients, and if not, how young should women be?

Increasingly, many women in their 40s and even 50s want children, having delayed childbearing to pursue careers. But older patients have higher rates of treatment complications (including preeclampsia, diabetes, preterm and very preterm delivery).[3,4] For women aged 44 and 45 who use their own eggs, birth rates are only 1.4% and 2.7%, respectively.[5] Many women who want to give birth have little choice but to purchase another woman's eggs.[6,7] Most donor egg recipients are 41 or older, and about a quarter are over 45. Among women using their own eggs, 9.1% were 41–42, 3.7% were 43–44, and 0.9% were 45 or older.[8]

The risks involved make these issues controversial. In 2004, the American Society for Reproductive Medicine (ASRM), the professional organization of infertility doctors, stated that "postmenopausal pregnancy should be discouraged" but that physicians should "carefully consider the specifics of each case."[9] In 2013, the ASRM expanded its guidelines to say that providers should implant embryos in women over 50 only after medical evaluation, discourage women over 55 from doing so, and counsel prospective parents about the risks.[10] The ASRM has not recommended upper age limits for women using their *own* eggs but has stated that physicians should develop "explicit" evidence-based policies and "may refuse to offer a treatment they see as futile or having a very poor prognosis."[11] This organization defines futility as less than a 1% likelihood of a live birth and "very poor prognosis" as less than a 5% chance.[12] The ASRM adds that in these cases, "[r]eferral information should be offered, if appropriate" and that "[c]are should not be provided solely for the financial benefit of the provider or center." Providers "may treat" such patients after assessment of risks and benefits and "fully inform[ing]"[13] patients with low odds of success. But key questions emerge concerning how providers view these issues, whether and when each of these scenarios occur, and whether these guidelines are followed and/or should be stronger, expanded, or more specific.

Other countries vary as to what age limits, if any, they permit. The British Human Fertilisation and Embryology Authority does not specify an upper age limit for treatment. Rather, clinics make their own determinations.[14] Australia bars IVF after the average age of natural menopause, "usually interpreted at 52 years of age."[15] Jurisdictions that publicly cover IVF costs also differ about maximum age limits. In 2010, Quebec decided to reimburse up to three IVF cycles but did not specify a maximum age. Older women with very poor prognoses consequently received treatment, prompting plans to alter the legislation and cap the age at 42, though still allowing older women to receive treatment if they pay out-of-pocket.[16]

Yet no prior research has examined how providers perceive these issues—what challenges, if any, they confront and how they respond. Though the ASRM has recommended that assisted reproductive technology (ART) providers develop policies concerning both age cutoffs and futility, I soon saw that clinicians in fact wrestle with how to view and make these decisions.

For women using their own or others' eggs, physicians vary in their cutoffs based on several types and levels of concern—from the patient's potential future regrets to the future child's welfare. Clinicians may follow clear age cutoffs, but these range. For women using *donors'* eggs, some doctors follow the ASRM's guidelines, accepting women until age 55, while other physicians set lower or higher ceilings. Providers also differ in how rigidly they establish such thresholds.

Occasionally, for women using their own eggs, infertility doctors, such as Henry, at a Midwest medical center, are cautious and conservative. He believes that physicians may "treat up to age 42, and then in rare circumstances 43, if the patient meets certain criteria that make us think she has a reasonable chance of pregnancy. With donor eggs, I will treat up through 47."

For women using their own or others' eggs, other doctors are more lenient, setting higher cutoffs. Edward, an Ohio reproductive endocrinology and infertility specialist (REI) in a hospital-affiliated private clinic, set a cutoff "at age 45 for a woman's eggs—essentially it goes to zero after that. . . . We give them a list of other clinics that might be willing to consider that."

"The controversy enters with donor eggs," Edward continued, "because it could be extended to almost any age. It gets down more to the health of the baby, and the mother and her life expectancy. The ethical guidelines used to stop at the natural age of menopause, but that's now been omitted from the ethical statement. Some data suggest you can go further. We've extended our age cutoff to 55 using donor eggs. Other clinics go beyond that"— highlighting how much doctors vary.

Providers vary in whether their limits are fixed or flexible. Peter, the West Coast IVF specialist, follows inviolable caps. "We have an upper age limit for women. We want them to deliver before they turn 55. So, they must be no older than 54. If they are over 50, they cannot have any medical problems. So, you can have medical problems and be 49, and we'll still do the transfer, but if you have any medical problems then you are ruled out at 50."

Clinicians who avoid such rigid thresholds then grapple with how exactly to decide case by case. "We discuss any patient over age 45," Marvin, a Massachusetts infertility specialist, said. Yet questions arise over what specific criteria to use. Physicians may not have clearly established cutoffs but feel "uncomfortable" treating certain patients beyond a certain age. "If the woman does carry the pregnancy, should there be some age limit?" Jill, an Illinois REI, wondered. "She can use donor eggs until she is 102, but it's a little hard for her to carry a pregnancy. We don't have an absolute age limit, but are more *comfortable* with somewhere around 50."

Recently, given technological advances, clinicians are also increasing their cutoffs but vary in how fast or slowly they do so, partly based on several subjective issues and uncertainties. "It's arbitrary," Valerie, the psychotherapist and "single-mom-by-choice," concluded. "50? 45? 55? Where do you stop it? I don't know."

At any age, women's health status also varies. Several factors can affect how old is "too old." "It depends on their health," Bill, a New England infertility doctor, said.

Yet, physicians differ in how they interpret the relevant evidence, and may be biased. Many physicians let patients try, even when the likelihood of success, based on assessments of the patient's reproductive functioning, is miniscule. "Some patients say 'I don't care if my FSH [follicle-stimulating hormone] is 20 and I'm 44. I want to try.' At an FSH of 10," which is relatively high, "a lot of clinics will go to donor egg," Bill continued. "We don't do that. We say, 'we may have to cancel the cycle based on the number of eggs. But if you want to try, we'll let you try. Some people . . . would rather have tried and failed, than not to have tried at all. We're not going to use the clinic's statistics [reporting requirements] to deny somebody what they want to try."

FSH levels indicate whether follicles are maturing and might thus release eggs. Doctors thus use these levels to predict the odds of success, but many women want to proceed, regardless.

Physicians might then provide treatment but seek to ensure that patients fully grasp the low odds. "If we give them adequate informed consent, tell them how dismal the pregnancy rates are, and they still want to do it, we'll do it," Calvin, a southern California REI, said.

Bill, Calvin, and other doctors who lack rigid age cutoffs can, however, then clash with patients. "The art of medicine is a judgment: whether or not you think this person really understands it," Roger, a New York REI, said. "We don't have a rule of thumb or cutoff, but strongly emphasize the evidence: that the chance of having a baby at 45 . . . is around 1%. Some patients say '$10,000 for a 1% chance? I should adopt.' But I have done it for patients. We do get crackpots who say, 'I feel like I'm 25.' I say, 'That's all well and good, except you are 45!' What you feel like and how many push-ups you can do has nothing to do with your ovarian age. I don't want to be accused of taking their money inappropriately."

While some doctors are flexible but require that patients fully comprehend these limited odds, other physicians are less circumspect. "Guidelines aren't always followed exactly," Steve, a Virginia private clinic REI, concluded. "Should they be more specific, or left as it is: up to the practitioner?"

Can Fathers Be Too Old?

Clinicians vary, too, in whether they consider the father's age, either by itself or in combination with the mother's age, and, if so, how. Many providers do not consider the father's age, arguing that such men could potentially have children on their own, if their female partner were fertile. "The 30-year-old woman with the 75-year-old man has existed throughout history," Steve argued.

Other providers reject older men but only if the female partner is also too old. "One couple came in: the woman was 64, and the guy was 78," Marvin, a Massachusetts REI, recounted. "The guy came in a wheelchair and had cancer and was dying, for Christ's sake! They wanted to use a donor egg. We did not treat them." Patients may request to have children without fully thinking through the long-term implications.

Usually, clinicians consider the father's age only in combination with the mother's. By weighing the father's age and establishing a maximum cutoff, providers increased the likelihood that at least one parent would be alive to raise the child. "Generally, their ages have to add up to under 100," Diane, an IVF nurse, said. "We came up with 100 because we don't want gender bias, but want a parent around to raise the kid."

Yet the thresholds for these combinations range. "Some clinics won't treat you if your combined age is more than 80," Nicholas, a New Jersey REI, said. "Or 90. Or 110. Increasing life expectancies can also raise the cutoff over time. Every couple of years, it goes up."

Providers confront questions, too, of whether this cutoff (like that for women) should be rigid and, if so, to what degree. Marvin, for instance, establishes soft, rather than "hard," lines: "How old is too old? A guideline of a combined age of 105 is not a *hard* line, but if you go over that line, you need a compelling reason. We're often very unsure of what decision to make, so we do the best we can."

Other clinicians set maximums for whichever partner is older. For Brenda, a Vermont infertility psychotherapist, "One partner must be under 55. For carrying the pregnancy yourself, the age cutoff is 50 because of obstetrical risks. At 55, you have a reasonable chance to see your child into adulthood. But . . . just because you're alive doesn't mean you're going to be a healthy, active parent."

Providers also face dilemmas of cutoffs for prospective mothers who use other women's eggs as well as wombs. Though not carrying the fetus, these prospective mothers will nonetheless, as older parents, be raising the child. "Ideally, if the mother is 55, and doesn't actually carry the pregnancy, she has to have a young husband," Jill felt strongly. "Just as when an old man has a young baby. The age *combination* is to make sure the kid has a parent."

How to Decide

"It isn't *my* family being created, *it's theirs*," said Ginger, an IVF clinic psychotherapist who underwent ART for her first child but adopted her second.

"I help that person move toward whatever decision they need to move toward. I don't judge. I just question." Providers often struggle with how exactly to make these decisions and how much to respect and follow patients' autonomy. Many providers feel that the patient alone, not the doctor, should decide—that the patient's autonomy is paramount—that they themselves would not do what the patient is doing but should not stand in the way.

Providers may feel that they don't know whether they themselves would use donor eggs. "I try to keep an open mind," Paulette, an IVF clinic psychotherapist, said, "because unless you are put in that position, you don't know what you would do. Who am I to say?" Psychotherapists, in particular, may aim to remain non-judgmental and non-directive.

Providers struggle with how much to weigh not only the parents' ages but the future child's rights and well-being—the odds that one or both parents will survive to raise the child through early adulthood. Parents who are 50 when the child is born have about a 15% likelihood of dying before the child is 15. Mothers and fathers who are 60 have a 20% and 30% chance, respectively, of doing so.[7] Advanced parental age itself can also hinder children.[18] Chronic disease or death of a parent can stress offspring at any age but especially adolescents, contributing to behavioral and mental health problems and substance abuse.[19] A child with elderly parents may also have to care of *them*, which can be stressful.[20] Parents using IVF after the age of 40 may have less physical energy to raise children.[21]

Alternatively, many clinicians weigh the child's welfare relatively little—essentially considering the mother's age alone. As an REI, Marvin, "probably *would* treat a 48-year-old woman with a 78-year-old man. . . . A 78-year-old guy could have started a relationship with a 50-year-old woman, and leaves her money. . . . We are pretty liberal about most of these things. The child doesn't take precedence [over the patient], but we need to consider that. If a kid is coming into an abusive relationship, or mental or drug abuse is involved, we have to look at them." Still, these decisions generally do not involve "abuse" but rather possible suboptimal future childhood experiences. In situations less extreme than clear abuse, more nuanced and difficult questions arise.

Whether to Use Other Women's Eggs

Countless patients resist using other women's eggs, seeing these as representing failure, defeat, and loss of a dream. Older women with less viable eggs wrestle with whether to accept or reject the need for others' oocytes. Occasionally, patients readily proceed to use these. "Donor eggs aren't a problem for me," Wendy, the 37-year-old Catholic secretary, told me. "I always really wanted to have kids. I always assumed I would." But she had ovarian cysts. Her husband had a low sperm count. She failed four IVF cycles at four different clinics and has already spent $80,000 on treatment. She has now found a new doctor and wants to try IVF one last time—all she can afford—using donor eggs. If it fails, she would adopt. "I will still love any child, whoever the parent is. If it comes from me, that's great. If it doesn't, that's OK too. It's not going to be any different."

Many patients at risk of transmitting dangerous mutations, for whom embryo screening has failed, also consider buying other people's sperm or eggs. Sally described a woman who "terminated two pregnancies for CF [cystic fibrosis]. Her choice then was to use a sperm donor. I can't picture myself doing that. But she and her husband did it. I would have gone through a lot more hell before I did that."

A woman with both infertility and risk of transmitting a severe mutation may also use others' eggs, rather than trying to screen embryos, if obtaining enough of her own eggs is hard. Paulette, an infertility clinic psychotherapist, described a patient with "two other children affected with a genetic disease. She's now 30-something. Do we do PGD [preimplantation genetic diagnosis]? She was on the border: she might get pregnant on her own, but will definitely have fewer eggs for PGD. But her FSH is borderline, so she's probably not going to get a lot of embryos. She said, 'If I'm spending money for IVF and then PGD, and may end up with nothing, I will increase my chances and do donor eggs.' And that's what she did. I don't know what *I* would have done if I were her."

Yet far more commonly, women seek a child biologically related to both members of a couple, who can serve as a concrete, physical blending of two individuals. Use of a gamete donor can clearly allow for such blending psychologically, but many prospective parents sense a difference. "You have to let go of something you grew up thinking you were going to have," Paulette, the psychotherapist, said. "Patients feel that with egg donation, they've hit the end of the line. A lot of women have undergone years of fertility treatments, inseminations, IVFs."

Patients' hopes and expectations can die hard. Brenda, the Vermont infertility psychotherapist, regularly sees women "who feel, 'I'm 44, and run 100 miles every day . . . I'm fine.' I tell them: 'But your ovarian reserve is nil. . . . Look at the blood test.' They say: 'I don't believe it. Look at me, I'm wonderful!'"

Commonly, patients focus only on successes, ignoring failures. "You always hear the story about someone who was blah-blah age, who had one good egg left, and it worked," Valerie, the psychotherapist and single-mother-by-choice, observed. "You don't hear the crash-and-burn stories."

Clinicians wrestle with what to do if patients do not accept these cutoffs. Providers may refer such patients to colleagues, but how often and whether these colleagues end up treating these patients are unknown.

Some patients who do not want to accept lowered success rates also dissemble about their year of birth. "Patients get around the guidelines by lying about their age," Diane, an IVF nurse, reported, "One woman snookered the physicians—she got in and was years older than she claimed. She gave birth in her early sixties!" Unfortunately, clinicians may fail to detect such deceptions.

Potential parents may also oppose using others' eggs because these trigger visceral feelings of "yuck." "Women have fantasies of a monster growing inside them, an alien," Diane continued. "That's normal. But if people don't *know* that these thoughts are normal, it can be very scary. A lot of women just close their eyes and do it. But others refuse or freak out. They haven't really processed it all—dealing with choosing the donor, and what if the donor doesn't work out. And most people do this in isolation."

Psychologists and bioethicists have described the "yuck response," a sense of disgust that can affect peoples' attitudes and decisions in other medical areas, such as abortion.[22] These strong emotions, some conservatives argue, are a legitimate basis for ethical decision-making.[23] But these quick visceral reactions can merely reflect personal distaste that requires careful and explicit scrutiny in the context of relevant ethical principles.[24] Historically, marriages between individuals of different religions or races, or of the same sex commonly provoked feelings of "yuck" but are now widely accepted, reflecting logical extensions of equal rights. As the Nobel Prize–winning psychologist Daniel Kahneman has demonstrated, individuals generally rely on fast intuitive thinking rather than slower careful analysis of data in making many decisions but may then be wrong.[25]

Other women fear that a foreign egg inside them will somehow infect or injure them. Even Ginger, an infertility clinic counselor who herself underwent ART, did not trust egg donors and therefore adopted her second child. "I said to my doctor, 'How do I know that the person who donated this egg doesn't have HIV?' He said, 'We don't think that the virus can live in an egg out of the environment.' I said, 'That's not good enough for me. I've got a kid and a husband who need me.'" Given that egg donors' medical problems could harm both the recipient and the child, egg recipients need to trust the donor. Yet establishing full confidence can be difficult.

Religion can also prompt patients to reject use of others' eggs. "My best friend and her husband are Christian," said Yvonne, who is Jewish, and whose husband is Catholic. For Yvonne's friends, donor eggs, "were not options because of their religious beliefs" that the baby would be from "another woman. It reflects unfaithfulness or infidelity."

Many women seek desperately, at all costs, to use their own eggs, rather than someone else's. For many women, the pursuit of a biological child can come to feel like an "addiction"—an intense biological impulse that seems to them beyond conscious control. "Women often get obsessed, having lost all reason," Diane, an IVF nurse, said. "They'll describe it in *addiction* terms: 'It's like heroin.' Most people feel 'Just one more time.' . . . It feels too terrible to contemplate what would happen if it doesn't work—not having done everything they could do.

"If they stop, or even pause, all those feelings of fear, failure and terrible loss will come crashing down. It's like gambling—they just throw good money after bad. Women say, 'I'm doing this until I'm in menopause, until the very last egg drops.' Or, 'Here: meet our $100,000 kids!' . . . *People don't want to get off the fertility train.* The only way of coping is: the next cycle, the next cycle." Hope beckons, as long as one can afford it.

Rather than deciding to cease altogether, patients may take respites that vary in duration and may continue indefinitely. "My body needed a break," Wendy, the Irish Catholic Chicago secretary, said after her fourth IVF cycle in five years resulted in miscarriage. "It was very hard . . . I didn't like being left to flounder."

Providers' Approaches

"When is enough enough?" Steve, the Virginia infertility doctor, asked. "Where do you draw the line? How do you tell people, and get them to stop if they want to do acupuncture, yoga? 'Why did you even come to my office?' They don't want to hear it. They get mad."

When ongoing treatment has little, if any, chance of success, physicians confront questions of not only *when* to stop pursuing a procedure but *how to weigh these competing considerations and decide.* Clinicians respond to these challenges in several ways, from agreeing and providing treatment to avoiding conflict by simply following patients' wishes, rather than refusing outright. Doctors may refer patients to colleagues who will fill such patients' requests. Yet, in the long run, such colleagues who agree to these patients' desires can generate problems. "That's the art of medicine," Steve reflected. "But it leads many practitioners to avoid saying 'No'—which devalues or discredits the whole process. I say, 'We won't do this here, but I know you can find somebody elsewhere who will do it.' Eventually, they'll find someone who'll do it. *It is a consumer-driven market."*

At times, for patients with low chances of success using their own eggs, he and other doctors will compromise, providing treatment but only to limited, proscribed extents, to help these adults avoid retrospective regrets. "For people with less than 5% chance at pregnancy, I will sometimes agree to a limited attempt," Steve explained. "Because part of being a doctor is also just caring for a patient. Doing two cycles with very low likelihood of success allows them to move on with their lives. Treatment is very low risk. Is that such a bad thing?"

Providers may gently, gradually explore and negotiate with patients the potential use of other women's eggs, recognizing that such acceptance can take time. "My doctor doesn't rush me into anything," Wendy, the Irish Catholic secretary, said. "He gives me options, and lets me think about it. I was dead set on just using my own eggs. He mentioned donor eggs, but didn't go further because he was trying to see my reaction."

Mental healthcare and other providers can also aid patients by reframing and resetting these hopes, which can be tricky. "Most patients say about using donor eggs, 'this isn't my first choice.' I say, 'Of course it's not,'" reported Ginger, an IVF clinic therapist who used ART for her first child and adopted her second. "You wanted a child who would be genetically connected to both of you. But it doesn't look like that's going to happen. So, let's look at what it's like to have a pregnancy with a donor egg.' When you use a donor egg, it's like there are three parents. Most people think they're just going to have a baby. But life unfolds however it does."

Other physicians, however, lead parents on, offering hope despite slim odds of success. "I have a problem when I see patients doing IVF after IVF after IVF, and the chances look really low," Helen, the Wisconsin psychotherapist, said. "Some doctors aren't really being responsible. Patients say, 'We made one egg last time. We're gonna do IVF with one egg.' I feel, 'uhhh, no.' One doctor in our town will tell women, 'It's enough! Stop! Move on!' Patients either love him or hate him for it. . . . He shoots from the hip, and tells it like it is. I've often sent patients to him, because he will say 'No.' More doctors should say 'No' or 'This is not good medical practice.'"

Who Decides Whether Women Use Their Own Eggs?

Providers also vary as to *who* should ultimately make these decisions—a patient, a provider, or a committee. Patients may look to the doctor to judge when to stop, but that discussion can be hard for providers, who grapple with the extent of their responsibility. "Patients expect that the doctor will tell them if they should move on," Helen said. "They want and expect the doctor to take responsibility. These days, doctors don't usually do that. They often think it's not their job—they can tell people the odds, but patients must make their own informed decisions."

In the end, many providers feel that these decisions belong to the patient. "It's not our position to make the decision for them," Steve, the infertility doctor, explained.

Yet, ultimately, given that ongoing interventions may fail and have risks, other providers feel that the patient should *not* make the final choice. These clinicians then confront questions, however, of who should do so. Clinics vary in how much of a formal decision-making process mechanism they use and whether they consult with a quality assurance or other committee. Hospital-based practices may get input from formal ethics committees. "Those decisions are not made by one person, but go to the hospital ethics committee," Diane, an IVF clinic psychologist, said. Age cutoffs, along with gender selection, can be "the major issues" that such committees confront and can be difficult.

Other providers may hesitate to consult an ethics committee and even avoid doing so, wary of that group's potential recommendations. "We were going to check a couple of things with the ethics committee, but thought these weren't going to go through comfortably," Paulette, the IVF clinic therapist, said. "So, we just backed off, and did our own review of it."

Providers often draw on their "comfort" or "gut feelings," recognizing that, though they are confronting fundamental moral dilemmas, they lack any formal ethical framework for doing so. At times, clinicians incorporate their perceptions of public opinion—how the public would respond—asking whether their decisions might evoke controversy. "If we think the general public will ask questions, then we think *we* should ask questions," Marvin said. "So, we discuss it. We're reasonably liberal, but discuss it, and decide what kinds of consent and psychological assistance the patient is going to need—the counseling and legal issues."

Physicians often flounder. "Just because we *can* do something, *should* we do it?" Nicholas, a New Jersey REI, asked. He avoided an explicit age cutoff policy, partly because older patients might then complain of age discrimination. "One clinic's front office received a phone call, and the caller said she'd like to come in for IVF treatment. The staff member said, 'Well, you're past our cutoff age—you're 48. The doctors won't treat you.' The caller was a judge, and filed a federal lawsuit for age discrimination. It's hard to make policies, especially if it's considered an ADA [Americans with Disabilities Act] issue. We try to stay out of that as much as possible."

To counter such potential claims of bias, he and other clinicians may give patients other, medically-based reasons, rather than age, for not offering treatment. "If a couple is 70 and 50, we don't want to do it. But we're not going to tell them it's based on age," Nicholas explained. "We find a way not to do it. We say, 'You're not going to get pregnant.' We counsel them, have them see a fetal medicine specialist, and say, 'Your only option is egg donation or a surrogate.' We try to wiggle out of it as much as possible." Presumably, given these legitimate medical reasons, accusations of age discrimination would be unfounded. Yet doctors still encounter questions of what exactly to say when declining to provide treatment based on age. Uncertainties persist. As Jill, the Illinois REI, sighed, "I'm not sure in my heart what the right answer is."

Still, despite these ambiguities, providers generally oppose any government regulations on this issue. Many providers remain ambivalent, recognizing the potential advantages of having a firm line but wary of regulations. "I'm torn," Valerie, the psychotherapist and single-mother-by-choice confessed. "I'm not big for regulating things, and it's hard to know how to educate people."

While guidelines thus state that clinicians should develop "evidence-based" policies, empirical facts alone do not answer these moral conflicts and uncertainties. When aware that age can limit fertility, patients differ on whether to accept, minimize, or deny this fact or lie about their birthdate.

The ASRM now states that clinicians can refuse to offer treatment if "they have a substantial, non-arbitrary basis" for thinking "that child-rearing will be inadequate" or "significant harm is likely."[26] The ASRM offers examples: "uncontrolled or untreated psychiatric illness, substance abuse, ongoing physical or emotional abuse." But providers wrestle with potentially less extreme situations—for example, when the probability is relatively high that a parent may die before the child is 18 years old.

John Robertson, the late University of Texas law professor and ASRM Ethics Committee member, argued that only imminent abuse and harm to the child should outweigh a patient's desire to have offspring since the child would still be better off than if he or she had not been born.[27] But children also have rights to an "open future."[28] In helping, through their direct decisions and actions, to create the infant, physicians arguably have certain professional duties and responsibilities, especially if enabling the birth of the child into a home that they strongly suspect to be deleterious. Physicians should at least carefully consider these issues.

Providers may fear accusations of age discrimination, yet such charges are not necessarily justified. The Age Discrimination Act of 1975 bans discrimination in institutions receiving federal financial aid[29]—such as hospital-affiliated clinics receiving federal assistance. But only about a third of IVF clinics are affiliated with such academic medical centers.[30] A physician's decision not to offer treatment because of an older patient's medical situation, based on that patient's lower likelihood of success, would presumably *not* constitute unfair discrimination. Nonetheless, despite ASRM recommendations that clinics develop "explicit" policies, some lawyers have advised infertility clinics *not* to have clear written policies about age cutoffs since doctors would then be legally obliged to follow these policies without any flexibility.

Many providers draw on their own "gut feelings" and "comfort," which may not reflect objective data. Unfamiliar situations may feel unacceptable or distasteful but in fact be ethically sound.

A doctor may refer older prospective patients to a colleague, rather than offering treatments; but such a colleague, if then treating these patients, may in fact be violating ASRM guidelines. Providers who decline to treat a patient because of age should carefully explain to the patient the relevant ASRM guidelines and the very low rates of success. Proceeding otherwise can fuel misunderstandings.

The ASRM states that clinicians should not provide futile treatment "solely" for their own financial benefit, yet some doctors may nonetheless do so. In its guidelines, the ASRM should, however, delete the word "solely." Providers may not always fully convey or highlight to older patients the very low chances of success and the full costs and risks involved. A vast continuum exists, however, between very loose current guidelines and formal government regulations. The ASRM could, for instance, address fathers' ages and encourage physicians to assess carefully patients' appreciation of these issues to ensure that the latter sufficiently comprehend these limitations.

How to Find Eggs

A good egg can be hard to find. Patients who opt for someone else's eggs must subsequently determine *whose* to use—family members', friends', or strangers'. Selecting and obtaining eggs from each of these types of individuals can pose challenges. Plans to use a particular donor can easily collapse.

Using Friends' or Family Members' Eggs

"The most difficult issues are the recruitment of good donors," Paulette, an IVF mental health professional who ran her clinic's donor egg program, said, "and the times when recipients don't get pregnant."

Unable to use their own eggs, patients may seek a sister's, in order to maintain a biological relationship with the future child. Yet creating a child through a family member's eggs can blur boundaries between these three individuals. Paulette described IVF patients whose sisters needed donor eggs. "We have a patient now going through donor egg who, several years back, wasn't ready to have a family of her own and donated her eggs to her sister. Her sister got pregnant and had a child. My patient was starting to feel she was getting to the point of having a child of her own, and her sister wanted another child. My patient said, 'OK. I'll donate to you and then worry about myself.' But my patient's FSH was elevated. Now she herself is going to need donor eggs. That can cause family friction: '*You took my only good eggs!*' "

Consequently, in a later case, Paulette refused such an arrangement. The patient "wanted to do a cycle, and donate half of the eggs to her sister and keep the other half for herself. The patient wanted her half of the eggs fertilized by her own husband. The eggs donated to her sister would be fertilized by the sister's husband. But what if my patient didn't get pregnant, and her sister did, and this sister had a child from my patient's eggs? We're trying to build, not *destroy* families."

Providers' prior clinical experiences can thus influence these decisions. Paulette and other clinicians might potentially consult ethics committees concerning not only *whether* to use other women's eggs but *whose*. Yet, partly afraid of the answer, many providers eschew such boards. "Instead of going to the ethics committee, we just decided that we're *not* going to do this," Paulette explained. "We told the patient that she would need to do an IVF on her own, have a baby, then do another cycle and donate to her sister. She came in, had a cycle, did get pregnant, and had a baby. She then did another cycle and those eggs went to her sister. Unfortunately, the sister didn't get pregnant. But at least we weren't trapped in that dilemma."

Patients may not fully appreciate all the possible difficulties. "It's very hard to figure out how you're going to feel about a situation when you've never been in it," Paulette observed. "People will say one thing, but feel differently when it happens. They say, '*Oh yeah, I understand*,' but might end up resenting [their] sibling." Confusion can arise, too, concerning who is related to whom. The child's biological mother would also be its aunt.

Giving eggs to family members can trigger additional complications since genetic tests on one member can reveal information about others, posing challenges for clinicians. Diane, the IVF nurse, described an early-menopausal patient "who needed donor eggs, and wanted to use her sister's. We tested them and they had a fragile X repeat in the gray zone." This mutation affects about 1 in 4,000 males and 1 in 8,000 females, varies in length and thus severity, and makes X chromosomes look "fragile."[31] It can cause intellectual disabilities, protruding ears, an elongated face, enlarged testicles in males, and early menopause in women. Diana wondered, "will the younger sister, in her mid-20s, also go through early menopause? We may be taking the younger sister's eggs, giving them to the older sister, and the younger sister will never be able to have children! The younger sister is engaged, and isn't willing to move up her wedding date, and try to get pregnant earlier. *What do we do?* We encouraged the younger sister to go through an IVF cycle and freeze her own embryos before she donates to her older sister. Basically: 'Do no harm.'"

Patients frequently attempt to use friends' eggs but then encounter further medical, social, psychological, and ethical hurdles. "Many doctors don't recommend using friends as egg donors," Francine, the Miami legal assistant, explained. "The donor might feel she's got some prior claim over the children."

Buying Strangers' Eggs

Walter and John, the gay couple described in Chapter 1, for instance, debated whether to buy eggs from a laid-back musician or an ambitious tennis star.

Countless prospective parents consider buying human eggs from strangers but then face dilemmas. Until the late 1990s, most patients accepted the first donor presented to them.[32] But approximately 10.5% of ART treatments in the US now use donor eggs,[33] and many foreigners travel here to purchase them, fueling a rapidly expanding and evolving market. In the US, both clinics and specialized agencies now broker eggs. Egg buyers usually try to match certain characteristics (ethnicity, physical appearance, intelligence, and interests).[34] In a study, mostly of students, the majority would choose tall, middle-class, Caucasian egg donors.[35]

The fact that women and agencies selling eggs earn relatively large fees—up to $50,000 for an egg—can create problems. As Helen stated, "It's big money." The ASRM has issued guidance about compensating, recruiting, and communicating risks and benefits to egg donors[36]; but doctors often fail to follow these recommendations. The organization states that "it would be prudent to limit donors to those who are 21 or older and have the emotional maturity to make such decisions."[37] Yet most clinicians actually advertise for younger donors.[38] In 2007, the organization recommended that donors do not supply eggs more than six times in their lifetime[39] since donation carries medical risks, including about a 1%–3% risk of OHSS,[40] and questions remain as to whether fertility drugs that donors must take increase risks of cancer.[41]

The ASRM suggested that payment should not depend on "the number or outcome of prior donation cycles, or the donor's ethnic or other personal characteristics—since payment is based on the donor's time, not the eggs themselves." Yet most clinics that advertise for eggs are seeking women who have donated before. The ASRM further stipulates still that "total payments to donors in excess of $5,000 require justification" and "sums above $10,000 are not appropriate."[42]

Outside of professional medicine, and unregulated by government or subject to professional codes of conduct, agencies operate as third-party companies.[43] Though agencies can agree voluntarily to follow the ASRM's guidelines in exchange for a listing on the organization's website as a professional endorsement, over 40% do not; and most that do so nonetheless in fact violate ASRM guidelines.[44,45,46,47,48] Among agency and clinic ads on Craigslist, 81% and 96%, respectively, have failed to comply with ARSM guidelines, including 85% of the agencies and clinics that were Society for Assisted Reproductive Technology (SART)–registered.[49] For instance, most egg donor agencies fail to follow the ASRM's guidelines against varying compensation based on donors' traits: 58.8% explicitly proclaim that they pay more for certain traits, and an additional 17.6% state that certain traits are "preferred" or "in demand." The most commonly mentioned compensated trait is prior donation success. While the ASRM recommends not using donors under 21, almost half of agency websites seek younger donors.[50]

The American Medical Association's guidelines require that all health websites present compensation amounts, but only 26% of egg agency websites mention psychological and/or emotional risks. None mention possible cancer risks. Most fail to mention short-term risks, and 88.7% do not mention possible dangers to future fertility.[51] Agencies ignore guidelines more than clinics do.[52]

Rene Almeling, a Yale sociologist, interviewed staff at two egg donation agencies who described encouraging donors to present "properly feminine profiles."[53] Agencies also try to persuade potential donors that donation is "natural" and "fulfilling," luring these young women through both monetary and non-monetary benefits that may inappropriately focus these individuals on personal gain, rather than potential medical harms.[54] These companies portray egg sellers in familiar, reassuring ways as altruistic "donors" to assuage buyers' fears about using strangers' eggs. These emotional appeals—trying to sell donors' personal, physical, or artistic characteristics and intellect, passion, sense of purpose, and general temperament or demeanor—can distract potential parents from appropriate assessments of relevant issues. These appeals can also promulgate misunderstandings that these various traits are somehow directly inherited, though non-genetic factors play major roles. Agency websites are thus placing commercialization over professionalism, promoting eugenic searches for "the best genes."

Yet strikingly, these concerns have failed to alter professional or other guidelines or policies. These issues are especially critical since the ASRM settled the Kamakahi case and eliminated caps on compensation. As compensation for eggs increases, more women will likely want to sell them.

Helen sees "market forces" changing the business. "Egg donor agencies are really taking over." Agencies have made more information about prospective donors available. "Clinics used to not show adult pictures of donors," Helen said. "But now, prospective parents can see photos. A lot of patients want to."

Variations in Egg Agency Quality

Agencies range widely in how well they evaluate, screen, and prepare potential egg sellers, and providers differ in how they view and whether and how often they use these companies—from often to rarely. Doctors can recruit egg sellers on their own and/or use these intermediary companies. A few clinicians have had good experiences with these brokers. "I haven't worked with that many agencies," Ginger, the IVF counselor who underwent ART for her first child and adopted her second, said. "I've liked the ones I know."

Many clinics use agencies to give patients more choices—frequently through websites. "A lot of people want more say about who their donor is," Ginger continued. "Before, when patients wanted to do donor egg, some clinics would find the egg donor. . . . The patient would have to either take her or leave her."

But other clinicians have had mixed or negative experiences with agencies. Many providers prefer to work only with a woman they know and have previously bought eggs from. "Most reputable clinics," Helen observed, "choose their own donors." Agencies can pose concerns regarding informed consent, potential harms, justice (possible eugenics), and approaches toward women selling their eggs—a vulnerable group. These for-profit businesses often fail to fully inform, educate, or prepare these women regarding the process and risks involved and list women with certain characteristics (attractive to many potential recipients) who are in fact unavailable. Financial profit commonly motivates these businesses, frequently lowering their standards for egg sellers and how well these companies assess these women.

Screening Egg Donors

"With the downturn of the economy, more people are responding to ads for donors," Paulette said. "I'm getting more applicants, but not more people passing through the screening. It hasn't increased my donor pool. It's still difficult to try and recruit quality donors—people who are educated, have a good family background, treatable medical issues, and no strong family history of disease, and are doing it for the right reason. They would not be doing egg donation without compensation." The Food and Drug Administration (FDA) and the ASRM require testing for various infectious diseases, such as HIV, hepatitis, syphilis, and gonorrhea. Yet agencies may barely screen donors for other medical or any psychological issues.

Clinicians can face difficulty collecting and assessing personal and family medical histories and deciding where to draw the line. "People do not get through [our] screening process," Paulette explained, "if they have a strong family history for anything—a first-degree relative with breast cancer, diabetes, colon cancer, family psych history, medications for depression. One first-degree relative with depression, medicated for a long time and not situational, would exclude them. A donor who was depressed and on medication in her teens because of her parents' divorce, and is not now on meds is OK. That would be disclosed to egg recipients, but probably not [exclude] her."

Yet would-be egg sellers do not always disclose all relevant information, exacerbating concerns. "People may not tell the truth," Paulette continued. "So, we go through the medical history, then talk. We have picked up stuff. One donor came in and had donated elsewhere, and in screening we discovered that her father was schizophrenic. That's an automatic 'No.'"

Since egg donor agencies do not always screen women adequately, many clinicians probe further, independently evaluating all potential donors. "We test the donors a lot," Paulette said. "An agency said that I reject donors that have donated elsewhere. We send our donors' profiles to a geneticist, and do extensive psychological testing. We regulate it ourselves. *I want to be able to sleep at night!* If I personally wouldn't feel comfortable using this person as a donor, then I'm not comfortable having a patient use her. The doctors, too, always say, 'It's just *a gut feeling.*'" Providers frequently come to rely on their intuitions about potential egg sellers.

Clinicians may also exercise heightened caution and thoroughly probe for how many times a potential egg donor has previously vended her eggs. "We limit their donation to six times," Jill said, "both because of the potential increased risks, as well as issues concerning having too many related children within the same community." Yet other providers may simply accept agencies' reports of a woman's past history of egg sales and other relevant behaviors. "It's difficult to be 100% sure that donors from agencies keep track of the number of cycles," Paulette reported. "One donor told me she's never donated before. But she knew too much about egg donation. She seemed to think that the injections were gonna be a piece of cake. I said, 'Something's not adding up here. What's going on?' She then told me she 'had a cycle done elsewhere,' and 'wasn't happy' with the way they treated her."

Consequently, Paulette and many other infertility health professionals now regularly inspect records of women who have sold eggs elsewhere. But agencies do not always agree to provide this crucial information. "I check records every time I can get them—60% of the time," Paulette added, "because we use a lot of agency donors. . . . I've gotten the records in every case—except when the program won't release them to me."

In not always monitoring, for instance, the maximum number of times a woman sells her eggs, agencies thus fail to follow a key current guideline. "The guidelines say egg donors are only supposed to donate six times, but a lot of agencies don't respect that, despite what it can do to women's bodies," Valerie, the psychotherapist and single-mother-by-choice averred. "I've read on the Internet and seen books by donors who say they don't tell agencies what they've been through. Nor do the agencies necessarily ask. Major medical centers ask. For-profit agencies may not."

Since women wanting to sell their eggs may dissemble, validating their information can be tricky. Suzanne, an IVF clinic psychotherapist, found it hard to "determine the veracity of what donors tell you about their history, and their family's genetics. There's no external monitoring—determination of whether what they say is true about whether they've donated elsewhere and how many times. There are no registries. So, donors could still go from place to place—which puts them at risk, due to hormonal exposure. There is also the risk of consanguinity of the offspring, since it's very populated here." These offspring could potentially intermarry.

Since egg sellers can lie, clinicians often administer psychological tests, such as the Minnesota Multiphasic Personality Inventory; but these questionnaires may not always detect problems. Various standardized tests can assess aspects of these women's personalities, providing some additional information, but in the end have certain limitations. "The question of whether the person is telling the truth always has to be in the back of your mind," Ginger, an IVF clinic therapist, said. Providers frequently find reasons to probe further. "I see things on evaluations from the agencies that lead me to question donors more," Ginger continued. "On one egg donor evaluation, everything looked great. The father had been dead for a number of years. 'What did he die of?' 'He was murdered.' I would want to get a good idea about what happened."

In the end, trust is crucial. "Ultimately, the prospective parents who receive the eggs have to trust the donor," Suzanne added. "Recipients trust whoever is doing the assessment to determine that this donor is reliable, has good genetic history, and is doing it for altruism, not compensation. Mental health professionals worry a little more. A lot of us have kids the age of the donors, and would not necessarily want our own kids doing this. It's a conflict. A fairly low percentage of donors, especially egg donors, are accepted."

Though psychotherapists screen these women, the purpose and thresholds of these evaluations can also be murky. "Some mental health professionals [MHPs] feel that our evaluation is to rule patients out," Helen explained. "Others feel very strongly, 'No. *We don't want to be gatekeepers!* This is *educational.*'"

With egg donor agencies, "anybody can sign up online," Helen continued. "There's minimal screening. People can lie. Donors come in, and are surprised at the depth of my questions and the amount of detail. And I'm only asking them standard questions! They'll say they did a psychological test online, which is really unethical, because a psychological test is supposed to be monitored in person. They'll say, 'Well, I met with somebody.' Or: 'met with someone in the hallway for 10 minutes' or 'The medical doctor met with me and said, 'everything's going to be fine.' But the doctor didn't check whether the donor understood what she would be undergoing. Then the donor met with a psychologist or social worker, and it was much briefer. Or there was nobody. Or they went to the clinic, and signed up. Who does quality control? It's not hard to be put 'on the books' as a potential donor. They get often flown to a different clinic, and by then, potential parents are invested in them."

Potential Harms to Egg Donors

Women, selling eggs primarily for money, may not sufficiently grasp the risks, especially OHSS. These women are not always even informed that these risks can occur.[55] Websites advertising for donors tend to present the benefits but not the potential hazards.[56,57,58,59] Unfortunately, no long-term study of egg donors has yet been conducted, and no plans yet exist to conduct one, to know how often donors experience harms or understand that these dangers can transpire. Providers do not want to pay for such research. Egg sellers themselves often want to get the money and not contemplate the long-term implications—the ongoing possible residual hazards to themselves and the fact that they have helped produce children that strangers will raise. In Europe, egg donors are not compensated more than limited amounts for direct expenses and are usually sisters or other close relatives; and data from them are available but have limited applicability to the US, where sellers are paid significantly more, attracting more, younger, and different women.

Clinicians vary in how well they educate donors. "Most women who donate their eggs would not do so without compensation," Jill, an Illinois infertility doctor, stated. "Because it really is a big deal; with anesthesia, most egg donors—certainly ours—are well counseled. We go through why they should and shouldn't do it. Fewer than 10% ultimately donate. Most don't really understand what it's going to entail. Those who do, understand very clearly that there is some risk. Parents say, 'I wouldn't want my daughter to donate'—to be exposed to that level of risk."

Commentators have argued that egg donors may later have difficulty conceiving children for themselves. These dangers may be relatively rare but have not been well examined through longitudinal long-term studies.[60] Uncertainties linger, and egg donors have not been studied over time. "There are unlikely short-term risks of hyperstimulating, and being very sick—liver, lung problems, risks from anesthesia, bleeding, and infection," Jill continued. "You could require a blood transfusion or hysterectomy, or lose your ovaries. I tell donors, 'The data thus far suggest that it is not an issue, but it will take 10–15 years before we have a definitive answer.'"

Donors may also suddenly learn for the first time that they have medical problems and may then face threats to confidentiality. Potential egg sellers anticipate earning dollars, not learning about their own genetic risks or diseases. "Donors are looking for money, but can't say that. They have to say they're altruistic, or they won't get chosen. But most are financially motivated," Helen added. Unfortunately, "potential donors get screened, and find out they have a genetic problem that's going to affect their own future, *and* they're not chosen as a donor, *and* don't get paid! Nobody says, 'Here, go to this nice mental health professional, and process how you just found out you've got this problem.' . . . That's a bum rap."

Many agencies may disseminate egg sellers' personal, potentially stigmatizing medical information, threatening confidentiality. Agencies' online details about these women can impact both potential parents and their future children. "We've seen confidentiality problems," Anne, a feminist-leaning northern California infertility doctor reported. "According to HIPAA [the Health Insurance Portability and Accountability Act], we can give information we find about the donor only to the donor, not to the agency. But the agency then calls the donor, and asks her a lot of questions. She's young and vulnerable and gives answers. The agency then gives out that information—which could be about venereal disease, genetics or drug testing—to the recipients. That is probably not illegal, because the agencies are not professionals, so they don't have to behave according to professional guidelines or ethics!"

US laws bar healthcare providers and institutions, but not egg agencies, from sharing with others (including clinicians, prospective parents, and the public at large) personal information about a woman selling her eggs.[61] Many agencies post photos and extensive information about egg donors on websites.

Eventually, offspring created through purchased eggs might also want to find their biological mother, using publicly available online information.

Knowledge about specific egg sellers can also fuel expectations that off-spring will possess these women's desired traits, raising expectations. "If patients see the adult picture," Helen explained, "it might distance them from the child . . . if the child does not sufficiently possess these anticipated, expensive, purchased traits."

The fact that egg agencies are involved with medical procedures but not operated by medical professionals can create problems. Still, regulation of agencies poses challenges. "I would find some way to put the brokers and agencies out of business," Anne, an REI who opposes sex selection, said. "But you'd have to figure out how to do that. I would put this aspect of medical care into the hands of medical professionals, and out of the hands of business people!"

Agencies thus increase supplies and choices of eggs but range in quality and do not always adequately screen women, assess or document numbers of prior egg donations, or provide relevant records to clinics. To purchase eggs, many foreigners travel to the US from countries that ban paying donors more than limited compensation for time and expenses.[62] Since the recent elimination of price caps, more women will also probably try to sell their eggs. But while some clinicians carefully screen potential egg sellers, other providers don't. Women who want to sell their eggs may not fully grasp the potential risks and may end up learning for the first time about medical problems but lack resources for assessment or treatment. Agencies can also foster both misunderstandings about genetics and eugenic-like notions.

Agencies should more fully inform potential donors about risks, including medical problems for which infertility clinics will not provide treatment. Clinics themselves should also give this information, to ensure that women selling eggs provide appropriate informed consent which, unfortunately, at many agencies, not all of these women do. Yet as the psychologist Daniel Kahneman has shown, knowledge that people initially receive about a situation "anchors" and establishes a mental framework that powerfully shapes how they weigh subsequent information and make decisions.[63]

Stronger guidelines or regulations can set and enforce higher standards for agencies regarding advertising and informed consent. The Centers for Disease Control and Prevention (CDC), the FDA, other agencies, and states could require that these companies get licensed or become certified, report data, and follow clear standards. The ASRM could mandate that agencies register in order to work with its members and adhere to guidelines fully.

Doctors should also vet agencies better to ensure that these companies conduct appropriate screening, carefully track each donor's prior number of donations, and provide this information to clinics. Agencies could report data to the SART or CDC annually, as ART clinics do, including yearly numbers of women selling their eggs, numbers of times each woman has previously done so, and any complications. Agencies should also only list as available those women who are indeed available.

Which Eggs to Buy

Patients and clinicians must decide not only *where* to buy eggs but *which* specific ones to purchase, presenting dilemmas. Prospective parents commonly seek eggs from sellers with a variety of characteristics, from certain height, hair and eye color to race and ethnicity, raising ethical questions. "One couple was having difficulty getting matched because she was 5'0" and he was 5'5" and they wanted a tall donor," Jill, the Illinois REI, reported. "But they didn't want to tell the child at any point about the donation. The wife said: 'It's been hard for us being short. Since we're able to pick, we want our child to be taller.' But they didn't want to tell their child: 'We'll worry about that when the time comes.' I said 'that's not going to fly. At some point, people are going to check blood type.' Well, we *can't* worry about that when the time comes. I didn't have a huge problem with it, but it was not ideal."

"A Japanese couple wanted a Japanese donor," Jill continued, "but there were virtually no Japanese egg donors." They ultimately said they would take a Chinese donor. "I don't mean to be disrespectful," Jill told them, "'but I'd think that they're not really all the same.' It turns out, there's an egg donor agency that specializes in Asian donors, and they got a Japanese donor."

Both patients and providers range widely, based partly on geography. Some patients are simply grateful to have a child. "Our city is not the same as New York," Jill continued. "Most of our couples using egg donation are so grateful to achieve a pregnancy. So, we don't get a lot of 'yuck' factor. I'm sure some people are willing to pay $25,000 for something they could never be"—for genetic traits they lack—"but that's not been *my* experience."

In larger cities and elsewhere, however, many patients place premiums on beauty and intelligence, raising concerns about possible eugenics. John, the Texas mechanic who had tried reversing his vasectomy and has a low sperm count, said, "Some sites are almost like manufacturing. Patients want a genetically perfect child. It's getting into sci-fi stuff. People want the Superman athlete sperm, and the 5-foot 8 gorgeous blonde, with massive boobs. 'We want [this man and that woman as donors]. That's the best baby. Splice them together. Give us what we need.' A lot of these sites have criteria to be an egg donor: college-educated, smart and beautiful. Not the average person out there just digging a ditch. They can also [exclude] certain things you don't want. People have babies naturally. The baby comes out, and you get what you get, and never know, and you deal with it. But if you're *paying* to get a child, and this technology is there, why not use it? You go car shopping. You don't just want something that has four wheels. You might want a Mercedes. That's why there's so many different kinds of cars. I don't see anything wrong with it."

Anonymity

Anonymous gamete donation, though banned in Great Britain in 2009, flourishes in the US. Most heterosexual egg buyers hide their use of other's gametes,[64] but most offspring would prefer to know.[65] Egg sellers are often willing to have future offspring contact them for medical, but not other reasons. "Donors, with very few exceptions, say, 'Yes, I am willing to be contacted,'" Brenda reported. "All donors say, 'If you need me for medical reasons, by all means contact me.' But there is not yet a central registry. Hopefully that will soon change."

A donor registry may help but requires resources. As Brenda explained, "England, Australia, Denmark and Sweden have government-run databases, but a lot of issues need to be answered: who funds it, and what it should contain. In the US, we don't trust the government to have that kind of information. Money is the biggest issue. We also don't want to stigmatize the kids." Still, voluntary registries have been established by offspring who were created through donors and seek biological kin.

Both anonymous and non-anonymous donors face certain challenges. "Known donor situations have been much more difficult," Diane said. "Some work really well: people with really good relationships, and close, respectful boundaries, who are open to processing things. In England and Canada, most donation is known because they don't allow people to pay. The staff ask, 'Do you have a sister?'"

Over time, US donor anonymity will become rarer. "Kids born through donors are getting older now and saying, 'We have a right to our heritage,'" Ginger said. "When asked, 'How do you feel about your child meeting you?' one sperm donor told me, 'When the child is 18, I would be delighted to meet him or her, and tell them who I am.' In the film *The Kids Are All Right*, the donor shows up. In some ways it was good. In other ways, it wasn't. That's probably realistic."

Disclosing Use of Donor Eggs

Many, if not most, patients do not want to disclose their use of donor gametes to their future children—or often even to friends, family members, and physicians. Studies have concluded that parents *should* inform their children in some way early on, partly to avoid later shock and feelings of betrayal if disclosure occurs.[66]

Yet prospective patients often fear stigma or discrimination from teachers and others. "With kids created from donor gametes, the world has a way to go," Ginger observed. "Parents commonly do not feel fully accepted and worry how it's going to play out in the community, at the kid's school: 'Can we tell our kid? It's not a secret, but it *is* private. How long can we do that? How do we help our kid talk about his beginnings?' Fortunately, ASRM recommends that you talk to your kid about his or her beginning. We're moving that way."

But doctors and patients commonly ignore this recommendation. Many REIs support donor gamete recipients *not* informing offspring. As another physician said, "I don't think it's a big deal if the parents don't want to tell them."

Due to fears or shame, many parents tell no one—not even the child's pediatrician. But such secrecy can cause problems. "A nurse with hip dysplasia, which is genetic, used an egg donor, and had twins," Jill reported.

"The doctors wanted to X-ray the babies to screen for hip dysplasia. The nurse didn't want to tell anybody, including the physicians, that these were not, biologically, her kids. So, she had them X-rayed! I expressed surprise. She said, 'I wasn't ready to tell!'"

Still, whom to tell and how to do so can be hard. Information can leak out. "You can't tell one person and think they're going to keep it a secret," Jill admonished. "Who knows who's going to say what when? The worst-case scenario is: the kid finds out from somebody else."

Psychotherapists can play important roles here, helping parents confront and address these disclosure dilemmas. Helen asks parents, ' "Who knows about the donation?' 'My best friend and my sister. No one else.' 'What about telling the child?' They'll say either, 'Yes. I'll tell them when they're old enough' or 'No.' I say, 'When do you think they're old enough?' They'll say '18.' I'll say, 'Let's think this through for a minute. Do you remember when you were 18? You're going off to college, and your parents sit you down and say what?'"

Freezing Eggs

"Should my daughter freeze her eggs?" a friend, Rachel, recently asked me. "She's 32, and not dating anyone now. I worry about her. She's working all the time as a junior partner in a law firm, and may not try to have kids for several years. By then, it may be too late. Eventually, I want grandkids! So, I've been telling her to freeze her eggs. She feels I'm bugging her."

Increasingly, younger fertile women who are not ready to have children but want to avoid future age-related infertility are freezing their own eggs. Apple, Facebook, and several other corporations pay to freeze their female employees' eggs[67,68] so that these women do not feel obligated to stop working to have children now. For several years, doctors have been freezing eggs and sperm for young people with cancer undergoing infertility-inducing chemotherapy and radiation. Egg freezing is now rapidly spreading for other young women, widely advertised, and encouraged. "Doctors are establishing egg banks," Steve, the infertility doctor, said.

Yet egg freezing is expensive. Costs vary somewhat by city but can be up to $18,000 per cycle, with women usually needing at least two cycles to obtain enough eggs. In addition, storage over 10 years requires around $10,000. Clinics charge at least $10,000 to fertilize the egg and transfer the resultant embryo into the womb.[69,70] But, for various reasons, these processes can fail. Egg freezing does not appear to harm birth rates in the short term, but the possible effects of freezing eggs for more than four years remain unknown.[71] Many women anticipate freezing their eggs for more than four years— perhaps one or two decades. The health of the resulting offspring at birth or later in life has not yet been examined. Moreover, at least one clinic recently lost thousands of stored eggs when its storage tank malfunctioned.[72]

Many providers also wonder whether frozen eggs are as good as fresh ones. "Frozen eggs are experimental," Paulette said. "We don't have great results with them." Providers thus wonder whether to provide this service and, if so, to whom and why. "Recently we've been struggling with egg freezing as a new thing," Henry, a Midwest REI, admitted. "Who should be offered that? Should we charge them for it? Should their insurance cover it? For doctors working at a medical center, almost everything else gets covered. But a single woman resident who's 40, without a significant other, wants to freeze her eggs for the future so she can get pregnant, and wants us to eat the cost. That's difficult. Should we even be doing it? At 40, the odds are small that these eggs are good. Are we giving her false hope? Do we bill it as 'infertility'?"

Generally, even for patients with cancer, insurance does not cover the costs. Consequently, Roger, a New York hospital-based infertility doctor, provided the service to a 23-year-old cancer patient gratis. "A student without insurance for IVF had myoblastic cancer. Her parents don't have money. She's about to face six months of horrendous chemotherapy that's going to make her menopausal. But she wants a child. So, I can freeze her eggs in advance. Yet no insurance billing code exists for that. I have to put down 'infertility.' Insurers, though, say that the definition of infertility is 'one year of unprotected intercourse.' She says, 'I have a boyfriend, but used the pill, and wasn't trying to get pregnant.' I petitioned the company, and wrote that this is extraordinary—have a heart here. But they refused."

Many private clinics are enlarging their egg freezing business for healthy women who wish to postpone childbearing and are willing to pay out-of-pocket; yet concerns emerge about whether these doctors may at times be overselling the procedure to increase their profit. "Entrepreneurial programs have to cover their overheads, and are overzealously marketing egg freezing," Calvin, an REI at a southern California hospital, stated. "Especially freezing 37- and 38-year-old women's eggs. We've all been doing very well freezing for egg donors, but they are only 25 and 26. Who knows if women at 37 or 38 are going to do as well? But a lot of IVF centers are pushing it, telling women in their late 30s, 'freeze your eggs now. They will decrease in quality in a few years. If we take them now, you can use them then.' Hospitals are a bit tougher on doctors doing that.

"If the doctors are SART members, they have to say that this is still 'experimental' and not market it as if it isn't," Calvin added. "The doctors need to check with a research ethics committee. But the doctors may just put a little disclaimer in the consent form. and go on their merry way." Patients may thus not fully grasp that the effectiveness of this procedure remains unknown.

At the East Coast academic hospital where Thomas works as an infertility doctor, egg freezing for future use perplexed the ethics committee. "We've talked recently: would we do fertility preservation to reproductively delay childbearing? We've struggled with these questions."

In sum, these new technologies for obtaining, storing, and using human eggs pose several dilemmas. Many patients seek strangers' eggs but must then decide whose, and where and how to find them, what criteria to use, and whether tell other people. Infertile couples who want a biologically-related child often end up feeling they should use strangers' eggs.

To avoid these scenarios, innumerable young women consider egg freezing but may not fully grasp the limitations—that the procedure may not always work. When my friend, Rachel, asked me if her daughter should undergo egg freezing, I said they should realize that the plan may fail. Rachel's daughter was still relatively young and could also potentially wait a couple of years. But Rachel's concerns illustrate how parents' and their young adult offspring's views and desires can clash.

Women, often 18–21, who sell their eggs enable many prospective parents to have children but face medical risks, without always fully grasping or appreciating these. The notion that creating their children may have harmed the biological mother disturbs many of these parents, who prefer to think that the donor and/or surrogate are "angels" who helped produce these children for wholly altruistic reasons. In fact, many agencies explicitly will tell an egg seller to say that she is donating for altruistic reasons when speaking to a potential egg buyer.[73] Even friends of mine who have used egg donors are surprised and disturbed by discussion of these risks, preferring to believe that donation is somehow risk-free. Women should be free to sell their eggs but should be informed about and understand the risks. Alas, such education and awareness do not always occur.

5 | Choosing Sperm

"I went for IQ," Valerie, a psychotherapist and single-mother-by-choice, said about how she picked a sperm donor. "My doctor recommended a particular sperm bank, but it didn't really have identity-release [non-anonymous] sperm, which I wanted. So, I just went online and found a sperm bank that emphasizes guys with doctorates. Eight profiles were in the ballpark. I ruled out two, based on looks. The company had 15-minute audiotapes, to hear the donor's voice. I ruled out anybody who sounded flat and schizoid. I went for somebody with the lowest the SAT scores of anybody in the group, who's a JD, not a PhD, but was funny and self-deprecating on the tape. So it was a *balance* of looks, IQ, and personality. I got most of what I was gunning for—which I realize is a genetic crap shoot, anyway. But it worked. I got a generally very nice-looking, easygoing child.

"He is turning four, and has some language delays. For the last year, I've been reading to him stories about donor sperm. But I'm not sure he really gets it yet. Not many good children's books are out there, and virtually none for single mothers. I haven't yet gone on the donor-sibling registry to touch base with some of the other kids related to my son. I don't know what to think about that, never having done it. It's unknown territory. I come from a small family, and never envisioned my son having 10 half-siblings. So, I've been reluctant. He's an easygoing kid, but is he going to start asking questions, and be hard to deal with? My guess is 'no,' but you never know."

Single women like Valerie and my college friend Abby, along with lesbian couples, and fathers with infertility problems or mutations contemplate buying sperm but then encounter questions. Prospective parents commonly consider or seek particular traits in gamete donors, though recognizing the limits of doing so. Still, amortized over the course of the child's lifetime, searches for such traits cost relatively little and may provide comfort, given the anxieties of raising a child. As with parents who use an egg donor, those who use a sperm donor also wrestle with whether to tell their eventual children.

Sperm is easier to obtain than eggs, and each year, thousands of US couples use so-called donor sperm. In 1952, sperm banks began[1] and gained notoriety with the 1980 creation of a Nobel sperm bank, which eventually closed due to difficulties processing such sperm and patients commonly wanting traits other than IQ alone. The media have described labs accidentally injecting women with the wrong sperm[2,3] and certain donors fathering as many as 150 children.[4] The American Society for Reproductive Medicine (ASRM) recommends that men not donate for more than 25 births in a geographic region of 800,000 people—to avoid half-siblings created from the same donor later accidentally meeting and having children.[5]

In heterosexual couples, many men resist addressing not only male infertility generally but specifically use of another man's sperm, seeing it as akin to being cuckolded. "The idea of raising somebody else's kid would just be weird to me—it's a 'man thing,'" said John, the Texas mechanic who reversed his vasectomy but still had trouble conceiving. The possibility of using a sperm donor bothered him. "It'd be like if my best friend came over, nailed my wife, and got her pregnant. I wouldn't want to raise his kid."

Many infertile men may thus delay this option. "They want another examination of their testes, or attempt to find sperm," Steve, the physician, explained. "I tell them I did a very thorough job and would not risk the viability of their testes. But some just don't want to stop. I don't know if these men go to other doctors." Little follow-up usually occurs.

Men and women can thus vary markedly in how they view and respond to potential donor gametes. "They have equally hard times using donor gametes, but it hits in different ways," Diane, an in vitro fertilization (IVF) nurse, said. "Two-thirds of the women using donor eggs are older, and have been through IVF, and see it coming. A provider may have planted the idea six or 12 months earlier. With men, it's usually much more sudden. Typically, they are younger and not expecting it, and have a Y-chromosome problem, or no sperm, and think they're not a man. They have to deal with that disappointment. It's disruptive."

Among prospective parents seeking sperm, otherwise-healthy straight men may encounter the most social and psychological difficulty. As Diane added, "A man who finds he has no sperm is different from a man who had testicular cancer in his teens, and is just glad to still be alive, and simply says 'bring it on,' or single women or lesbian couples where it's fun looking for 'Mr. Right.' Men with no sperm feel terrible—that they've disappointed their wives—and face identity issues and a sense of mourning. Our culture is not very kind. We socialize men about what's important—being a 'fully biological parent.' Men are expected to do this easily, according to what they tell us in high school."

Potential sperm recipients face decisions of whose sperm to use. Just as some women donate to their sisters, some men ask their brothers, though tensions can ensue. "Men bring in their brothers to stand in for them in the gene pool," Diane continued. "How well it works depends on the brother and their relationship. Sometimes we don't go forward with it. Other times, it's perfect. When brothers are rivals, or competitive, we don't want to participate. When a brother volunteers because his brother's been struggling, it works fine. When the brothers say this is a secret they're going to 'take to the grave,' and that they don't want to tell the rest of the family, I'll say 'No.' These secrets are too powerful."

Providers thus need to assess the familial dynamics carefully. "You don't want it to come up and bite you later on," Diane amplified. "I'll sit down with them, alone and together, and listen to them talk about their family. I'll say to the potential donor: is there any reason you want us to reject you for medical reasons—so you don't have to do this? I'll just say 'we can't take your brother as a donor.' Some will tell me that they don't want to do it, but don't want their brother or sister to get mad at them. In that case, we reject them. That happens all the time. Some tell me privately that their sibling is 'crazy' for pursuing the possibility."

Choosing Strategies

"They pick sperm donors who resemble them," Diane explained. "Somebody who's as close a match as possible to the man they're replacing, so that the child grows up looking like he belongs in the family." Infertile men without brothers as potential sperm donors must then consider strangers and generally select based on physical appearance.

Yet finding sperm donors can prove puzzling. Sperm banks offer vast choices, bewildering patients with questions of how much information to seek. More information about potential donors generally raises costs. "The whole process with sperm donors has been terrible," Wendy, the Irish Catholic secretary who had failed four IVF cycles, said. "I've been left to do it on my own. Initially, I was all gung-ho. My gynecologist was going to do IUI [intrauterine insemination] with donor sperm. The nurse said, 'I've never done it before. I don't know what to tell you.' She asked the doctor and said, 'The sperm donor just has to be CMV [cytomegalovirus]–negative.' I said,

'What is that?' I went on websites and looked myself. I put in the qualities I wanted to match: blond hair, blue eyes, height, and CMV-negative. That's all I could do. But with time, I've learned what to look for: there's sperm for $200 a vial instead of $600 a bottle, because they don't have enough information on the donors to charge you $600. With the more expensive sperm, you get pictures and more health and family history—about the parents', grandparents', and siblings' occupations and hobbies. The $200 sperm has no picture. It's anonymous, and it doesn't go in depth. They tell you if he has brothers and sisters or a child, and the family medical history. Is the extra information worth it? No. I don't need to know what the grandmother and grandfather did for a living."

Potential sperm recipients may seek and obtain external input, but it can vary and be confusing. Wendy's psychiatrist asked her, ' "What are his hobbies? You should be looking for his hobbies as well. Does he wear glasses?' I wouldn't have thought about that."

Ultimately, however, questions surface about the trustworthiness of potential donors' self-reports. Wendy wondered: "Is he going to say he is 'a bookworm and sits home and reads all day,' or, 'No, I go fishing and hiking.' They don't screen these people very well," As Suzanne, a San Francisco IVF mental health professional, commented, "With sperm donors, no real kind of assessment is generally done."

Anonymity of Sperm Donors

Though banned by the United Kingdom in 2009,[6] anonymous sperm donation flourishes in the United States; but should it? As the progeny age, many want to know who their biological fathers are. Wide third-party donations also pose questions of relatedness among unknown half-siblings. Adult offspring of sperm donors have voluntarily established national registries, seeking half-siblings and their shared sperm donors. "We desperately need a donor registry in this country," Brenda, the Vermont psychotherapist, urged. "We used to just worry about the patients getting pregnant. Nobody thought about the kids. But now, the kids aren't kids anymore. They're growing up!"

Voluntary online donor-sibling listings now exist that parents and eventually their offspring can search for offspring from the same donor but that also generate challenges and ambivalence. Parents must decide when and what to tell offspring about this possibility. Individuals may search a registry only if they know they were created through donor sperm. They then need to determine how to relate to newfound relatives. DNA databases also now let half-siblings find each other.

A donor registry could in fact include and periodically update other data as well, such as donors' genetic information. Anne, a northern California reproductive endocrinology and infertility specialist who opposed sex selection, "would definitely have a national 'donor gamete' registry, including sperm donors, to keep track of genetic disease. In 2006, one man who sold his sperm was found to have transmitted a serious genetic disease to at least five children. But the sperm bank does not know his whereabouts, is unable to contact him, and does not know how many other children he sired who now have the disease.[7,8] "Still," Anne continued, "no one now tracks pregnancies or the children's medical problems. We're very careful that nobody gets an infection from a donated gamete. But we don't care so much whether there's a genetic illness in a child, or the donor later develops a disease."

In short, many prospective parents seek sperm from others but face intricate choices that both resemble and differ from challenges in seeking eggs. Both sperm and egg recipients wrestle with what to look for in a donor— from medical history to physical appearance or other behavioral traits that are not necessarily fully inherited. Parents also confront dilemmas of whether to disclose the donation to others, including to the future offspring. The ASRM says, "while ultimately the choice of recipient parents, disclosure to donor-conceived persons of the use of donor gametes or embryos in their conception is strongly encouraged."[9] But US sperm donation is usually anonymous, with no option for future contact with the biological father. A minority of donors allow for such possible future communication, to be mutually decided at a later date; but that may not occur.

Recipients vary widely in responding to these challenges. Infertile men grapple with raising "another man's" child and not feeling "macho." Single-mothers-by-choice and lesbians have no option but to disclose the fact that donation occurred. Yet even these women can wrestle with when and what to disclose and whether to maintain contact with the father. Some lesbians do not want the child, before age 18, to know the father's identity or to meet him. Many heterosexual couples decide *never* to reveal the donation to the child. Yet psychological studies demonstrate that it is better for offspring to know, ideally earlier, to avoid later feelings of being betrayed and deceived. The biological father's medical history may also be critical in the child's future medical care. Moreover, at age 18, on reaching adulthood, children arguably have an ethical right to know.

6 | Choosing Embryos to Avoid Disease

"It's a miracle!" Sally, the Maryland website manager, proclaimed. "I tried PGD [preimplantation genetic diagnosis] and got pregnant with my three-year-old daughter! She is a CF [cystic fibrosis] carrier, but does not have the disease." Previously, Sally had aborted two pregnancies because the fetuses had this disorder, for which she and her husband are both carriers. "When I got pregnant again naturally, and had to terminate it, I thought: 'I don't think I can keep doing this.'" Her third cycle, using PGD, produced her daughter.

"Unfortunately, IVF [in vitro fertilization] is made for infertile women, and is much different for women doing IVF for PGD," she continued. "We don't often have very good results. Technically, I *can* get pregnant—I've been pregnant four times. But from my first two cycles, using PGD, I had no embryos to put back. The IVF got 25 eggs, yet only three fertilized. I thought, 'This is a lot of effort for basically a 0% chance of having a child.'" Embryo screening involves risks and often fails.

Still, the alternative—having to abort a fetus with a severe mutation—can be worse. The challenges of each of these paths can lead potential parents to vacillate. "The back and forth was difficult," Sally recalled. "This is so horrible: I can only do IVF/PGD. Then I would do IVF/PGD, and think: 'This is so horrible. I have to try naturally again.' I tried naturally and terminated again, and thought, 'I can't keep doing this.' On each path, I reached a crisis point, had no other recourse, and took the other route."

When her first child was born, using PGD, without CF, she "wanted another child, and didn't want to go through IVF/PGD if I didn't have to. So, I tried it again naturally, and am now six months pregnant. This time, the baby is just a carrier." She and many other prospective parents thus flip strategies over time, especially for a first versus a second child, gambling differently after having had affected fetuses and/or children. "A lot of women who are carriers—mostly for CF—find me on different online discussion boards and ask me for advice, trying to decide between options," Sally added. "I tell women that they're probably a lot more resilient than they think. The first time, wanting a baby so badly, having to go through an abortion is really horrible. It pretty much devastates you. Then somehow, you get through it. Terminating was horrible. I suffered, but didn't think twice about it."

Prospective parents can now choose not only eggs and sperm but embryos. Previously, couples at risk of transmitting dangerous mutations to their children could opt to get pregnant and then test and abort the fetus, if needed. Now, PGD lets these prospective parents select a mutation-free embryo *before* pregnancy occurs, and avoid abortion. "I can virtually reassure them that, as long as there are no major reproductive problems, they will eventually have healthy children," Jennifer, a Northwest medical center doctor, said. "It's just a matter of how long it takes. There's no reason for them *not* to want a healthy family."

Patients can also reject embryos with certain genes (e.g., for CF or Down syndrome) and/or seek genes explicitly associated *with* certain characteristics (e.g., sex). Routinely, doctors now screen embryos to choose children's sex. Yet these processes can prove arduous and raise moral questions.

Increasingly, researchers are also using advanced whole-genome sequencing to identify additional markers associated with other diseases. Clustered regularly interspaced short palindromic repeats (CRISPR) also allows for insertion or deletion of particular genes in embryos. In upcoming years, doctors will probably be able to choose or create embryos with genes for characteristics such as blond hair, blue eyes, height or perfect pitch.

Almost all IVF clinics perform PGD, mostly for large chromosomal abnormalities, X-linked diseases that affect boys more than girls (58%), non-medical sex selection (42%), adult-onset disease (28%), and selecting for disabilities (3%).[1] The British government accepts PGD for approximately 250 conditions[2] and is considering 19 more.[3]

But among US internists, I found out that only 7% feel qualified to answer patient questions about PGD; most are uncertain whether they'd refer patients for it for various disorders (e.g., CF, Huntington's disease [HD], breast cancer).[4] Most neurologists and psychiatrists would refer patients for PGD for HD and Tay-Sachs; but for CF, about one-quarter were unsure.[5]

Generally, patients and providers support PGD, but disagreements emerge.[6] In one study, only 24% of patients with hereditary cancers were aware of the procedure; but once it was explained, 72% felt it should be offered and 43% would consider using it, though 29% remained uncertain.[7] Most women at risk of breast cancer did not know about it but, once told, felt it was acceptable,[8] though they would not undergo it themselves since it requires IVF and they themselves, as embryos, would have been rejected.[9,10,11]

Many prospective parents who are at risk of transmitting a fatal genetic disease find the alternatives to PGD worse—fetal testing followed by abortion, pending the results. In a study of families at risk for two such diseases (Von Hippel-Lindau and Li-Fraumeni syndromes), 35% viewed PGD positively, and most were not concerned about potential disadvantages, including unknown long-term effects, low success rate, and misdiagnosis. Still, among couples undergoing PGD for these two diseases, only 56% would use PGD again; 12% would instead use prenatal diagnosis followed by abortion, and 6% would do no tests.[12] These couples perceive advantages such as having unaffected embryos, avoiding abortion, and decreasing miscarriages, but patients also confront financial[13] and psychological strains.[14,15]

Many clinics screen embryos not to detect the presence of particular mutations associated with disease in the future child but to check that all embryos have the overall correct number and type of 46 chromosomes—large bundles of thousands of genes each. In 1996, two doctors hypothesized that screening for the gross numbers of chromosomes in this way would not only prevent disease in the offspring of older women, but also increase live birth rates and reduce miscarriages more broadly.[16] Such screening, known as preimplantation genetic screening (PGS), spread rapidly for the purpose of increasing births and decreasing miscarriages, despite the lack of evidence to support the use of the procedure for these two reasons, raising costs and physicians' profits. Eventually, data showed that it actually lowered the odds of pregnancy in many women, who ended up with fewer embryos to use.[17]

How then do patients and providers decide whether to screen embryos and, if so, for what, and how to respond to these limitations? In fact, providers and patients struggle with several quandaries: for which conditions to use PGD, how to decide, how many times to keep trying PGD if it fails, whether to select embryos to create so-called savior siblings who can donate organs to existing children with severe disease, what to do if patients want to screen embryos for two genes instead of one, whether patients should always be told the results of genetic tests on embryos, whether to test the fetus to confirm the results, and what to do with leftover embryos. Insurers and patients wrestle with whether to pay for this procedure and, if so, how much. These questions may appear disparate but are in fact closely linked. Patients' and providers' understandings and attitudes can shape these decisions. This chapter explores the uses of this technology to avoid medical disease. The next chapter examines selection of non-medical traits, such as sex.

Whether to Test Embryos and, if so, for Which Diseases

Many prospective parents learn that they are at risk of transmitting a lethal mutation only after a child develops the disorder. They must then decide whether to try to prevent the disease in future offspring by screening embryos, even though doing so requires IVF, which they otherwise do not need.

Patients and providers grapple with not only *whether*, in general, to screen embryos[18,19,20,21,22] but *when* exactly and for *which* diseases to do so. PGD can be used for disorders that range from fatal to non-fatal and from childhood- to adult-onset. The assay of embryos, rather than adults, poses different risks, benefits, and moral issues,[23,24,25] affecting the creation of future human beings who cannot consent. Many providers and patients readily support PGD for lethal and childhood-onset disorders but are often ambivalent about disorders such as breast cancer that would not cause symptoms for decades, if ever, and vary in severity, heritability, penetrance (the likelihood that a mutation, if present, will in fact cause symptoms), and treatability. Clinicians and patients generally support PGD for HD but are less unanimous regarding breast cancer, which is treatable and only affects adults. Physicians commonly test embryos for genes associated with breast cancer (BRCA) only if the patient's family history is extensive.

"What would you do or not do PGD for?" Sam, who works as a patient advocate for a Midwest patient organization, asked. "We want to do our best for our kids. But diseases may have some genetic predisposition, though it's not that high, and the disease is not bad. Everyone is going to die of *something*."

Individual providers and patients may disagree about whether and how much to screen embryos for particular diseases—how predictive a gene should be, how severe the symptoms, and whether the gene needs to cause a disease, not just a disability. Sam, for instance, worked with a patient with Walker-Warburg syndrome who "felt judged by the geneticist. Some patients are deaf, have different color of eyes, and different conditions that not everyone calls disabilities. We had long discussions: should we do PGD? It's not life-threatening. Most PGD we do is for more severe disease. The patient decided to do it, but felt the doctors were questioning his motive." In particular situations, providers and patients may thus support PGD to differing degrees.

Decisions about whether to undergo PGD can be tough because a particular mutation can manifest itself differently among people. "A patient had ankylosing spondylitis, a dominant genetic disease, with highly variable penetrance," Anne, the northern California fertility specialist, explained. In this disorder, a serious form of arthritis, spinal vertebrae fuse together, causing immobility, pain, and discomfort. No cure exists, but cases can vary in severity. "Only 20% of people with the [gene] have the disease, which ranges from crippling to fairly mild arthritis and maybe eye problems. That's a very difficult borderline decision-making area. This is not Tay-Sachs, where you die in infancy. We *all* have something. *Just because we can identify it, doesn't always mean we need to eliminate it!*"

Patients may thus struggle with not only risks of the mutation being present but with its potentially variable expressions. "Patients tend to assume," Anne continued, "that what they've seen is what's going to happen. One husband had retinoblastoma, a dominant disease, and is blind, and was extremely comfortable with a child being blind. The husband has a very good life. He's very productive. So, the couple did not do any testing with their first child, who was positive, but who unfortunately, developed a cancer, needed several brain surgeries, and was on chemotherapy and very ill." In the husband, the penetrance (i.e., the likelihood that the mutation will cause symptoms) "was different than in the child, which they hadn't anticipated. So, for the second child, we did PGD. Their decision was based on cancer, not blindness. They successfully had unaffected embryos—and a baby."

Reproductive endocrinology and infertility specialists (REIs) also use PGD for CF, which is becoming more treatable, raising further dilemmas. "Kids with CF used to die in their early 20s," Karen, the IVF provider who underwent infertility treatment herself, said. "Now, they live into their early 40s to 50s."

Yet, both infertility providers and patients vary on whether and how much they understand these genetic variations and treatment advances. Some providers seek to educate patients very carefully. "We get asked about Gaucher's disease," Jennifer, the Northwest physician, reported, "because it's a very common autosomal recessive condition in the Jewish community— one in 11 to 12 people are carriers. Their knee-jerk reaction is, 'This is going to be a terrible disease.' Couples may have difficulty making these decisions. [They] suddenly learn that they are both carriers for a recessive disorder, which they had not anticipated.

"We educate them. Other providers don't always appreciate that there are different types—type I, II, and III. Some are awful—the kids are mentally retarded, and deteriorate and die before their fifth birthday. It's horrible. Others have bone crises, and pain similar to sickle cell. But in the Jewish community, the most common manifestation is individuals living to be 80, never knowing [that they have the mutation], and dying blissfully ignorant. Some individuals end up having a big spleen, but not much worse. Other individuals are more severely affected, but it's a chronic disease. Enzyme replacement treatment is now available. Patients are happy, productive, active. But patients think they want to do PGD for Gaucher's disease. We can make some predictions about how severely someone would be affected, depending on their genetics: 'Even though no one in your family has Gaucher's, it's likely to be like *this*. These are treatment options.'"

Social and even political factors can play key roles. Due to past injustices, screening embryos for sickle-cell anemia, for instance, causes controversy, given wariness in some African American communities toward the medical system. Edward, an Ohio infertility doctor, sees "resistance, because a lot of people will have sickle-cell disease and live long productive lives. But [PGD] should be offered."

Another limitation of PGD is that many eggs are needed, yet diseases for which PGD is used may also lower the number of usable eggs. "Women who are fragile-X carriers, which can involve premature ovarian failure, face complicated IVF/PGD," Jennifer observed, "because they may get very few, or poor-quality eggs. We do not know that until we actually start the process."

Preventing all serious genetic conditions through PGD is impossible. Patients may test embryos for one mutation, only to have a child born with a different problem. "One couple did PGD for spinal muscular atrophy," Jennifer reported. "I always recommend a confirmatory prenatal diagnosis, once the couple's pregnant. We did that, and found that the child had Klinefelter syndrome"—caused by an extra chromosome in males and leading to various bodily symptoms. "The couple decided to continue the pregnancy anyway. That child is doing really well, and they wouldn't have had it any other way. . . . So, these procedures are very unpredictable."

When couples confront two genetic diseases, screening for both is extremely difficult. But deciding between two disorders can be excruciating. "One couple had two recessive conditions: CF and another disease that results in infantile death," Jennifer explained. "The couple decided *against* PGD. But theoretically, we can do PGD for both of those conditions simultaneously. Another couple are both carriers for a lethal recessive condition, and lost three children to it. We recently found out that the husband had myotonic dystrophy—a dominant condition. We talked about deciding between conditions. Admittedly, I have definitely been trying to push toward: don't worry about the myotonic dystrophy . . . an adult-onset condition. The other condition is lethal."

Psychologically, these situations can be especially hard. Couples may fear they will never have children. "One couple lost their newborn baby to one disease, and the next baby had a *completely different* condition," Jennifer continued. "They've gone from one genetic disorder to the next. It's helpful to have a larger perspective—knowing they'll eventually get there. Patients learn as they go, and often modify their decisions based on their experiences." Based on changes in their experiences, views, finances, or partners, prospective parents frequently shift over time.

When PGD fails, patients and providers confront additional dilemmas about whether to keep trying and for how long. Over time, many such patients fluctuate, attempting PGD and then perhaps trying to conceive naturally. Even women at risk of transmitting serious mutations for a disease may flip-flop between IVF/PGD and natural conception, gambling on the odds of having a healthy child. "We've had everything," Jennifer noted, "from couples who have been very thoughtful, to those who decide on either prenatal testing or IVF for one cycle, and had a baby, and then, on the next cycle, do the other procedure—couples who've gotten pregnant the old-fashioned way, naturally, and then gone through one or more terminations, and decided it was too traumatizing. Couples have gone through multiple IVF cycles and just couldn't stomach it anymore—the whole emotional roller coaster. They had a goal of having a baby by a certain time, and clearly weren't going to meet that, jumped tracks, and went for a more efficient way. There's no predictable path.

"Couples may try one cycle, and depending on how that goes—not just financially, but whether it's successful or how frustrating or difficult it was—go back and forth, even for recessive conditions. One couple ended up with five out of five embryos affected. So, they had to rethink: 'Are we going to do a second cycle?'"

If PGD fails, many couples decide, given their increasing age, to accept the risk of undergoing an abortion, rather than the possibility of childlessness. "A lot of couples start at 25, and think that's early, so that it's going to be easy," Jennifer explained. "But they go through a couple of cycles, and end up with no healthy embryos. We expected that this would be the perfect solution for them. They now start to reconsider. They might never have considered terminating a pregnancy, but now decide to go the natural route, and undergo amnio [amniocentesis] or chorionic villus sampling [CVS] and terminate if they have to. They think: 'That's not so bad after all, because otherwise I'm never going to have a baby. I only have a couple of reproductive years. I've got to start quickly.'"

"When PGD doesn't work it's extremely difficult," Anne, an REI, confided. Over time, numerous patients weigh the chances of PGD or IVF failure against the risks of having a diseased child. When such failures occur, patients may, depending on the severity of the disease, opt to try to conceive naturally and produce a mutation-free child. "Patients sometimes misunderstand how many eggs and healthy embryos they're going to make," Anne continued. "Two patients went through expensive IVF and PGD, didn't have a baby, and then conceived without IVF. One patient with breast cancer just abandoned PGD and conceived naturally. A patient tried PGD for Reiter's syndrome"—reactive arthritis associated with a particular gene. "We had transferred unaffected embryos, but eventually, after three cycles, she transferred the healthiest looking *affected* embryos."

Over these arduous, multiple-staged journeys, patients may reappraise the pros and cons of PGD versus alternatives, including natural conception with or without confirmatory testing, using gametes from known or unknown donors, or remaining childlessness. Paulette, the psychotherapist, described a patient from whom, "we got no eggs" and "who then used her sister's eggs, and had a boy. But at circumcision, he bled heavily, and was found to have hemophilia. The family was tested. Nobody had it. The sister had developed a spontaneous mutation. When the patient wanted a sibling for her son, we did PGD on all the frozen embryos. One embryo was female, and therefore unaffected. We transferred it. But she didn't get pregnant. The [patient's] sister agreed to do another cycle, and we PGD'd all the embryos, and she delivered a second time—and the child was OK."

Given these unpredictabilities, prospective parents at risk for a serious disease may eschew embryo testing and instead use a stranger's gametes. "Patients who've had diseases they *could* screen for have chosen donor sperm," Diane, an IVF nurse, said. "*That* seems pretty radical. But they've seen too many family members suffer with this disease: 'Let's forget about it! Just take it out of the gene pool!'" One patient had "watched sisters and cousins suffer with a disease that required multiple surgeries. The bones of the head fuse together, so the brain gets crushed." The prospective father, "didn't want to risk it. He would have had to put his wife through IVF and PGD, getting through all these gates, one at a time, hoping he had embryos at the end."

Given these conflicting considerations, patients commonly harbor doubts, even after making decisions. "My last miscarriage was due to Down, not my translocation," Yvonne, the Philadelphia child psychotherapist, said. Two of her chromosomes had switched parts. In her cells, these parts were "balanced," such that all the DNA was still present but on different chromosomes. She was at risk, however, of having children with "unbalanced" chromosomal translocations, where some DNA might end up missing, causing severe symptoms. "That's weighed on me: what if I passed on the translocation? I am very grateful my daughter is translocation-free.

But at times I've thought about just passing on the burden." By the time her daughter grows up, "maybe technology will be improved."

Factors Affecting Embryo Testing

Provider Characteristics

In these decisions, clinicians vary based on several factors, such as how much and for what diseases they have heretofore performed PGD. Overall, the amount, types, and success of embryo testing have been increasing. "We've used PGD for CF, sickle cell, and muscular dystrophy," Edward, an Ohio hospital-affiliated REI, said. "BRCA has not come up. We've debated it, but not done it."

Over time, providers may change how they perceive and weigh the risks and trade-offs involved. "PGD has great promise for the future in diagnosing embryos," Peter, a West Coast IVF doctor, said. "We'll get better at it, and less invasive." Many physicians come to alter how they weigh the benefits and risks. "Initially, I thought the removal of the cell was very traumatic, since pregnancy rates decrease after PGD," Peter continued. "But it's actually less traumatic than I thought. So, you sacrifice pregnancy rate. But sometimes, the information is worth it."

With time, many providers alter their embryo testing practices and decisions. Due to testing errors, certain clinicians no longer assess embryos for mutations. Sued over a misdiagnosis, Peter now screens embryos not for disease but *solely* for sex. "We're only doing PGD for gender selection now," he explained. "We stopped doing single-gene defects because we were sued for a misdiagnosis. It was absolutely outrageous. An infertility patient in her early 40s needed IVF. We discovered that she was the carrier of an unusual sex-linked metabolic disease. We did single-gene detection and had a misdiagnosis. She ended up with a defective male. She had signed an informed consent that said very clearly: 'I understand that there's a 5% chance of misdiagnosis, and that I will expect to do amnio, at which point I will have the opportunity to terminate the pregnancy.' But she decided *not* to terminate the pregnancy, had the baby boy, and then sued us for *wrongful life*. The lawsuit took many years. Eventually, we were found not liable. But the experience was so incredibly painful that we decided: this wasn't worth it. The technology is now at a stage where a lot of mistakes are going to be made. As long as there's a possibility of a mistake, somebody's going to sue. This is not a major part of our practice, and was the most traumatic, disruptive thing that ever occurred. My embryologist simply refuses to do it."

Providers' Understandings of Genetics

Clinicians themselves range considerably in their understanding of the particular diseases assayed and the genetics involved. Infertility providers may lack sufficient knowledge of rapidly advancing genetic discoveries. "Many IVF providers have difficulty distinguishing among genetic disorders," Jennifer, the Northwest physician, said. "They may just not have the experience. Many IVF doctors recognize their knowledge deficits, and fill those gaps—spending more time doing it, or partnering with doctors who know these diseases." But the swiftly burgeoning amount of information leads to knowledge gaps. "I've had patients come in with diseases I've never heard of," Edward, an Ohio infertility doctor, admitted.

Clinicians' PGD experience, knowledge, and skills will surely increase, shifting their views of risks and benefits. Still, given inherent limitations, Peter thought that having some providers specialize in screening embryos would be helpful. "If we ran a PGD clinic, and were going to be at the forefront of knowledge, always analyzing, doing quality-control, publishing on it, I would be less reluctant to do it. But we were doing one case here and there. *That's* where you get into trouble."

Patients' Attitudes and Understandings

Patients, too, often have difficulty grasping the potential complexities of testing embryos. Initially, they may be wary due to inaccurate or unrealistic fears about removing cells from embryos. Clinicians also differ in how they discuss PGD with patients, potentially framing the procedure in helpful ways—explaining how embryos split naturally in the case of twins. "Initially, I feared whether PGD causes any other effects, because it's so unnatural," Francine, now planning to try experimental treatment in Mexico, confessed. "You're taking out cells. That sounds bad! I felt: 'The baby's not going to form normally. It's going to miscarry.' Then, a doctor said, 'When you have identical twins, cells split, and you start with half of each, and it goes from there.' *I never realized that!* That gave me a lot of comfort. You're taking out these cells at *that* stage. That analogy of twins is important for women to remember. It put it into a natural context."

Other patients like the idea of screening embryos but fail to grasp key details: "If you're already going through IVF and they can tell you what [the baby] may have down the road, why not do it?" asked Isabelle, now pregnant after undergoing PGD for Emanuel syndrome. Yet she and others may find the procedure appealing in the abstract in ways that do not reflect the reality. "You can know what to look for, and prepare for it," Isabelle continued. "Instead of having prostate cancer at 65, you can start treating it at 60," and if you could know in advance that your son will be born deaf, "it may be easier." Prospective parents may thus mistakenly see embryo testing as providing more than it does—guaranteed, complete and reliable information to help them and their offspring plan for future, adult-onset disease.

Many patients need careful education about the diseases for which PGD should be used. "One family with polycystic kidney disease, a dominant condition, thought they needed to select against that," Jennifer explained. "Once they started understanding more about the disease—all the variations, and everything involved—they decided they should just have kids and not worry about this."

Prospective parents may also have difficulty comprehending the results' potential inconclusiveness, bewildered by the lack of clarity. Francine described a friend, who "had five embryos. Three were normal. Two: they weren't able to tell. I asked the [doctor], 'How often do you get an inconclusive result?' . . . It's frustrating if I'm going to spend all this money, and maybe not get answers."

Patients' views and choices regarding PGD can be shaped, too, by prior experience with the particular illness confronted—whether relatives or prior children have been affected. "A lot of women have had kids die from genetic diseases, or know these diseases' effects," Sally explained. "They're coming from a worst-case scenario." If they've seen severe symptoms, "they are just not very concerned about the procedure causing problems."

Patients may also be uncertain or decline embryo testing because of moral or religious views—discomfort "playing God." Certain patients wrestle with such concerns, though probably less than with abortion. "The most difficult issues," Brenda, the Vermont psychotherapist, said, "are, 'can we do it?' and then, 'should we do it?' We're manipulating so much, should we manipulate more? Some people just say: 'bring it on.' There's zero conflict: 'This is great. It provides information.'" She feels that chromosome testing is "pretty clear"—that it's not "fair to bring a child into the world with [chromosomal abnormalities]. But with a Down diagnosis on amnio, patients say: 'What the heck? . . . We're already this far along in the pregnancy, [let's just have the child].' Testing embryos simplifies the situation a lot, so people don't struggle with it as much."

Broader social attitudes also mold these decisions. Embryo screening is increasing as more patients become aware of it, often making it less "scary." "The younger generation is very willing to think about PGD, and deal with these issues, whereas earlier generations just didn't talk about these things," Jennifer observed. "Fewer people today think it's freaky and weird."

Cost Limitations

"Many patients," Jennifer thought, "want to do PGD, but face financial barriers." Unfortunately, high cost and lack of insurance coverage block many patients from testing embryos. Insurance coverage for IVF overall tends to remain limited,[26] but additional problems arise regarding insurance coverage for PGD. Jennifer estimates that only about 40% of couples at risk of transmitting dangerous mutations can afford PGD. A few countries cover the procedure in certain cases,[27,28] but the amounts and the diseases covered are changing and not always clear. In 2015, only 10 of the 27 European Union countries reimbursed PGD, of which four reimbursed it only in public clinics and 11 allowed it but did not reimburse it.[29] In Australia, PGD is generally not covered.[30] Insurers that cover PGD may do so only for certain diseases. IVF, required for embryo testing, hikes up costs even more. Insurers may not cover PGD, simply because they do not cover IVF. Yet the rationales for these two procedures differ—creating a child versus preventing severe disease.

High PGD costs lead countless parents to transmit dangerous mutations to offspring, when doing so could have been prevented. Physicians often push insurers to cover PGD—arguing that refusal to cover embryo screening for serious diseases is short-sighted. But these efforts generally fail. Patients may consequently decline the procedure or pay out-of-pocket, unsure whether they will eventually be reimbursed.

US insurers as a whole do not appear to be changing with time. "Over the last 10 years, for every company that takes it on, another one will drop it," Calvin observed. "So, it's been fairly stable. There's been no increased coverage."

These companies may calculate that since patients frequently shift insurers over time, a competitor will probably be responsible in the future. "Every time we have a PGD," Calvin explained, "we write to the insurance company that this procedure prevents the baby from being very sick. But that argument usually doesn't help, because patients change insurance so often that the chances are that when the baby's born, another insurer will be covering them."

Insurers may also see IVF and PGD as "elective"—unessential and hence unnecessary. "Insurance companies think, 'If we cover IVF, we'll have to cover cosmetic surgery,'" Calvin continued. "They equate the two, which is crazy."

In addition, insurers may not want to be the first company to cover PGD since doing so might attract disproportionate numbers of patients seeking the procedure. As Calvin reported, "Some insurers say, 'If we cover it, there'll be adverse selection because those who need it will come to us.'" An insurance company that *did* reimburse it might then face prohibitive costs. Calvin felt that this argument was not, however, wholly valid, "since most people don't have any choice of where they go for insurance."

Insurance can be not only limited but also inconsistent and illogical, with a few companies reimbursing only part of the overall expenses involved (e.g., IVF but not the PGD or vice versa). "One of our large insurers covers PGD but, if the patient is not infertile, won't cover the IVF," Thomas, an REI at a large East Coast academic hospital, explained. "So, if a patient has a single-gene defect or recurrent miscarriages covered, the IVF, which is the larger expense, isn't covered."

Insurers may argue that since they cannot reimburse all expenses, they must make trade-offs between embryo screening and interventions for all other conditions. "With some insurers, it's a zero-sum game," Thomas continued. "They say they only have a certain amount of money, so if they spend it on PGD, they're going to spend less on something else. They don't argue that it's a bad thing medically, only that they'd have to take dollars from another pot."

Insurers may also not cover PGD because the procedure and the necessary IVF are more expensive than testing the fetus through amnio or CVS and then performing an abortion if a serious mutation exists. Thomas reported that "insurers argue that 'the patient will terminate anyway,'" so they think 'should we cover PGD for $15,000, or should that patient have a pregnancy termination for a couple of hundred dollars?' Two CF carriers have a one-in-four chance of having a baby with CF. They try on their own, roll the dice, and do an amnio. If it's CF-positive, they have only two options: termination or delivery."

Providers thought that insurers would never publicly state this preference to have patients undergo abortions but that public awareness of this preference might motivate change. "Insurance companies can never say that it's cheaper to just have the couple terminate," Thomas explained. "But it might be good to make them realize that *they're advocating termination!* That would be uncomfortable to say—and a very powerful argument on our behalf."

Insurers may also focus only on their own short-term bottom line, rather than on longer-term projections that the healthcare system as a whole could save money by paying for PGD, rather than for a severely ill child's lifetime of care. "Insurance, generally, is very shortsighted," Tim, an infertility doctor in a private practice affiliated with a hospital, said. "They're interested in covering the lives they have to cover [now], with very little thought for the future. So, they'll do some crazy things, like not cover PGD for CF. Yet once a couple has a baby with CF, the insurer must cover all the costs associated with that. Insurers are thinking short-term: cover this delivery, and hopefully get that off their books, and make it somebody else's problem, so that *they* are not responsible." Alternatively, insurers could view embryo testing costs as preventative and less than the lifetime care expenses for a child with serious medical problems. "Would you rather have another child with a million-dollar hospital stay, or prevent it ahead of time?" asked Cathy, whose daughter had died at five months of a chromosomal abnormality (trisomy 22), which she now desperately seeks to avoid in a future child.

For-profit insurers may simply dismiss these long-term costs. "No insurance company thinks it's worth it to save them money 20 years down the road, much less 10," Anne, the northern California REI who opposed sex selection, concluded, frustrated.

Patients are then forced to grapple with profound moral, not only financial, tensions. Isabelle underwent an abortion because of Emanuel syndrome and was now pregnant again through PGD, though unsure whether her insurance would cover the procedure. "The company thought: Get pregnant on your own, and if the fetus turns out to have the problem, have an abortion." She felt that insurance employees are "paper-pushers and don't necessarily look beyond the simple business, bookkeeping end of it." They make reimbursement decisions but may not fully grasp the types, natural courses, or severities of inheritable diseases or the trade-offs and larger issues involved.

Insurers may also worry that if they cover PGD, demand for it would mount. "Insurance companies don't want to open the floodgates," Jennifer felt. "So, it's very easy for them to say, 'Your employer doesn't cover this. So, we have no choice.' Or, 'You have $10,000 for IVF. You can use *that* for PGD.'"

If insurers did cover PGD, they would also have to decide for which mutations to do so, raising challenges of how to weigh particular diseases. "Insurers have difficulty discriminating between very serious and less serious genetic conditions or genetic enhancement," Jennifer continued. "To them, genetics is all the same. They don't have a way of coming up with gradations about how severe it is, even with physicians' letters about medical necessity. Each of us carries a lot of heritable conditions we could pass onto children. To insurers, it's a slippery slope: they don't want to expose themselves to a potentially huge cost."

These corporations may also not cover PGD because they feel it is not accurate enough. "PGD doesn't always work, and has false negatives," Henry, a Midwest IVF doctor, said. "And *then*, insurers would have spent a bunch of money for nothing."

Providers Advocating for Patients

"We argue," Thomas, the East Coast academic hospital-based physician, said, "that they should cover PGD—that they then won't have to care for a baby with CF. But they tend to see it as: they would have to pay right *now*." Given these obstacles, many clinicians nonetheless try to persuade insurers, articulating the above rationales for reimbursement, but unfortunately, generally fail. As a doctor, Henry has "gotten into debates" with these corporations. "One of my colleagues had some luck in one case. But so far *we've* not had much luck."

Rarely, these companies make exceptions—for instance with patients who, because of strong conservative religious beliefs, would never terminate an affected pregnancy. "A few Orthodox Jewish couples appealed insurance decisions," Sam, the patient advocate, reported. "But you usually have to *scare* the insurer. The doctor wrote that this couple will *never* terminate a pregnancy. 'It's either PGD or they're going to have another sick child.' The insurer paid for it. But that's not the norm."

Providers can expend considerable amounts of time trying to assist patients in getting PGD covered, but succeed only partially, at best. Anne, an REI in a large practice, has "only once successfully challenged an insurer to cover it—a patient with Reiter's syndrome. . . . She had IVF coverage for infertility, but not for genetic diagnosis. . . . She had a lot of resources—intellectual and educational."

Ironically, insurance company employees' misunderstanding of these issues occasionally leads them to cover PGD because they fail to comprehend it. Yet that reason leaves providers wary. "One state has a mandate that if you have 50 employees, they get IVF insurance for up to four cycles," Jill, the Illinois REI, said. "Most of that insurance will cover PGD, even if patients aren't infertile—but *not* because the insurers know [the] difference."

Providers who have failed to convince insurers can feel exasperated. "It was a waste of time and effort," Bill, a New England REI, sighed. "After I got beaten up a few times, I realized there's the 'party line.'"

Doctors may understand and appreciate insurers' priority rationales and limitations but remain frustrated. "Insurance is a business," Nicholas, a New Jersey private practice infertility doctor, acknowledged, "if they pay for everything, they're not going to be in business." Still, he and others felt that these corporations could make more longer-term, cost-effective decisions.

Patients Forgoing Embryo Testing or Paying Out-of-Pocket

Many patients must thus make excruciating decisions, weighing the costs of disease prevention through embryo screening against the risks of a child developing the disease, which can vary depending on the particular mutation's penetrance and lethality. Patients may strain to pay for PGD when confronting a highly penetrant and lethal mutation but be unsure about less severe mutations. Regarding ankylosing spondylitis, which Anne mentioned earlier, "One couple would probably have done PGD, but didn't have insurance for it, and so decided not to. They had a baby who wasn't tested. So, they don't know if the child has the disease." Uncertainty shadows them.

Patients must commonly pay PGD costs themselves out-of-pocket up front and afterward try to convince insurers to cover the amount. "We had to pay and hope we got reimbursed," Isabelle said. "We figured if this whole process cost us $4,000, it would be worth it. But most people have to pay 100% out-of-pocket."

Thus, only after their child is born do many couples learn whether their insurer will in fact offset any of the costs. "The most frustrating part of the whole IVF process was definitely just not knowing," Isabelle added. "Not only are you going through infertility issues, but you spend hours on the phone with insurance companies for months. They ended up covering the PGD, but only six to eight months later!"

To try to obtain coverage, patients may even move to other states. After aborting a fetus with CF, Sally and her husband felt they had little other option if they wanted a healthy child who was biologically theirs. "We moved out of state to get insurance for IVF/PGD." Yet their two cycles there failed. Decisions to relocate to different regions for ARTs can thus require major gambles that can fail.

Even after covering high costs for a prior sick child with a lethal mutation, insurers may remain short-sighted with a particular couple. "One child died of myotonic dystrophy, and the parents went through a lot of money as he was dying, covered by the insurer. But the insurer would still not cover PGD," Nicholas, the New Jersey REI, said. "The couple was fighting, and ended up very bitter, paying out-of-pocket."

Needs to Confirm Embryo Tests

Embryo tests can err—though overall, relatively rarely[31]—and providers generally understand these limitations and seek to confirm results through amnio or CVS.[32] Patients screening embryos should thus also perform confirmatory testing of the fetus using amnio or CVS but often refuse. PGD can be inaccurate, in part because an embryo's cells may not all be genetically identical but rather vary, creating mosaicism.[33] Amnio and CVS require, however, sticking a needle into the amniotic sac holding the fetus, and can harm the pregnancy, triggering miscarriage in about 1% of cases, though this rate appears to be decreasing over time.[34] Perhaps as a result, only 37% of patients becoming pregnant after embryo screening undergo this follow-up testing, of whom 40% found it more stressful than the PGD itself.[35] Both providers and patients therefore face dilemmas. "CVS is advisable for every couple, but only 50% or fewer do it," Jill said. "CVS has its own risks: 1% or 2% misdiagnosis, and potential miscarriage."

Patients Refusing Confirmatory Testing

Many prospective parents shun fetal testing because they oppose abortion and would never terminate the pregnancy based on the results. Patients may test an embryo but not a fetus because they would refuse abortion. Yet, lack of confirmatory testing can produce children with devastating diseases. Henry described two cases where the embryo tests were normal but the child was nonetheless born with the disease. "In one case, we were trying to avoid Fanconi's anemia, while trying to do an HLA [human leukocyte antigen] match for a child that was affected. The child they delivered in fact *had* Fanconi's anemia. In another case, the child had an unclear rare storage disease," in which the body fails to store properly enzymes that digest proteins and other chemicals, leading to "prolonged degeneration and deafness. In both cases, we counseled the parents extensively that we can have a false diagnosis, and do amnio to make sure the PGD is accurate. In both cases, they refused. The parents said, 'We'll just take whatever happens. After all this, if it's God's will that I have another child that's affected, I'm just going to have to deal with that.' And that's what happened."

Even for women not undergoing IVF and PGD, decisions of whether to undergo amnio or CVS can be fraught. Karen, the California physician whose son, born through IVF, was deaf, recognized the harsh trade-offs involved for Down syndrome or CF. "The mothers have been pregnant for a while, and will have felt the baby move inside of them. That's hard. The fetus looks perfectly developed, but is tiny. It would be really tough to make a decision, say for CF. Few people with CF have normal lives."

Women who test embryos may decline amnio or CVS because they miscomprehend the test's purpose and the potential errors of PGD, and hence the need for this follow-up testing. Possible miscarriage also frightens many patients who have lost prior pregnancies. "I didn't understand," Cathy, whose daughter died of trisomy-22, said. "They suggested amnio or CVS, but I thought that testing was the whole point of PGD! I didn't quite understand that it's not 100%. Amnio or CVS have a chance of miscarriage. So, I thought, *I'm not going to bother with it.*" She also opposed abortion, which may have fueled her wariness of fetal testing that might lead to recommendations to terminate the pregnancy. Her physician, discussing embryo screening, presumably told her about the need for follow-up fetal testing, but she seemed not to remember that.

Yet while some women refuse abortion and thus fetal testing, others undergo this testing, not to terminate the pregnancy but instead to prepare for having an ill child. These patients test the fetus not to potentially end the pregnancy, based on the results, but only to ready themselves psychologically to have a child with a disability. "I'm not for abortion, but you can *prepare yourself* to deal with things," Cathy explained. "Another mom had amnio and found trisomy-13, but didn't undergo an abortion. If [my husband and I] get that information, we would deal with it"—not have an abortion, but mentally prepare to have an ill child.

Women also avoid confirmatory testing of fetuses because they fear resultant miscarriage. Especially for women who had prior miscarriages and/ or difficulty conceiving, such pregnancy losses can be devastating, given the high stakes. Women who have screened embryos, "because they've had a lot of miscarriages tend not to do amnio or CVS," Sally, who herself had aborted two fetuses due to CF, said. "If an embryo stuck this time, it's probably a good one. They feel they're more at risk for having a miscarriage. You don't want to take on even that 1 out of 100 or 1 out of 400 chance. I'm sure I'd be the same way."

Other women struggle to balance their wariness of abortion and hence of fetal testing against the severity of the specific mutation they might transmit. "It depends on the genetic disease and whether you would terminate," Sally added. "One woman did IVF/PGD [for a disease], got pregnant with twins, and decided *not* to do amnio or CVS. One baby ended up being misdiagnosed. The other lived for a little while, and then died from the disease. With CF, the child would probably live for a time, which in ways, is worse. I don't know."

"I was going to terminate if the amnio was CF-positive," Sally said. "Most people would terminate. But every situation is different. A lot of people try to prevent a disease, but won't necessarily terminate a pregnancy for that disease."

Waiting for amnio and CVS results can itself also be stressful. "The genetic counselor had mistakenly told me that I would get the amnio results back in four days," Sally reported. "It ended up being four weeks! I almost had a nervous breakdown."

Creating "Savior Siblings"

Doctors and patients can also choose embryos so that the future child can donate stem cells or other tissues to an existing child with a severe disease who needs blood or other human tissue transplants. Embryo screening can thus be used to design a new child—a so-called savior sibling—to be an immunological "match." This procedure, in which doctors select embryos with certain immune system markers, known as Human Leukocyte Antigens (HLAs), is known as PGD-HLA and has been used for an array of disorders, including sickle-cell anemia, Fanconi anemia (FA),[36] Diamond-Blackfan anemia[37] several types of leukemia[38] and beta-thalassemia.[39,40] For 70% of patients with these diseases, no existing HLA-compatible donors are unavailable.[41,42,43,44,45] Hence, HLA-matched "savior siblings" can provide umbilical cord blood or stem cells for transplant. Among 137 IVF clinics in 2006, 24% had used PGD-HLA.[46] As IVF, PGD and genomic testing continue to expand, demand for this procedure will surely continue to rise.

Yet the procedure does not directly benefit the future child, who may in fact be burdened with the need to regularly donate tissue. Providers thus have to consider the rights and well-being of this future, unborn infant and weigh these concerns against the potential benefits to the current, sick child. These future risks and benefits are inherently difficult to assess and balance, partly because they impact different individuals, one of whom is not yet born and thus cannot speak for him- or herself.

Further logistical obstacles to PGD-HLA emerge as well. In one study, only 12% of embryos tested were both free of the disease-causing mutation and HLA-compatible, and only 18% of couples who attempted to create a "savior sib" succeeded.[47] Other limitations include implantation rates of only approximately 29%, tissue graft rejection, ill siblings dying before the transplantation occurs, and misdiagnosis. In one study of two networks of parents with FA children, fewer than 35% were offered PGD and only 70% were aware that PGD-HLA was an option.[48]

Critics have expressed concerns that parents may view "savior siblings" as merely a means to an end, rather than as valuable human beings in and of themselves[49] and that the procedure should be treated as experimental research, not clinical care.[50] The UK government has permitted PGD-HLA but reviews each case individually, and the embryos must be at risk of the disease as well.[51]

But parents may in fact view such children lovingly, rather than instrumentally.[52,53] Proponents have pushed to end the debate about whether such "designer babies" should be permitted, suggesting instead guidelines regarding PGD-HLA use. These advocates aver that clinicians should discuss this procedure with families only when matched donors are unavailable, when the mother is still of reproductive age, and when the ill child has only been diagnosed within the prior few weeks and is expected to survive at least a year.[54]

Still, many questions persist about how infertility providers in fact view and make these decisions; what difficulties, if any, these clinicians encounter; and whether they discuss these issues with patients and, if so, how. Strikingly, no studies have probed these questions.

Providers and patients, I soon found, face ethical and logistical challenges concerning whether to try to create "savior siblings" and, if so, how much so and how to decide. While many doctors have not yet confronted these issues, others have done so. Henry, the infertility specialist at a Midwest medical center, for example, has performed "a couple of cases for HLA-matching for bone marrow transplant in a sibling."

Nonetheless, providers who have not yet themselves performed PGD-HLA are aware of colleagues who have faced the possibility. "HLA-typing has not come up in my practice, but has in many other practices," Joe said.

Providers may be uneasy and uncertain about the procedure since it creates but discards viable embryos and seeks to prevent problems not in the parent or future offspring but in an existing child. "HLA-typing with marrow donation to a living child is difficult because the couple doesn't have a disease," Henry explained. "They can conceive on their own. You're creating embryos, knowing that you're going to discard a bunch, simply because they aren't an HLA match—not because they are defective in any way. . . . In our business, we commonly discard embryos, but don't like the idea of creating embryos for no reason. In this case, there's a reason, but it's a little bit different."

"I had a sad situation of a little boy, now nine, who had thalassemia major," Roger, a New York hospital-based infertility doctor, stated. "The boy's wealthy parents wanted to do PGD to avoid transmitting the disease, and wanted to do HLA matching so the new child could transfuse their son. The mother had already been through the procedure five or six times at another clinic and failed. The odds of success were less than 1% that the parents were going to get what they wanted. The parents said, 'Money isn't a problem,' so we tried it, but kept failing, going on IVF cycle 10 or 11. We could not get through to them that we needed a plan for closure." Such ongoing failures can disturb both providers and patients.

Unsure whether to start a twelfth cycle for PGD-HLA, Roger "went to the ethics committee. One member said, 'You are aiding in this woman's suicide.' I said, 'That's a vicious thing to say. I'm not that kind of person. I'm not handing this lady a gun.' They said, 'Yes, you are. You have no idea how these drugs are going to affect her. She may get ovarian cancer and die because you've stimulated the hell out of her. There's no literature to say this is safe.' I said, 'There is a difference between a woman who kills herself by standing on the railroad tracks, waiting for the train to hit her, and the woman who notices her two-year-old playing on the tracks, and runs to grab him to throw him off the other side, knowing that she's not going to make it.' That is a form of killing herself, but isn't suicide—the motivation is different. She's not trying to die. She wants to be the mother of that kid, but thinks the future kid is more important than she is. If you had a nine-year-old with thalassemia major getting transfused every week, who can't run around with the other kids, at what point would you say 'we're going to stop?' So, we did it two more times. But it didn't work. We then got the parents to stop. We were honest enough to tell the ethics committee that we couldn't answer this. They said, 'You have to weigh these things.' They never give you the answer." Consulting ethics committees can potentially help but poses challenges, especially since a single "correct answer" may be elusive.

Physicians face conundrums of *how many times* to attempt to create savior siblings since success rates per cycle are relatively low. Parents seek to select both *against* the mutation causing the disease as well as *for* a certain immunological HLA match. Yet using PGD to screen for more than one target at a time is hard.

Parents must decide how much they want a savior sib versus merely a healthy child. Jill described dilemmas of how to proceed in a case when HLA-typing failed: "We did not have success, and ultimately the kid had a transplant with a donor from a cord blood bank. Then, the couple wanted another child, not just to create a match, and ultimately delivered a healthy, non-matching sibling. Thus far, the ill kid who had the transplant is doing well."

Physicians may assist such couples if and only if the prospective parents would have had another child anyway, but how to assess such "intention" is unclear. Couples may insist that they are not planning to have a child *merely* to create a donor, but their desires can be difficult to gauge since the parents may be conflicted, wanting a new child to "rescue" their existing child. Providers and parents may also clash on these issues. In a study of parents of 74 children with FA, 89% valued most the health of their FA-affected child, and only 20% were concerned about the ethics of choosing a child's characteristics.[55] Yet, providers may express concerns about the unborn child and the discarding of good embryos. Prospective parents may thus not answer questions about their motives truthfully. As Nicholas said about a sickle-cell anemia case, "I would not feel good if a couple were going to create a life *just* to get bone marrow. But if a couple wants to have more kids, and this is an added benefit, that's fine."

Proponents argue that the benefits to the current child outweigh the burdens to the future offspring, justifying the procedure. Yet unborn offspring arguably have rights to an "open future," to make their *own* decisions about their lives as much as possible.[56] While the future children may feel grateful to be able to help their ill siblings, that may not always be the case.

The long-term psychological development of such savior siblings as they grow and can themselves convey views also remains unknown—for example, whether they perceive their experiences positively or negatively. Research is urgently needed.

Several recommendations have been proposed to routinize PGD-HLA,[57] but additional guidelines are needed, such as requiring robust informed consent to ensure that prospective parents fully understand the risks, limitations, low success rates, costs, and ethical concerns that savior siblings may be used as a means to an end. The possibility of such instrumentality need not bar PGD-HLA, but doctors should discuss it with prospective parents. Professional guidelines could address how clinicians should proceed if PGD-HLA fails—the number of cycles to do and the potential value of consulting ethics committees. Guidelines should address, too, whether providers should ever encourage parents *not* to pursue PGD-HLA and instead have a healthy "replacement child" who is not HLA-matched. Arguably, providers should at times consider doing so—especially when efforts at such HLA-matching have failed.

Testing Embryos Without Revealing the Results

"The most challenging cases are non-disclosing protocols," Jennifer, the Northwest medical center physician, said, "when the patient is non-disclosing, but *I can see symptoms!*" When prospective parents at risk for a serious mutation such as HD want to ensure that their child does not receive the mutation but they do not wish to know their own mutation status, doctors can perform such non-disclosing PGD. Since HD, a fatal neurological and psychiatric disease caused by an autosomal dominant mutation, lacks treatment, most individuals at risk for it do not want to know whether they will have the mutation, seeing it as a "death sentence."[58] Depending on the country, only around 3%–21% of at-risk adults get tested for HD—only 3%–5% even in Sweden, for instance, which has universal health coverage.[59]

In non-disclosing PGD, the physician transfers into the womb only mutation-free embryos and does not tell the patient whether any embryos in fact had the mutation. Individuals at risk for HD can thus have a mutation-free child, without having to learn whether they themselves have the defective gene. Such embryo testing without disclosure of results, though performed, particularly for HD,[60,61] remains controversial.

A geneticist, Ayelet Erez, and her colleagues oppose this procedure, arguing that the parents' right to remain ignorant of their HD mutation status has limits.[62] These authors describe a woman seeking PGD for HD since her husband did not wish to know whether he had the mutation, though his father had it. Yet due to testing done during a prior pregnancy, the woman's clinician knew that the husband was mutation-free. These critics argue that PGD, if the clinician agrees to perform it, would be a "sham," subjecting the women to IVF, which carries risks, and PGD when the parents could instead have conceived naturally. Moreover, if all the embryos are mutation-positive, clinicians "could feel compelled" to still perform an embryo transfer, though not in fact transferring any.[63]

Critics have also asserted that doctors should instead perform so-called exclusion testing—checking the affected prospective grandparent, if he or she is alive, and the fetus, to determine whether the future child has inherited one of this grandparent's fourth chromosomes, which might contain the HD genetic defect.[64] Yet such exclusion testing has limitations. It requires, for example, discarding half the healthy embryos. Moreover, genetic recombination can occur, in which bits of DNA switch between chromosomes. Hence, an affected grandparent's fourth chromosome, even if the fetus inherited it, may not in fact contain the mutation. Nonetheless, IVF and PGD would still be required, even though these procedures may not then be necessary.[65] Given the limitations of both non-disclosing PGD and exclusion testing for HD, the Dutch government has banned both.[66]

But questions arise of how providers and patients view these issues—what they decide and what challenges they face in weighing the pros and cons.

Disclosure Dilemmas for Providers and Patients

Non-disclosing PGD, I found, can put clinicians in awkward positions, especially when they know the patient's mutation status, while the patient wants to remain ignorant. "Patients are supposed to be presymptomatic or asymptomatic, but *I know!*" Jennifer elaborated about cases involving HD. "I'm not a neurologist, but part of my obligation is to know whether symptoms now may affect how safely the parents will raise a child. With one patient, I could see the symptoms progressing. She could not sign the consent form because she was so severely affected. Up until the very end, the husband was in denial, saying he didn't see any symptoms. They did several cycles of IVF, and never took home a baby. They decided to conceive naturally, and asked me to do a non-disclosing prenatal diagnosis. They still wanted to be in the dark as to whether, if we terminated the pregnancy, it was due to Down syndrome or HD. I went through all of this very carefully with them and agreed—reluctantly—that we would do non-disclosing prenatal diagnosis. I said the odds were '50 to 1' that if I told them to terminate, it would be because of HD, not Down syndrome. They got pregnant, and ended up with an affected fetus, and terminated it. I didn't tell them it was HD. They got pregnant again—naturally—and the baby was unaffected. But by then, the mother was very symptomatic. The family finally admitted she was symptomatic—but unfortunately, only when the baby was a few months old, and had an infection and trouble breathing. The mother literally didn't know how to take care of her baby and called 911. When they came, the mother had a psychotic break and became belligerent, hostile and aggressive. The family realized then that she was not fit to care for the child. That was very difficult for me, because we were trying to guide them as much as possible, without telling them what everyone around them was saying. *That* type of situation is the most difficult for me—when I know what the patients don't know, and I'm trying to think about their best interest, but allow them to have their autonomy, not telling them information they don't want."

"The hardest decision, which just tears me," Jennifer continued, "was a patient doing an HD non-disclosing PGD elsewhere, who then came to me for prenatal diagnosis and help having a child. The other IVF program did not really appreciate that a non-disclosing protocol meant that *I* also don't know the patient's genetic status—so that it's not influencing how I counsel patients. This IVF program sent me the patient's genetic test results. So, I then went from *not* knowing to knowing. The patient assumed I didn't know; but I *did* know, and found it challenging to be completely even-handed."

Clinicians who strongly suspect or know the patient's gene status face quandaries concerning both the content and process of interactions—whether to be neutral when they have information that the patient does not want. "If they don't know their odds, and I do, it becomes more difficult for me to be truly non-directive," Jennifer explained. "I can't disassociate the rest of me. It's hard for me to objectively step back and say: 'Did I really present that even-handedly, or did I favor one approach more than another because I had more information?'"

Other physicians have performed non-disclosing PGD when they knew that the at-risk parent did not in fact carry the mutation, and that the procedure was therefore unnecessary—"a charade."

Doctors and patients also face questions of how to hide the test results in the electronic health record, given possible discrimination. "Patients worry about what's going to be in the record," Jennifer explained. "How do we do things under the radar?"

Provider and Patient Responses to Testing Embryos Without Disclosure

In responding to these challenges about non-disclosing PGD, physicians vary from avoiding the procedure to remaining uncertain or performing it in differing ways. Many doctors remain torn. "Non-disclosing HD is a weird one for me," Peter, the West Coast REI who now uses PGD only for sex selection, confessed. "I understand it. I've heard all the arguments: there's discrimination. But I just don't know. Misdiagnosis is possible."

Both providers and patients are often unsure. "A couple talked about 'non-disclosing' PGD for HD, but did not come back for treatment," Edward, the Ohio REI, reported. "We're still thinking about it." Based on hope, age, finances, attitudes, risk tolerance, and reproductive history, patients may also debate these decisions. "Many couples will be fluid," Jennifer added. "They go through non-disclosing PGD, but after many unsuccessful cycles, say 'Maybe we should just do this the old-fashioned way instead.'"

Patients may also want to pursue variations of non-disclosing PGD. Clinics differ in whether and how they offer these. Certain providers and patients mistakenly think the procedure entails revealing the results to the spouse but not to the at-risk prospective parent. Jennifer described "a man at-risk for HD who didn't want to test himself, and wanted us to disclose the results to his wife, but not him. If an embryo came back positive, she would know that he was HD-positive, but he would still be blissfully ignorant. That was just preposterous! I said: 'You need to think about what you're saying, and why. And why this doesn't really make sense for the two of you.'"

Given these complexities, clinicians may offer non-disclosure testing but require robust counseling and informed consent to ensure that patients fully grasp all the potential limitations. Other providers may neither readily agree nor decline to perform the procedure but rather consider these issues carefully, counseling patients and/or recommending external medical or neurological consultations. Providers face questions regarding the extent of their responsibility to the at-risk patient, especially if symptoms appear—whether to encourage the affected parent to get a diagnostic evaluation. Jennifer offers "a neurological examination and genetic testing. I'll ask their spouses what they think. If they ask me what to do, I always suggest a formal neurological exam."

Providers and patients may also consider alternatives, such as using donor gametes. Peter described a couple who underwent four cycles of non-disclosing PGD for HD. "They had no pregnancies, so decided to use donor sperm. She got pregnant and delivered."

Thus, while critics have argued that non-disclosing PGD could lead to sham transfers of embryos and should therefore be banned, this possibility presents only one of several potential outcomes and, arguably, should not be the basis for universally prohibiting this procedure's use. Informed consent can specify that no embryos may be available for transfer—that is, that none may be viable and mutation-negative and thus transferred. Thus, "sham" transfers need not occur. Frequently in IVF, no embryos are available for transfer, for various reasons unrelated to PGD, such as none appearing healthy under the microscope.

Moreover, risk–benefit assessments of whether to ban all non-disclosing PGD must include the fact that if it were prohibited, at-risk patients who wanted a mutation-free child would have to undergo testing themselves. But 50% of these patients would learn they have the mutation and have increased risks of severe depression and suicide. HD patients kill themselves at around 10 times the national average.[67] The relatively small self-selected minority of at-risk individuals who undergo testing (about 5% in the United States)[68,69] has presumably undergone extensive counseling and sees benefits of testing but still has increased rates of depression and suicide.[70,71] Tested individuals may not all have mental health problems but do not represent most at-risk individuals, constituting instead only a small, self-selected minority who feel they will be able to handle knowledge that they possess the lethal mutation.

Questions therefore arise of whether these extraordinary risks of depression and suicide outweigh the potential harms of IVF to the mother. If this procedure were banned, an at-risk woman (or man) who wants a child without HD would have to abort fetuses with the mutation. After thorough informed consent, she may instead decide to accept the risk of IVF and PGD—knowing that these procedures may not in fact be necessary—because she wants to avoid the mental anguish and suicidality that she might experience if she learned she had the HD mutation. Such a decision by her seems reasonable.

Additionally, given the principle of autonomy, to ban her from being able to make this choice about her own body and instead to compel her to have an abortion if she wishes to have an HD-free child does not appear entirely justifiable.

Though non-disclosing PGD might lead to cases in which IVF/PGD is performed if patients are mutation-negative (with a "sham" transfer of embryos), this potential scenario needs to be viewed not in insolation but as only one of several possible scenarios. Debates about whether to prohibit non-disclosing PGD should thus consider *all* of the risks and benefits involved with *the full range* of potential cases—not only one subset of patients. Clearly, thorough informed consent is imperative, to ensure that patients grasp all of the pros and cons. Ultimately, some providers may not want to offer this procedure; but the complete set of issues involved should at least be considered before universally outlawing the procedure.

If the practice is permitted, insurance coverage should also be considered so that the procedure is available not only to wealthy patients who can pay out-of-pocket. In addition, professional organizations should develop guidelines and "best practices," addressing whether, when, and how to perform the procedure and what issues clinicians should discuss with patients.

Leftover Embryos

As Jim showed me in his office, REIs store thousands of barrels of unused extra frozen embryos. Patients can keep these embryos frozen, discard them, or donate them to other prospective parents or to research. These options each pose challenges. Many of these clumps of cells were rejected because of suspected problems with their genes or physical appearance. Yet embryos, even if of poor quality, contain both parents' genes and represent potential future offspring, and few parents like destroying them.[72] What, however, should then be done with these cells? Their ongoing permanent storage can be expensive. "There is maintenance," Steve, the Virginia REI, said, "having cryo-tanks and liquid nitrogen—resources, time."

Among patients with frozen embryos, 54% have said they would be very likely to use them for reproduction, 21% would donate them for research, and less than 7% would choose other options, such as discarding them or donating them to other patients. In various studies of IVF users, those who report donating embryos for research trust the medical system and view science positively,[73] while others view embryos as persons and cite risks and lack of information about the specific research projects. Patients may also not know about donation for research or be given this option.

A few states have laws regarding the disposition of frozen gametes or embryos—usually in case couples divorce or one member dies. Louisiana bars destruction of embryos, seeing them as possessing the same rights as people. Florida requires that prospective parents and the provider sign an agreement before treatment regarding future embryo and gamete disposition.[74]

Many providers and patients struggle with what to do. A few patients do readily discard them. John, the Texas mechanic who underwent a vasectomy, when asked by his infertility doctor about disposition of leftover embryos, replied, "I don't care. Do as you wish with them. Donate them for stem cell studies? Fine. A lot of people believe that as soon as you have an embryo, it's a person. You start getting into religious beliefs. For me, if it helps doctors to further research, saving people's lives and helping people have children, I'm all for it."

But embryo disposition disturbs other patients. "Whether to dispose of extra embryos, or continue to pay for them to be frozen is a hard decision," observed Ginger, the IVF clinic therapist who underwent IVF herself and adopted her second child. "One couple had a frozen embryo, but ended up adopting, as opposed to trying to get pregnant again. They thought they would sign away the embryo. But it turned out to be a very emotional experience—especially for the woman. So, they decided, 'Let's keep the embryo, and continue paying for its storage.' The decision is a little bit easier if you've already created your family, and have the number of kids you want. No one I know has donated their leftover embryos to another couple."

When initiating treatment, clinicians should address embryo storage with patients but may still encounter quandaries. "I discuss storage of embryos up front at our initial visit," Henry said, "the possibility that they will have extra embryos, and that several options exist of what to do with them. Most couples have no ethical dilemmas, and just say, 'I'll discard them and it won't bother me.' Other couples have thought it through, and say 'That's a problem.' We then limit the number of eggs we inseminate. But an advantage of IVF is the selection process: you're usually eliminating some, which get left over."

But leftover embryos may have poorer quality—which is why they were not chosen—or possess traits deemed undesirable. "What do we do with the *other-gendered* embryos?" Brenda, the Vermont psychotherapist, asked about cases where patients choose to select embryos based on sex.

Unfortunately, leftover embryos and their creators can also get judged. After Karen used IVF and had a son born deaf, no one wanted her remaining embryos, which disappointed and saddened her. "I don't know what I would have done had I known, before implantation, that he would be deaf. It's very hard to know what decision you're going to make until you're actually in that position. I used to make fun of deaf people. When they talked, they didn't sound very intelligent. I thought they should sign, rather than talk, because signing looked very elegant."

Karen now feels that prospective parents and providers preemptively and unfairly judge embryos. Whether her son's mild disability is hereditary is unclear. "They don't know if it's in fact genetic. He is otherwise completely normal. My mother may have a similar, though much milder trait. I have no idea whether it is genetic or not. The actual gene has not yet been discovered. There's not a lot of research on whether it's genetic."

The fact that PGD requires discarding unused embryos troubles many patients morally. "My son is perfect," Karen continued. "Well, not perfect—he's the biggest pain in the ass right now—but he's a great kid. He's handsome, smart, funny, reads well and is very perceptive. After his diagnosis, I worked very hard with him. Had he been born in a different family, it would have been different." Prospective parents who reject such embryos are doing themselves a disservice, she feels. "I know what it feels like not to be able have a kid when you want one—how hard it is to be infertile, how desperate you are to have a child. So, it's ridiculous that people are prejudiced against an embryo for that diagnosis. They are *losing out.*"

She wanted to give the embryos to another couple who desired a child, but doing so would entail freezing them, for which she would have to pay. Karen's husband thought the expense was too high, highlighting how much members of a couple can disagree and raising questions about how much these possibilities for future life are worth. "I never intended to use the extra embryos," she explained, "and would have loved to give them away—to give a child to someone. I wanted to just keep the embryos frozen. But my husband didn't want to spend the money anymore—a few hundred dollars a year."

Providers may avoid arranging for embryo disposition and donation due to fears about litigation, government regulations, and the religious right. "The government treats every egg as if it's a kidney," Diane said, "making everyone jump through hoops, testing. If a woman wants to give embryos she made with donor eggs to her sister, we can't transfer them, unless the sperm came from the woman's husband. . . . It makes problems for people wanting to donate leftover embryos to other couples. They can't, because of the FDA and donor organ laws. Couples are certainly willing to give embryos, but clinics are terrified to touch them, because the FDA could shut them down. We all believe embryos should be given for free, so there's no money in it. We don't want the Christian right dealing with it; yet they have the money and impetus to lobby. You have to get your legal department involved. It's tedious, time-consuming, and expensive. For most clinics, it's just an altruistic sideline."

Fears of potential religious and hence political opposition can also hinder providers from arranging for embryo donation for stem cell research. "We've never been hassled by the religious right," Bill, a New England REI, said. "They may create problems once we start donating discarded embryos for research, but so far they have left us alone."

Policies or regulations could help. "It would be terrific if the profession had clear guidelines," Steve remarked. "Maybe orphaned embryos should go to a federal facility for a period of time, and then get destroyed or done whatever with. But it's hard to change government."

Clinics may consequently keep embryos indefinitely, despite patients giving permission to discard them. As Jill, the Illinois REI, said, "Theoretically, we can legally destroy unclaimed embryos after three to five years, if the patients don't give us a disposition on the consent, or they fail to respond to letters after a certain point in time, or fail to pay their bill, and we're unable to find them. But we don't destroy them because patients could come back and say, 'Where are my embryos? I never got the letters you sent.'"

Moreover, providers feel that the costs of storing embryos are relatively small and can be built into IVF fees. "That's not the biggest issue," Jill continued. "Because it's not *that* expensive." Still, such costs then get included in clinic fees that thereby rise.

Many clinics retain embryos because patients may sign permission for disposal but remain ambivalent, and the moral status of embryos is controversial. "Embryos are stacking up," Diane, an IVF nurse, said. "Everybody's afraid of throwing out leftover ones because of Christian right/government concerns. You have to have a signature saying to discard them. But most patients are so ambivalent about it that they aren't signing to discard them, but don't want to pay for it, either. They're an incredible research resource, but nobody's willing to use them because of fears the government will shut them down. It's a problem."

Given these complexities, clinics, especially at academic medical centers, may consult a formal ethics committee. Henry's ethics committee was formed "when we developed one of the first embryo donation programs. We were wrestling with: What options should we give parents? Is this donation or adoption? What priorities should we set for *who* gets the embryos?" Yet many clinics lack such committees and may be unsure how to resolve these conundrums.

In short, embryo testing can benefit countless patients but poses questions, including for which diseases to use it (from mild to severe childhood- to adult-onset diseases and future or existing children), whether to confirm the results via later fetal testing, how many times to attempt the procedure, who should know the results, who should decide, how much to pay for it, and what to do with leftovers. Doctors can select embryos for ever-widening numbers of genes that each raise related but also differing issues. Individual providers' and patients' views and experiences shape these choices. While some patients test embryos to avoid needs to abort fetuses with serious mutations, other potential parents decline both PGD and abortion, not wanting to "play God" and feeling that children born with certain diseases can still live many years, even if sick or disabled, or not have symptoms for decades, if ever, and eventually have improved treatment options. In addition, embryo testing carries risks due to IVF and can be inaccurate.[75]

Given possible PGD errors, providers routinely recommend confirmation through fetal testing once the pregnancy occurs, but some prospective parents decline, concerned less about PGD errors than about abortion and possible miscarriage due to amnio or CVS.

For many women, abortion is stressful to undergo, but pursuing it because of a mutation can add difficulties. Most abortions occur because women do not want to have the child, yet women who terminate a fetus because of a mutation in fact *want* to have the child but feel that the mutation is too grave to warrant life. These women must balance complex competing sets of risks, benefits, and moral values.

The American College of Genetics and Genomics has issued policies concerning genetic testing of potential parents, fetuses, newborns, children, and adults[76,77,78] but not embryos, and should consider doing so. Especially with increasing sophistication of genetic tests, such recommendations should address whether doctors should routinely check embryos for certain genes and, if so, which; what genetic information REIs should give to potential parents; and how providers will discuss and patients will understand these issues. Currently, patients and providers are often ill-equipped to confront these challenges.

Given that insurance policies impede access to care both generally[79] and regarding PGD, forcing countless patients to make suboptimal choices, insurers should reconsider refusals to reimburse these expenses. These companies should see PGD as a form of disease prevention and adopt longer-term perspectives, not merely assume that a child born with severe disease will probably be another insurer's responsibility. Insurers may simply not have thought through all these issues and instead just find refusal easier.

Larger questions surface of whether PGD should always be covered and, if so, who should decide and how. Ideally, insurance should include essential healthcare services, but dilemmas arise of whether PGD fits that category, given limited healthcare resources overall. The US and many other countries lack constitutional rights to health or to have a child. But I would posit that PGD should be covered, if the prospective parents want it, to prevent severe disease. By preventing severe untreatable disease, the procedure can advance the welfare of the future child and decrease society's overall health costs. Still, gray areas emerge regarding where to draw the line—how to define "serious disease" versus disability or milder treatable conditions. Breast cancer, for instance, is treatable and only afflicts adults. For such diseases, prospective parents certainly have rights to choose PGD; but to argue that they have rights to insurance coverage for the procedure is harder. Countries vary widely in what prenatal genomic tests they cover. In only 36 and 29 states, respectively, does Medicaid covers genetic testing and counseling regarding fetuses.[80] Louisiana requires insurers to cover screening only for cleft palate/lip. Alabama requires insurers to cover "medically necessary" prenatal fetal testing, especially if the mother is 35 or older. California requires coverage of disorders in cases of "high-risk pregnancies."[81] Given rising health costs, rights to have treatment available and to have insurance cover it clearly differ.

Ideally, insurers should be required to cover PGD for serious infant or childhood disorders, given the long-term benefits. Doctors should also ensure that patients understand the costs, in order to reduce financial uncertainties as much as possible. Enhanced understanding of these issues may also help insurers and their employees who make these decisions appreciate the long-term public health and industry-wide advantages of coverage. Given these challenges, physicians and their patients may benefit from doctors subspecializing in PGD (to ensure possession of optimal expertise) or otherwise avoid performing it.

Even when patients agree to discard or donate embryos, dilemmas can emerge. Some patients simply stop paying for storage. Though doctors may try to address, in advance, options concerning leftover embryos, patients may not always know beforehand how they will later feel. Physicians may seek compromises, storing embryos indefinitely; but costs then mount. Though doctors can build these expenses into clinic fees, treatment prices then increase. Better informed consent can help.

These choices prompt broader questions, too, concerning the moral status of embryos. These cells should be treated with respect, but how much so? While the religious right might argue that these cells should all be donated to other patients, these embryos were often rejected for transfer because their quality was significantly less. Providers, too, can feel uneasy about these quandaries, seeing them as highly personal patient decisions.

Improved education of providers and patients about these issues is vital. Clinicians (including REIs, obstetrician/gynecologists, geneticists, nurses, and genetic counselors) need to understand and be prepared to address these complexities with prospective parents. Pediatricians, internists, and other physicians who genetically test children or adults should consider the reproductive implications for family members and be able to discuss with patients embryo testing. Patients should understand the realistic odds of success and the ways phenotypic expressions may vary.

As we will see, additional predicaments surface concerning selection of genes associated with not only disease but with socially desired characteristics, such as the child's sex.

7 | "Family Balancing"
CHOOSING SEX AND OTHER TRAITS

"The biggest ethical issue is gender selection for 'family balancing,'" Thomas, an infertility doctor at a large East Coast academic hospital, declared. "For one case, we had about two years of ethics committee meetings, and after a long period reached a consensus. . . . Some places do it, but we don't. It's non-negotiable. When ART [assisted reproductive technology] started here, the program was set up through the hospital ethics committee. The committee rapidly decided that it did not have expertise in this area, and suggested an offshoot—an ethics committee specifically designated to look at assisted reproductive issues. The committee specifically looks at things like end-of-life decisions. It's like a court. On at least two occasions, once for gender selection using PGD [preimplantation genetic diagnosis], our ART ethics committee had an impasse, and took it to the hospital ethics committee, which also had an impasse and sent it back to us. But those discussions were helpful, and we then reached a solution.

"Our ethics committee includes community people, ethicists, physicians, geneticists, psychiatrists, social workers, and a judge. The hospital administration was involved. Whatever outcome we reach, we've thought it through extraordinarily carefully, and audited all of these processes, and collected minutes for all those meetings. You can disagree with the conclusion, but it would be hard to disagree with the *process*. That gives us both a level of discussion and safety, instead of 'I'm a single practitioner doing IVF in a rural area. I'll make it up, and decide by myself that I'm doing this.'

"We use 'the newspaper headline approach' to ethical decision making: How comfortable would we be if we were on the front page of the local newspaper, with the headline that our program decided to do elective sex selection or not? I'm comfortable, if the reporter puts into the article that we at least vetted the process very completely.

"The downside is time: my time and other people's time. That process can take three or four months. This couple is waiting to do IVF. That's a long time. There's a long learning curve. The chair of our committee is now one of the hospital attorneys, and is relatively new in that role—about a year—kind of a novice." The length of this decision-making process can vary, depending partly on how new versus familiar the dilemma is for the committee's members.

Doctors and patients confront questions of so-called positive selection—to choose embryos for various social traits—most commonly regarding sex but also deafness and dwarfism. Yet these requests can pose ethical conundrums. Many physicians routinely screen embryos for sex to prevent the transmission of serious sex-associated diseases, including disorders for which specific, highly predictive genetic markers have not yet been identified. But every year in China, India, and elsewhere, far more boys than girls are born, due usually to aborting female fetuses,[1,2,3] raising concerns about a possible "slippery slope" toward eugenics.

While Karen found that no one wanted her leftover embryos because she had had a son born deaf, other, deaf-perspective parents may indeed want a deaf child. The philosopher Michael Sandel opens his book *Against Perfection* by describing a deaf lesbian couple choosing a sperm donor with a long family history of hearing loss, in order to have such a child.[4] The media have also described dwarf parents seeking to have children with dwarfism.[5] Among IVF clinics in 2006, 3% have used PGD to select for a disability.[6] Yet the deaf community itself appears split. At a British "Deaf Nation" conference in 1997, 55% of 87 individuals with hearing loss felt that such genetic testing would cause more harm than good, 46% thought it might devalue deaf people, 29% would prefer not to have deaf offspring, and 16% would consider undergoing fetal testing.[7] Among hearing children of deaf adults, 46% felt that deafness was a distinct culture, not a disability; 72% had no preference regarding having deaf or hearing children; and 60% felt that individuals should not select for or against deafness.[8]

Individuals with achondroplasia (ACH), a hereditary form of dwarfism, were more likely than their average-height relatives to be interested, if pregnant, in fetal testing followed by possible abortion (62% vs. 28%).[9] Among both groups combined, 40% would consider abortion for homozygous ACH, which is fatal in infants. Respondents who saw dwarfism as a disorder (more than a trait) and were less educated were more likely to see knowledge of fetal diagnosis as important.[10] Most of these respondents would refuse abortion, which may make the option of embryo testing appealing.

Such non-medical selection of embryos raises controversies about possible eugenics—choosing and promoting certain *non*-medical characteristics over others. In the early twentieth century, eugenics spread in the United States as many White Anglo-Saxon Protestants felt threatened by large waves of Catholic, Jewish, and other immigrants and consequently sought to purify the country's genes. In the Holocaust, the Nazis drove these American initiatives to horrific ends, trying to purify the genes of Germany—killing Jews, gypsies, gay men, and others to eliminate "bad" genes.[11]

Today, clinicians face quandaries about how to respond to potential parents' requests to choose future children's sex and other non-medical, social characteristics; how to decide; whether to ever say no and, if so, when; and whether to develop explicit policies or decide case by case.

Clinicians range in their decision-making processes—in how *systematic* they are and who is included (e.g., whether and how many outsiders participate). As Thomas suggested in using "the newspaper headline approach," clinicians may seek various criteria for decision-making but can vary widely in outcomes.

"Family Balancing"? Whether to Choose Sex

Professional organizations have proffered mixed opinions about choosing a future child's sex. The American College of Obstetricians and Gynecologists has opposed sex selection for personal, non-medical reasons[12] but stated, "Because a patient is entitled to obtain personal medical information about the sex of her fetus, it will sometimes be impossible for healthcare professionals to avoid unwitting participation in sex selection."[13] The American Society for Reproductive Medicine (ASRM) Ethics Committee, however, wrote in 2004 that while sex selection for social reasons "should be discouraged,"[14] sperm sorting, if found to be safe and effective, may be used for "gender variety in a family."[15] Sperm determine the future child's sex. Eggs all contain a single X—or so-called female—chromosome. If the sperm contains an X chromosome, the child will be a girl. Sperm with a Y chromosome produce a boy. X chromosomes are larger and hence heavier and can thus be sorted by machines, though with only around 90% accuracy[16] and with potential damage to the sperm. This process also requires relatively large quantities of sperm and is hence of limited utility.

In 2015, the ASRM committee revised its opinion, lacking "consensus," but concluding that providers are "under no ethical obligation to provide or refuse nonmedical sex selection" and "are encouraged to develop and make available their policies and to accommodate employees' decisions about whether or not to participate in such treatment."[17]

Opponents of social sex selection argue that it could be discriminatory (disrespecting embryos and children of the other sex), has unknown long-term risks to mothers and children, creates a slippery slope toward choosing additional socially-desirable traits, and is opposed by most of the public.[18,19,20,21] In the US, 8% of the general population said they would use sex selection if it cost $2,500, 74% opposed it, 18% were undecided, and 18% would use it if it involved only taking a medication.[22] Yet the public may be becoming more in favor of it as well. Of US women seeking fertility care, 41% in 2002 would select their next child's sex if it were free[23] and 49% in 2007 would do so, of whom 56% currently had no children.[24] In response, sex selection proponents argue that it should be permitted in order to respect patient autonomy and reproductive liberty and to help women who are under pressure to have sons. These advocates argue that the practice need not produce gender discrimination since preferences for males and females are roughly equal in the US.[25,26,27,28,29,30,31]

Certain ethnic groups strongly prefer males over females overall, but other populations have more mixed views. In one survey, Caucasian couples were split in their gender preferences, while Chinese, Indian, and Middle Eastern couples desired more males.[32] In another investigation, patients seeking gender selection, overall, and especially Africans, Asian Indian, and Chinese, preferred males, while Caucasians and Hispanics preferred females.[33] Female Indian immigrants to the US seeking gender selection feel pressure to do so, often facing physical or verbal abuse otherwise. Forty percent had terminated pregnancies because the fetuses were female.[34] Women with only daughters generally want sons; those with only sons prefer daughters.[35] In one study of 18 couples considering gender selection, 78% preferred boys, feeling that their preferences were ethical, due to their desires for gender balance and reproductive rights, though expressing some ambivalence about destroying embryos of the opposite gender and about disclosing their decision to family members.[36] Still, only two of these 18 couples ultimately utilized PGD. Why the others declined is unknown—possibly finances.[37]

Studies of how providers in fact view and make these decisions remain limited and have shown heterogeneous results. Among US clinics providing PGD, 42% have used it for non-medical sex selection.[38] On their websites, 18% of private infertility clinics advertised social sex selection, but no hospital or university-based clinics did so.[39] Physicians in other fields remain largely uncertain. Few internists, neurologists, or psychiatrists (5.2%, 7.6%, and 11.5%, respectively) would refer patients to PGD for family balancing, and large numbers (45.4%, 25.5%, and 23.5%, respectively) are unsure.[40,41] Among 15 sex selection technology physicians, all vouchsafed non-medical sex selection, while primary care providers appeared to oppose it.[42] In the US, interviews with 19 PGD service providers found that they supported this procedure for medical reasons but varied concerning sex selection.[43] In at least one English clinic, staff have faced challenges balancing patients' autonomy against the possible negative social effects.[44] Given the lack of consensus, the ASRM states that clinics should make their own policies.[45]

Yet while prior studies and guidelines have focused on whether social sex selection should be allowed or banned,[46,47] the practice is expanding, raising crucial questions of not only *whether* it should be performed but *when* exactly—for which specific patients—and how these decisions should or do get made. Though the ASRM has recommended that clinics develop policies related to social sex selection, many questions emerge about how exactly providers and patients make these decisions.

Requests for Social Sex Selection

The patients I've spoken with generally want to screen embryos to eliminate disease more than to seek behavioral traits, but providers receive both types of requests. While critics have denounced embryo selection for traits, most providers are far less cynical. Patients "ask because they want to give their child the healthiest start—not for a negative reason. Not 'I don't want a child with *that*' but 'I don't want to *burden* my child with that,'" Brenda, the Vermont psychotherapist, reported. "With some, it is a bit of narcissism: 'Look, I need my child to be a varsity player who can also attend Julliard, while . . .' But that's few and far between. In general, people want the best for their kids."

Reasons For and Against Social Sex Selection

The notion of choosing embryos for non-medical traits, rather than leaving it up to nature, struck some providers as problematic. "Sex is not a disease!" Henry, the Minnesota reproductive endocrinology and infertility specialist (REI), argued. He thus refused to select embryos for this trait. "You really can't design kids," said Ginger, the IVF counselor who underwent ART for her first child and decided to adopt her second rather than use donor gametes. Desires to plan one's offspring's characteristics too much may founder.

Clinicians vary in whether they select sex and, if so, when and how they define "family balancing" and how they decide. IVF and embryo screening can endanger both mother and embryo, making some providers wary. "As long as PGD is invasive and damages the embryos to some extent," Edward, the Ohio physician, argued, "it should *not* be used for completely 'elective' situations—sex selection, or other genetic engineering—eye color, hair color. Unfortunately, some doctors talk about PGD and PGS [preimplantation genetic screening] as if these procedures have zero downsides for the patient or for the embryos. But that's incorrect."

While Thomas, Henry, and Edward hesitate, other REIs readily fill such patient requests. "I don't have a problem if a couple is doing family balancing, and would like to have a girl or a boy," Nicholas, a New Jersey private infertility doctor, said. He then tests embryos and only uses those of the desired sex. But other doctors vehemently oppose such selection.

Many clinicians normalize selection of social traits and dismiss objections that such selection may veer toward eugenics. "The eugenics arguments are very specious," argued Peter, a West Coast REI who was sued over a PGD misdiagnosis and now uses the technique only for sex selection. "We all practice eugenics in our choice of mate! Patients who would like to practice eugenics can go to the sperm bank, and get sperm from Michael Jordan, so they can have a basketball player!"

Peter thus adopts a much lower threshold, agreeing to select sex for a second child but not a first. He has "a tough time when the *first* child's gender is being selected. But if someone has a boy, and now wants a girl because they only want two kids, I have a hard time blocking that. I get it!"

Yet choosing a future spouse with whom to spend the rest of one's life raises similar, but also very different, issues. People usually select mates based on whom they fall in love with and want to eat, sleep, and spend time with over many years. Moreover, not all couples want or have children. Prospective parents generally buy eggs or sperm based on the presumed presence of genes for traits they would like to transmit to their offspring. But such transmission may not occur, leading to disappointment and difficulties for both the parent and child.

Most physicians here supported sex selection, often asserting, too, that parents request boys and girls equally. Nicholas, for instance, gets "requests for just as many girls as boys—unlike China or India where everything is 'boy.' *That* would be scarier. I think it's fine to allow patients to decide." Yet many parents here request girls because they have had a boy with autism or another disease that occurs more in boys and that they wish to avoid in a future child.

Physicians also contend that gender selection is easier to oppose in the abstract than when seeing the details of particular cases. "People who aren't close to this, and don't see these patients across the table, may not understand it," Calvin, the Los Angeles REI, argued. "When you look at patients from afar and are not involved with them, the first knee-jerk reaction is not to do it. Personally, I don't have any problems with gender selection for social reasons, but our hospital will not let us. I would have no problem, particularly if the patient is going through IVF anyway. Even if the patient didn't, I wouldn't have a problem for family planning and balancing." Still, "not being able to do it doesn't bother" him. He and other doctors might be willing to perform social sex selection but decline because their clinic's policy opposes it. Providers vary in not only the nature but the *strength* of their preferences pro or con.

Yet, especially in the current, highly competitive professional marketplace, doctors may be torn, supporting a "free-market" business approach: if a patient wants the service, doctors should have the right to provide it (though they may remain somewhat uncomfortable). Economic motivations may sway physicians. "Financially, should I just let it go through?" Steve, an infertility provider in a private group practice, wondered. "Because otherwise, she's going to go down the street and do it."

He and some others try to balance financial against other motives. "As providers, we always like to stay objective," he continued. "But as a human being, it's hard to stay objective on those topics. To have somebody casually say, 'We're just going to disregard those embryos' doesn't feel right. At our physicians' meeting we talk about the 'ick' factor. If it feels *icky*, we don't want to do it."

As a compromise, Jim, Steve, and other providers have transferred into a woman one embryo of each gender, concomitantly. This approach can nonetheless generate controversy among staff. "They're still 'leaving it up to God' but at least know they had a male embryo transferred," Steve explained. "But some staff in my practice say we're being ridiculous, because if the parents wanted a boy, and only have a girl, she's going to have a 'horrible life.'" Many clinicians remain uncomfortable or uncertain, generating conflicts within and between providers. Personal feelings can collide with understandings of professional duties.

Challenges in Defining "Family Balancing"

Doctors who perform sex selection for "family balancing" must decide how exactly to define it: whether to set any limits and, if so, where. Clinicians frequently draw certain boundaries, but these vary. Some physicians cite instances where they think the practice is clearly permissible, such as for relatively major, but not minor, gender imbalances in a family—for example, five children of the same sex. "When somebody has five boys and wants a girl," Roger, a New York hospital-based IVF doctor, said, "a part of me says that's not unreasonable or crazy. They are not bad people."

Other providers, however, struggle with precisely how high a threshold to set, generally requiring that a family be relatively more imbalanced. "The procedure is definitely questionable when somebody comes in and wants a boy as a first or second baby," Diane, an IVF nurse, said. "After that, it gets gray."

Many clinics seek to navigate these controversies by claiming that they are screening embryos for reasons *other* than sex—such as checking for chromosomal abnormalities—and simultaneously also test for sex. "Facilities say they're screening for aneuploidy," Diane reported, "and 'Oh, by the way it's a girl' or 'it's a boy.' So, the excuse was aneuploidy. The *reality* was that it was gender selection. . . . Everybody knows the couple came in asking for gender selection."

Even within one clinic, doctors' views may vary, impeding a unified formal policy. Physicians may refuse to perform the practice but refer patients to colleagues in their office. "I just say I don't do it," Anne, the northern California REI in a large clinic, reported. "Patients could go to another doctor in my practice. I feel I have to tell them that. It would be much easier for me if we had a policy against doing it. I could just say, 'I'm so sorry. I wish we could do it. But we have a policy against it.' Since we don't have that, it puts me in a very difficult position. I have to say, 'no, I don't believe in this, so you can see one of my colleagues and do it.' That's very hard." Explicit policies can benefit programs, bolstering clinicians who may want to refuse these requests.

Within a practice, doctors can therefore differ about whether and where to draw lines. "I counsel against it," Marvin, a Massachusetts REI, said. "One other partner feels like me. But other partners disagree." Their clinic has thus tempered use of such selection: "We set the bar reasonably high."

Personal, professional, institutional, geographic, and other factors can shape physicians' attitudes. "It's done routinely in California," Edward, the Ohio REI, said, "but tends *not* to be done electively a lot on the East Coast." Among clinicians overall, women (such as Jill, Jennifer, Karen, Anne, and Diane), concerned about possible preferences for males, tended to be warier than men (such as Nicholas, Peter, Calvin, Steve, and Roger).

Given these tensions, many providers seek various strategies, parameters, or compromises. Clinicians may practice sperm sorting, seeing it as less egregious than PGD, even though it is less effective. Compared to embryos, sperm is less constituent of life itself and thus has less moral status per se. Consequently, some providers see selection of sperm that will produce boys rather than girls as less egregious than choosing embryos for the same purpose. Other clinicians agree to select embryos for sex but only if the patient is undergoing IVF anyway and without promises of necessarily obtaining the desired sex. Jill has "done PGD for gender selection once or twice, but with the understanding that if they don't have any healthy embryos of the gender they want, yet *do* have healthy embryos of the other gender, we will use them."

Given these conflicts, some providers remain uncertain about their beliefs. Jill is "not convinced there is a 'right' answer."

Challenges in the Decision-Making *Process*

Providers differ widely concerning not only the content but the *process* of these decisions. Larger hospital and university-affiliated clinics appear more likely to consult formal external ethics committees than do smaller, freestanding private clinics. Yet, these committees vary widely in whom they include and how they work—from formal to informal and from internal to external.

As Thomas suggested, a hospital ethics committee, in confronting this issue for the first time, can undergo a long and intricate process, including not only fertility providers but others as well. Such committees generally appear to discourage social sex selection. When Joe's university-affiliated clinic had a patient with three or four girls who wanted a boy, "We dealt with that through the ethics committee, to make sure we were doing the right thing. In general, we are not proponents of sex selections."

Ethics committees may also question sex selection for ostensible medical, rather than elective, social reasons. Thomas has "done PGD for gender selection for autism, but that issue also went to the ethics committee" since it affects boys more than girls.

Clinics in academic medical institutions can differ from others in having not only more external oversight committees but other, indirect interactions that can serve as de facto ethical reviews. "Affiliation with a university or hospital can only have a beneficial effect," Thomas explained. "A lot more peer review and thought go into one's practice. Doctors have to stand up and present their own data, and justify why they did certain things. *That* makes it much more difficult to be entrepreneurial. We present our data all the time, not necessarily to an ethics committee, but in front of our faculty. Residents rotate through our program, so they're always seeing exactly what's happening. . . . It's a check and balance to know what's going on, as opposed to just doing it and advertising it without any oversight."

But how do these ethics committees themselves operate—which cases do clinicians present, how exactly do these boards then decide, and what criteria and standards do they use? "If the parents have four boys and want a girl," Roger brings it "to our ethics committee and we talk about it." They use, " 'the *sniff* test': is this totally comfortable?" Committees may thus base these decisions on intuitive, subjective factors, rather than explicit ethical analysis per se.

Unaffiliated with academic medical centers, private and/or smaller clinics commonly lack formal ethics panels and instead simply have some or all of the staff discuss difficult cases. These conversations range in scope, length, inclusiveness, and formality, and whether they set precedent or formal policy. Physicians and staff may have no formal ethics training. "If you want to go outside the rules, we have the nurses, the nurse managers, the doctors, the lab director, and the psychiatrist sit down and talk about it," Steve said. "But we can't expend a tremendous amount of effort on one patient, trying to assuage [his or her] cultural reality." These decisions may hence be made case by case, rather than through broader formal policy.

The amount of oversight and feedback can increase with the size of the office. Doctors working individually or in very small practices may receive even less scrutiny. "On these sticky topics, most people in a reasonably-sized practice figure it out on their own," Steve continued. "We're four doctors. The one- or two-doctor shops may have less interprofessional oversight." Clinics can thus differ regarding not only whether they are university-affiliated or private but large or small.

In other programs, the decision-making process can mostly involve consultation with a mental health professional. "Family-balancing couples go to psych and make a decision," Steve reported. Yet at other private offices, these processes can remain vague or unclear—even to staff. Mental health providers (MHPs), for instance, may not be involved in these deliberations at all or even know how decisions are made. When I asked Ginger, the infertility psychotherapist and patient, about it, she replied that, though she had been working at an IVF clinic for several years, "I don't know how they make those decisions. That's a good question."

Clear policies, if developed and established, can offer more clarity and less confusion. "If it's a larger clinic, and they have an ethics review board and/or policies and procedures, they're probably better off," Helen, the Wisconsin infertility psychotherapist, said. "If you're a smaller clinic, all sorts of things come up. Ethical issues are really big in infertility."

Opposition to Regulations

While many countries explicitly prohibit social sex selection, most US providers perform it and oppose any regulatory restrictions. Nonetheless, some clinicians, especially women, remained ambivalent and wondered whether more external input might help. "I usually like having fewer policies," Ginger said. "Yet, some things beg for oversight. I don't know how I feel about the little boys getting let go, and the little girls given a chance at life."

Ultimately, however, even providers wary of sex selection tended to oppose legal bans. Rather, they supported the current looser, essentially unenforced guidelines. Many patients, too, held this view. Even if personally against social sex selection, they were often leery of government bans. Sally, for example, the Maryland website manager who underwent two abortions because the fetus had cystic fibrosis (CF), does not "agree with such laws" and similarly opposes laws against abortion. "Every case is different. Just because someone's using a technology in what I believe is the wrong way, doesn't mean it should be against the law. If autism is in the family, and the parents are worried that a boy might be more autistic than a girl, so they'd rather have a girl, having a law is a bad idea."

Selecting for Deafness and Dwarfism

Selection of embryos not for sex but for other socially desirable characteristics raises even larger concerns about eugenics and "designer babies." Many patients feel torn, as did Sally, who had terminated two pregnancies due to CF and had undergone three IVF cycles to produce her daughter, who is a CF carrier but lacks the disease. Genetic screening had worked for her, but she sensed that it could potentially be abused: "Maybe you'll be able to make a smarter or prettier child. I don't know if that's right or wrong. In some ways it's kind of wrong. But it does sort of make sense that the next step in human evolution is: we're going to be able to start affecting genetics directly. I'm not sure where that's going to take us. But it does seem like the inevitable next step."

Most of these clinicians knew of patients' requests for a non-medical trait other than sex, though none had yet personally received any inquiries. Still, they expect such inquiries to increase and are unsure how they will respond, given the underlying ethical quandaries. Bill was the only provider who had had such a request in his practice. "We've actually had a couple who wanted a deaf child because they were both deaf. Everybody was astonished. The patient wasn't mine. The couple doesn't consider deafness a disability."

Other clinicians had heard of such requests elsewhere that had not necessarily been granted. "We have *not* had positive selection for disabilities come up, but it *has* come up in other centers," Edward, the Ohio REI, said: "Two parents have a form of dwarfism, or two deaf parents want to create a deaf child. We have not confronted that, but know it's an issue."

No other providers, however, had received such inquiries for positive selection for non-medical traits other than sex; and some clinicians even thought it might be apocryphal. "I have not seen requests for positive selection for deafness or dwarfism," Brenda, the Vermont infertility psychotherapist, reported. "I keep wondering if that's more of an urban myth."

Nonetheless, providers anticipate more such requests. "I'm sure it will come up in the next five years," Steve, the Virginia REI, predicted. "People are going to start to use gene screening technology to sweeten the pot for their child. It's a slippery slope." CRISPR will surely further these quandaries.

Such desires present logistical challenges as well. "There are deaf couples who would love to have a child who shared their condition," Karen, the California physician who had leftover embryos, following the birth of her deaf son, said. "But how do you get them together with a frozen embryo?"

Difficulties of Deciding About Such Requests

These clinicians all sensed inherent challenges in such cases, given varying views and assumptions about what in fact constitutes a disability (versus a disease), and whether all patient requests for embryo screening should be granted. Hereditary deafness may cause disabilities but not threaten life itself. "With hearing loss, I've had to email many geneticists and get their opinions," Sam, the patient advocate, said. "It was horrible because the question becomes: what is and is not a disease?"

Providers may try to learn from individuals living with the condition. "The deaf community tells you that deafness is OK," Sam continued. "Yet people with dominant deafness tell me, 'I'm not dying of it; but it's a very, very inconvenient life for me.' So, patients helped us with that." He concluded that he could, ethically, aid a prospective parent wanting to select for this trait.

Ultimately, he and many others felt they should not criticize a patient asking for such selection because they cannot experience first-hand the dilemmas that the prospective parent faces. "Personally, I think it's crazy," Sam concluded, "but I'm not deaf so cannot judge them."

Given these uncertainties, clinicians might also require that such patients consult a psychotherapist. Regarding "deaf parents wanting a deaf child," Calvin, the Los Angeles REI, said, "I suppose that with appropriate psychological screening, we would probably do it. We have a psychologist in our office."

Physicians such as Thomas, at academic medical centers with formal ethics committees, would refer such cases to these boards. "We would probably run that by the ethics committee here and get a final verdict," Henry, the Midwest medical center infertility doctor, added, "because you're running into a gray area."

Doctors were generally open to the possibility of treating such patients but only under certain provisos or extenuating circumstances. Providers might stipulate, for instance, that the patients be infertile and not seeking a deaf child as the primary reason for IVF—"that they have a disease *and* are infertile," Henry declared.

Clinicians have thus often heard about but rarely, if ever, personally seen such requests for positive selection for traits such as deafness and dwarfism and are unsure how they would proceed, recognizing the inherent moral dilemmas.

Blond Hair, Blue Eyes, and Other Traits in the Future?

In the near future, genes for other non-medical "desired traits" will certainly be identified and requested. "If we did come up with a strong candidate for a gene for athletic prowess, I suppose people might come in for that," Henry said. "But it hasn't happened yet." Individual doctors have, in fact, occasionally suggested to the media that they could "design babies" for these features, but such selection is not yet technologically possible.

Nonetheless, providers and patients are concerned about selection of other "socially desirable" non-medical characteristics. Researchers are identifying genes for hair and eye color and will no doubt soon be able to select embryos with genes for certain hues. Such possible "elective genetic engineering" in the future sparks worries. Edward, for instance, felt that "if a doctor out on the edge said, 'we can do this elective genetic engineering for eye and hair color,' they'd get more requests for it—patients coming in thinking, 'I want to have a blond, blue-eyed, 6'3" son, then a brown-haired, brown-eyed 5'6" daughter.'"

Yet, choosing genes based on physical appearance may lead to eliminating other, important traits. As Peter, who now screens embryos only for sex selection, warned, "in the future, if there's going to be the potential for selecting eye color or hair color, that's going to be tough. Not so much because it's a slippery slope, but because if you choose eye color or other genes, are you going to start eliminating some desirable qualities from the population?"

Patients may seek other personal, social, or cultural traits for which no direct genetic marker exist. "A woman only wanted a Harvard egg," a California REI told me. "No other egg would do. But I said no because we would have to monitor her in another city. I offered an egg donor from another top school, but the patient refused."

A friend of mine, Judy, wanted a tall, smart, dark-haired Jewish sperm donor. She got one—from Yale Law school. Her eventual son was very bright but ended up being very socially awkward, not getting along with people and having no friends. "I forgot to ask about the sperm donor's personality," she told me. A problem with trying to choose certain traits is taking other characteristics for granted and forgetting to seek them. Tiresias, the blind prophet, asked the gods for immortality but neglected to ask, too, for youth. He survived—as a decrepit, old man.

As we move into the future, debates will increasingly arise about whether to select embryos based on additional non-medical, socially desirable traits and, if so, which, who should decide, and how. Though the media have discussed dilemmas concerning possibilities of positive selection for deafness or dwarfism,[48] thus far, such cases remain rare—"urban myth" more than daily reality for providers and patients. Still, doctors remain uncertain and troubled, not yet having fully thought through the complex pros and cons. Providers wrestle with whether deafness and dwarfism are "diseases," "disabilities," or "normal" and whether such selection might implicitly restrict a child's "open future"[49] and ability to make key choices for him- or herself. Providers may be wary but still avoid judging such a patient, not knowing what *they* would do in similar circumstances. But, in helping to create a child, they arguably have a degree of moral responsibility and thus need to consider the child's best interests, which at times may outweigh the prospective parents' rights to make such choices.

The future parents' autonomy can clash with their future child's best interests and broader social justice. Providers tended not to discuss eugenics-related issues explicitly but to speak of dilemmas regarding disabilities versus disease. Importantly, numerous individuals now living with these conditions feel that they do not have a "disease" and that unfair stigma exists. Many clinicians may weigh parents' desires for seeking certain offspring, more than potential harms to the child. At times, providers also wrestle with concerns about possible future requests for athleticism or other traits, and about eugenics.

Providers vary in how they would decide and might consult MHPs, formal ethics committees, or relevant communities and/or consider extenuating circumstances—for example, seeing the practice as acceptable if the patient were undergoing IVF for other reasons.

Hank Greely, the Stanford law professor, has suggested that PGD may eventually become the norm in reproduction.[50] Yet the data here suggest that providers may vary considerably in how they make these decisions. More consensus is needed on how clinicians should proceed in weighing these conflicts.

Based on the responses of the men and women here, full social consensus will probably be elusive, resulting in providers and prospective parents making their own decisions, driven by their own levels of comfort. In the essentially unregulated US infertility market, I expect that many providers will profit from such procedures and offer them, despite these moral questions.

As we will see, providers and patients today face quandaries of whether to use these technologies to affect not only which genes to transmit to children but also how many such children to have.

8 | "Two Kids for the Price of One?"

CHOOSING TWINS

"I struggled with whether to put two embryos in, in case one doesn't make it," Isabelle, the pregnant Connecticut office manager, told me. At 29, she was fertile but at risk of transmitting a chromosomal abnormality, and hence tested her embryos. "But, what if both embryos take? What if we put in only one, and it doesn't work? We then have to do it all over again. And how do you choose *which* one to put in?"

She wrestled with these decisions. "The clinic doesn't actually tell you how many eggs or embryos you have until you are ready to do the transfer. The doctors said they'd call us that morning. But they don't tell us the actual number until we're on the gurney. . . . So, you don't have time to really think how many you want to put in or freeze. If you have three, is it worth freezing just one, or freezing two and putting one in?"

In using these technologies, patients and providers grapple with not only *which genes* to give offspring but *how many* offspring to produce. Isabelle, Jane, and many other patients and providers seek twins to save time and money. Consequently, since the late 1970s in the United States, Europe, and many other countries,[1,2] IVF has generated an "epidemic of twins,"[3] with rates of twin and other higher-order multiple births doubling. "Octomom" brought worldwide attention to the potential problems of multiple births due to IVF.[4] Since then, the numbers of triplets and higher-order multiple births have decreased slightly, but very high rates of twins continue.

Yet when transferring more than one embryo, the risks of complications rise significantly. Among infants born to mothers through IVF, twins are 12 times more likely than singletons to be born prematurely, 16 times more likely to have low birth weight, and about five times more likely to have breathing problems or jaundice.[5] Mothers of twins are about 2.5 times more likely to have preeclampsia, over eight times more likely to have premature rupture of membranes, and four times more likely to require a cesarean section.[6]

Several reproductive endocrinology and infertility specialists (REIs) have supported transferring in a cycle, in otherwise healthy women, two embryos simultaneously (known as double embryo transfer, or DET), rather than one embryo (so-called single embryo transfer, or SET), arguing that twins are desirable and that the reported risks are exaggerated.[7] Among studies published from 1995 to 2008,[8] DET had about twice the pregnancy and birth rates of SET (since the two embryos are being transferred).

Yet since around 2008, when many of these studies were conducted, SET success has increased substantially, while higher complications with DET persist. A more recent meta-analysis[9] found that DET had five times more preterm births than SET and that the birth rates from two consequential SETs versus one fresh DET were not significantly different (38% vs. 42%). In one Italian study, twin pregnancies had 31.8 times more perinatal complications.[10] For women over 40, transferring three rather than two embryos does not increase the birth rate.[11] Moreover, the overall costs per live birth for DET and SET are roughly equal, given neonatal intensive-care unit (NICU) and other expenses,[12] yet twins can cost even more, given the long-term complications.[13,14]

Increasingly, commentators have argued that SET should become the standard of care[15] and can be achieved by altering insurance coverage, developing appropriate educational materials, and including patients' spouses in these discussions.[16,17,18] Various countries including the United Kingdom,[19] Australia, New Zealand,[20] and Canada[21] have successfully reduced the numbers of embryos transferred. As of 2013, 17 European countries had limited how many embryos doctors can transfer.[22] The British Human Fertilisation and Embryology Authority has lowered the incidence of multiple births from 24% in 2009 to 15% in 2010 and to 10% in 2012.[23] Belgium reduced the multiple pregnancy rate from 27% in 2003 to 11% in 2010 by reimbursing expenses only for transfers of limited numbers of embryos.[24]

Yet, many countries still permit wide physician discretion. The American Society for Reproductive Medicine (ASRM) states that "physicians should be encouraged to counsel good-prognosis patients to accept" SET and that a doctor's judgment is important for selecting the best candidates.[25] For patients younger than 35 with a good prognosis, "providers should only transfer a single embryo, and not more than two embryos."[26] For patients aged 38–40 and 41–42, the ASRM states no more than three and five embryos, respectively, should be transferred. Patients who have failed two or more IVF cycles or have a less favorable prognosis can receive an additional embryo. For patients aged over 43, there is no limit.[27]

Since 1998, the US rates of twin and multiple births have begun to decrease slightly—as providers transfer three or more embryos less frequently[28]—but remain high in numerous countries. In 2013, 17% of all Canadian births were twins.[29] In 2011, 19.4% of all European births were multiples.[30] Transfers of three and four embryos, respectively, still accounted for 49.4% and 9.9% of all cycles using IVF and intracytoplasmic sperm injection (ICSI) in Greece, 34.6% and 4.5% in Italy, 26.2% and 3.1% in Hungary, and 40.7% and 7.9% in Bulgaria.[31] As a result, twin and triplet deliveries accounted for 18.6% of IVF births in Europe as a whole, including 41.5% in Greece, 28.2% in Romania, 22.0% in Moldova, and 26.5% in Serbia.[32] In the US in 2016, 61.1% of transfers of embryos using fresh non-donor eggs involved more than one embryo, 11.6% involved more than three embryos, 18.8% of the resulting live births were twins, and 0.6% (114 births) were triplets or more. The average number of embryos for women under 35 was around 1.5, and 15.8% of pregnancies resulted in multiple-infant births.[33] The published *Society of Assisted Reproductive Technology Clinic Summary Report*[34] lists the percent of cycles, retrievals, and transfers resulting in live births for each clinic but unfortunately does not separate these data by the number of embryos transferred, presumably because doing so would lower the success rate. For singletons, the success rates would presumably be less.

Most patients prefer to pursue twins than singletons,[35,36] and only 34% of IVF doctors routinely discuss SET and its benefits with patients.[37]

Some observers have claimed that patients' preference for twins results from limitations in insurance coverage, but the data to support this claim are mixed. In a 2011 US study, states that mandated some IVF insurance coverage had only slightly lower rates of twins (26% vs. 28%) and triplets (3.4% vs. 3.9%), and the absolute differences were minimal (2% and 0.5%, respectively).[38] Moreover, in Denmark, which reimburses up to three IVF cycles, 59% of patients still nonetheless preferred twins, wanting less physical and psychological stress from IVF, as well as siblings for their offspring, not because of economics.[39] Gay male couples frequently prefer twins so that each member will have a biological child.[40]

Patients' opposition to SET may result from clinicians not presenting the potential dangers. In one survey, most patients wanted twins, but 42% did not remember being counseled about the risks of multiple births. The median length of the counseling, they recalled, was only five minutes![41]

More insurance coverage for IVF would help many patients[42] but faces obstacles. Countries such as Japan, Italy, and Israel that want to boost their population pay for more infertility treatment, through nationalized, single-payer health systems, than do other nations. In almost every country, however, insurance to cover unlimited fertility treatment for all those who seek it, including heterosexual couples, single men and women, and same-sex couples would cost enormous amounts. Given competing health needs, sufficient political will may not exist to mandate IVF coverage for everyone. Physicians can also easily avoid multiple births by simply transferring fewer embryos.

Hence, critical questions remain about why many IVF providers continue to produce twin births despite the risks, how providers and patients perceive and discuss these risks and decisions, and how and why patients often fail to recall these conversations.

Both providers and patients, I soon saw, struggle with how many embryos to transfer. The relevant ASRM guidelines, though generally followed by clinics, allow considerable flexibility and contain ambiguities. Decisions about how many embryos to transfer are dyadic and dynamic. Several factors affect whether and how clinicians address these issues. Many doctors end up transferring additional embryos, due to both their own and their patients' attitudes.

Patients' Perspectives

"I never feared multiple pregnancies, because we just wanted a baby," Roxanne, the Michigan marketer undergoing infertility treatment for eight years, said. "We *should* have been thinking, but weren't: what are we going to do with two or three? We thought, 'Whatever it will be, will be.'" Given that the odds are less than 50% that any one embryo transferred will become an infant, many prospective parents, especially those who have previously failed to get pregnant, are eager simply to have a baby, and thus ignore the risks involved. Patients may recognize the conflicting pros and cons but remain unsure and transfer additional embryos.

In part, as Isabelle described about only learning on the gurney how many embryos she had available, doctors may not know in advance the number of viable embryos and may consequently offer patients scant time to make these decisions.

Patients may also fail to grasp fully the dangers, and instead focus only on the potential benefit—having a child. "Most women don't understand the risks," Valerie, the psychotherapist and single-mother-by-choice, observed, "A lot of women want to get it over with. Most patients are pretty much in denial: 'It won't happen to me.'"

Eager to increase their success rates, providers may simply not communicate these potential dangers to patients effectively. "Clinics want to keep up their numbers," Valerie added. "Most patients are not informed of the risks of having twins. The only therapy patient I've had who was fully informed of those risks was herself a doctor. And she agonized for a long time about what would happen if both of the transferred embryos took. The field is moving toward single embryo transfers. But I'm not sure if patients are adequately counseled."

Given strong competing desires, anxieties, and stresses, patients may dismiss, minimize, or deny the hazards of multiple births to themselves and their offspring and rationalize these decisions. "Patients are very fickle because they're desperate, looking for magic guarantees," Helen, the Wisconsin psychotherapist, observed. "Most people don't understand the statistics. Some patients look at pregnancy statistics as opposed to 'take-home baby' statistics. Statistics can also be moved around. Statistics on live births are inflated by the number of multiple pregnancies." Indeed, the relative odds are not straightforward, and patients may thus miscomprehend them or be misled.

Misunderstandings partly reflect low education about these complications among patients and the public at large. "Forty percent of women say they would rather have quadruplets than have nothing," Calvin, the southern California REI, said, "because the public has no idea how devastating prematurity is. In part, the popular press is not reporting the problem of multiple births well." The media can in fact promote inaccurate understandings of the odds and dangers. Stories about cases such as "Octomom" can omit important details and follow-up. "Even with the octuplets, six are doing fine, breathing and having normal blood pressure," Calvin added. "The press doesn't talk about what they're going to look like later."

Both financial and non-financial factors can heighten patients' desires for twins. Amanda, the high school administrator, underwent six intrauterine inseminations and could afford only one IVF cycle. She described a 35-year-old friend who had IVF and "had twins, and paid 100% out-of-pocket. If I were in her place, I would also be more likely to transfer two. I wouldn't want to have to go through this all over again, *and* lay out more money."

Many providers are acutely aware that patients, to reduce expenses, may want twins, rather than two separate pregnancies. "Patients should have insurance that would help us provide care to everybody," Jill, the Illinois infertility doctor, said. "Then, we would have greater ability to do single-embryo transfers."

Physicians' Perspectives

Patients' desires to save money can sway physicians. "It's hard to argue against people wanting to transfer two because of cost," Henry concluded. But doctors should also consider the future child's health. In fact, in trying to estimate the odds of success per embryo for any one patient, providers must balance several competing sets of statistics, along with their own experiences with past patients. Paulette, the psychotherapist, said that her clinic transfers two because, "Usually only one takes. . . . We have less than 20% twins. If our rate of twins were higher, we probably would go to single embryo transfer. Transferring three? Extremely rare. We might consider it on frozen embryos. If the couple has only three frozen ones, we'll probably transfer three. If we had a lot of recipients get pregnant with twins, the doctors would feel more comfortable doing single-embryo transfers." Still, questions arise of what rate of twins would be high enough to prompt transferring only single embryos.

Importantly, not all providers feel that the data sufficiently demonstrate the dangers of multiple births. "Triplets are a little gray," Jill, the Illinois REI, said about the risks when a patient, pregnant with triplets, refuses to reduce the number of fetuses. About a patient who ultimately refused to reduce from three fetuses, Jill added, "If the data were *compelling*—if she had quadruplets—I'd say, 'You really do need to get this done.' But for triplets, the data are not compelling enough for me to say that." Providers may thus draw on their own perceptions of the evidence, especially given a patient's strong desire for more than one child.

Yet the data *are* very strong, given the relatively high rate of complications that twins and triplets face. Many providers may be unaware of this evidence or may not fully want to accept it.

Some clinicians fully recognize the dangers of transferring more than one embryo—"even twins are dicey," Ginger, the IVF counselor who used ART for her first child and adopted her second, said. Providers who have witnessed close up the physical harms of multiple births appear more convinced. Valerie, for instance, has "seen first-hand what could happen: I know people who went through IVF and ended up with extremely premature infants in the NICU, born at 25 weeks. Another set of twins are now 15 [years old], but one's extremely disabled." Psychologically, risks are more "salient" to individuals who have seen or experienced them first-hand.[43]

Unfortunately, however, doctors who feel the data are insufficient may fail to mention or highlight these risks to patients. In Calvin's program, some physicians "just tell the patient, 'You have two [embryos],' and put back two, and don't even talk to them about one. Maybe twins aren't bad: we're overexaggerating it. But it's still an admirable goal to try to cut down the amount of twins."

The risks of complications due to twin pregnancies are approximately 40% for any one patient but consequently become of considerable concern from a broader public health perspective. Providers may, however, fully appreciate these larger medical and social costs only when they themselves have witnessed or experienced them.

Doctors and patients may transfer extra embryos, too, to avoid later retrospective regret, frustration, and anger if no pregnancy ensues. "If we put two in a 37-year-old, and she doesn't get pregnant," Nicholas, a New Jersey REI in private practice, admitted, "then you say, 'Maybe I should have put in *three* instead.' "

Given growing competition for patients, physicians may also agree to transfer additional embryos because doing so can help boost the clinic's success rates. "Part of the motivation for some physicians is to get their numbers up of 'take-home babies' per cycle," Valerie said.

"People are moving toward a single embryo transfer, but not so quickly," Helen concurred. "It's very competitive. Big money." Henry transfers just single embryos but can do so only because no other clinic exists within hundreds of miles. "The other nearest IVF center is three or four hours away from us. In contrast, if you have a clinic down the hall, and you say, 'No I'm not going to do that,' patients say, 'Fine,' and go there. It's monetarily and competition-driven."

In addition, providers may "game" the statistics, "cherry-picking the best patients, to try to have the best results," Nicholas said. "It's hard to find a private practitioner who is not intently aware of the statistics." Doctors may manipulate the data they submit to the Centers for Disease Control and Prevention (CDC) and that thus get published, carefully selecting or rejecting certain patients or only treating those with very high-quality embryos.

Currently, providers report their pregnancy rates without specifying how many embryos they transferred per IVF cycle, fueling insertions of extra embryos. "The reporting system disincentivizes transferring fewer embryos," Calvin elaborated. "We look mostly at pregnancy rates, rather than implantation rates. So, in many programs, if a patient wants to put more than one embryo back, nobody will argue with them. If the guidelines say 'one,' a lot of patients want to put back two. Doctors say, 'Fine.' But if we were looking at implantation rates, rather than pregnancy rates, [it would] be better if you put back one; and you might be a little bit more persuasive with your patients who are trying to decide between one and two."

Improvements to current reporting requirements could reduce the numbers of embryos transferred but face significant hurdles. "I've been pushing for change for years," Calvin said. "Implantation rates are very 'down in the depths' of the reporting system. I'm kind of radical: you shouldn't report pregnancy rates at all, because people are going to look at pregnancy rates because they understand it. If you reported implantation rates without pregnancy rates, they'd have to look at *that*. . . . They'd be able to judge a program better, because implantation rate is a better judge. But I don't think that's going to come to be."

The quality of reported data can also range widely. "The more data you get, the less accurate they are," Calvin added. He is very "gung-ho about getting the data, and putting them in appropriately. But a lot of programs see it as another task to do, and just put in whatever data are the fastest to put in. The more you do that, the more problems you're going to have. You end up with numbers, but they may not be accurate. . . . What do they mean?"

Not all doctors even report their statistics, since doing so is in fact optional, rather than mandatory, perpetuating problems. The "Octodoc," for instance, did not list his success rates. "He doesn't report to SART [Society for Assisted Reproductive Technologies]. He doesn't have to," Steve, the Virginia REI, explained. "Doctors can choose not to. Transferring double embryos skews success rate data. Those clinics publish better rates, but are not more successful. That's double-speak right there!"

Clinics may also try to report better rates by creating separate, non-reported "research" protocols. "One particular institution has two protocols," Steve said, "including a 'research protocol' where they put through their high FSH [follicle-stimulating hormone] patients, and don't treat other women who have a FSH greater than 10."

To boost their reported statistics, providers may also simply refuse to care for certain patients whom other physicians might accept. "Doctors differ somewhat in turning away patients who have a reduced prognosis for pregnancy," Jill said. "We get patients who have been turned away elsewhere for things like elevated FSH. The other provider doesn't want to treat them. I feel, depending on their prognosis, that for patients with a 10% likelihood of pregnancy, as long as they're adequately counseled, treatment may not be unreasonable."

How to Decide

Doctors who disagree with prospective parents' desires to transfer extra embryos have to determine whether to follow or oppose these patients' autonomy. Steve described "the 'how-many-embryos-to-transfer?' conversation with a 42-year-old who wants six embryos transferred because she gets the math." He says, " 'I don't want to do six,' or, 'No, I won't do that.' She says, 'They're *my* embryos!' I say, 'Well, I have to follow our professional society guidelines, and I won't deviate from *no more than five.*' "

Yet other providers grapple with these conflicts, uncertain. Henry and his colleagues "wrestle with the competing ethical areas of patient autonomy versus the physician's social responsibility to do what's right for society. The patient says, 'I want two embryos because I want twins.' You say, 'Well, twins are not a good outcome.' They have premature delivery. . . . 'Well, *I* want them anyway!' "

Providers vary widely in whether to let the patient decide and, if so, when. "Doctors used to make all the calls: this is the way we do it," Henry reflected. "Lately, it seems that the patient gets whatever she wants. Doctors throw up their hands and say, 'Well, that's what she wanted.' That's ridiculous! It shouldn't remove the responsibility from the physician. At what point should a physician say, where excessive healthcare costs are involved, 'No, we're not doing that'? Healthcare costs are increasing hugely because of the attitude: we can do it. We get paid to do it. But no one asks *should* we do it." Many providers thus see these dilemmas as fundamentally ethical—how to balance patients' individual desires against broader societal costs. Unfortunately, wider social benefits and harms usually get ignored.

Who Decides? Physicians versus Committees

Providers differ in whether they make such difficult decisions on their own or consult other professionals, such as a quality assurance (QA) or other committee. "If anything looks exceptional, we bring it to QA . . . to discuss," Marvin, a Massachusetts REI, explained. "Our patients sign a consent form about how many embryos to transfer. It hurts our pregnancy rates, but we don't treat our pregnancy rate. We treat our patients! . . .We need to get our twin and triplet rates down." Transferring one embryo and freezing and then later transferring another one was, he felt, "almost as good" as transferring "two at one time."

The type of clinic can shape these decision-making processes. Academic medical centers and hospitals can impose additional implicit and explicit pressures. The hospital where Nicholas consults "had a meeting of the medical executive directors with 500 people, and they had me talk . . . about the impact of IVF on the hospital. I thought there was going to be a lynching. I'm independent, but on staff at the hospital. I'm not there very often. I do outpatient things and am not tied by the hospital, or run by them." Hospital environments, in which physicians periodically present their work to colleagues in specialties other than infertility, can provide a form of oversight.

Changing Guidelines or Policies?

Since relatively high rates of both twin and higher-order multiple births persist, additional guidelines or regulations can potentially help but have both pros and cons. Clinicians tend to oppose stricter mandates that would diminish their autonomy. In deciding how many embryos to transfer, clinical judgment can be important, but simply following patients' autonomy can have limitations. "Unfortunately, some doctors don't exercise appropriate judgment, or follow the guidelines," Jill said, and "act like cowboys and cowgirls, making the rest of us look bad."

Yet though providers may not all follow guidelines, even Jill opposed more stringent approaches, feeling that current guidelines, suffice: "The question becomes, 'What exactly is a guideline? Should doctors be punished for not adhering? Should there be rules, rather than guidelines?' No! We're taking care of *individual patients!*" But, while preferring current, looser guidelines—rather than regulations—she nonetheless questions their scope and implications. Types of guidelines span a broad continuum, based on their implementation, enforcement and penalties.

Other providers feel that improved reporting policies would be beneficial but elusive, due to complex federal bureaucracies. "The CDC basically says, 'We want you to report pregnancy rates, and decrease the multiple pregnancy rates,' but these two goals conflict," Calvin said. "It's very hard to change it. . . . We've done very well minimizing triplets. But [the CDC is] starting to see that the report is not going to minimize twin deliveries. Still, changes at the CDC have to be approved by the Office of Management and Budget."

Despite needs for enhanced policy implementation and enforcement, SART's regulatory power may simply be too restricted. The punishment of losing SART membership fails to incentivize all physicians. "The only thing SART has to threaten doctors with is non-membership, which some doctors would probably welcome, because then they wouldn't have to pay, or register their results," Henry said. "By law, you'd still have to register with the CDC. But until we can create brand-name recognition—that a 'SART-approved clinic' is in fact superior, and that clinics don't become a member simply by paying dues—no one really cares."

Generally, clinicians favor stronger professional self-regulation, not government involvement. "It should start with self-regulation," Henry said. "There have been improvements over time. If there isn't continued improvement, it's time for more regulation—as in Scandinavian and other European countries." The threat of government regulation, if physician self-regulation fails, could potentially enhance adherence but has thus far fallen short.

Government regulations could potentially mandate transfer of only one embryo per cycle but may also face opposition from financially constrained patients, unless insurance coverage increases. To reduce the incidence of twins and other multiple births, government policies may thus also need to expand reimbursement. "If the government would mandate that insurance companies pay for IVF, I would be the first to say: only do elective single embryo transfers," Sam, a patient advocate, stated. But currently, "the patient is paying for it."

Reducing the Number of Fetuses: Partial Abortions

"When the doctor transferred three embryos, we basically had to sign that if the pregnancy had more than three fetuses, we would reduce," Roxanne, the Michigan marketer, said. After three miscarriages, she underwent three IVFs over eight years to have her son. Now, she was trying to have a second child. "We didn't really think it through at the time. I just thought, 'Let's do this.' I don't know that I would have been able to say, 'Sure, go ahead and take one of the fetuses.' Because when you do that, you could lose the whole pregnancy."

If more than one embryo is transferred and two, three, or four fetuses consequently develop, clinicians and patients face quandaries of whether to abort all but one of the fetuses. "With all we had been through, to lose a fetus because of an elective reduction would have been terrible," Roxanne explained. "At the time, it was such a far-off possibility that it really wasn't in my mind. It also wasn't in my mind that I should worry about having multiple children. Just getting *one* child is difficult. You're so desperate to have a child that you think, 'Hey, two, three, four would be just as good.'"

Partial abortions that reduce triplets and twin pregnancies to singletons improve results[44] and lower complications.[45,46] Such reduction can induce miscarriage, but spontaneous twin pregnancies miscarry at about the same rate.[47,48]

The number of patients who undergo such partial abortions remains unknown. In 2011, 22 European nations had 343 fetal reductions[49]; yet several of these countries did not provide data, and whether these reductions were for twin or higher multiple pregnancies is unclear. Before embryo transfer, 77% of 36 Oregon IVF patients would, hypothetically, consider reduction; 67% would consider reducing triplets, while only 6.5% would consider reducing twins.[50] In India, among 186 IVF/ICSI patients, 83% were counseled about reduction, but 61% didn't know the procedure was invasive and risked miscarriage. Eventually, 42 reduced, of whom 40% initially said they were unwilling to do so.[51]

How do patients and providers in fact perceive and make these tough decisions? I found that they wrestle with these questions. Many physicians in fact transfer extra embryos, assuming that patients will later reduce, if needed. Yet many patients say "yes" beforehand and later renege.

Extra Embryos, if Patients Agree to Later Partial Abortions

When transferring more than two embryos, providers often discuss with patients, beforehand, possible needs for later reductions. "We have a long checklist of complicated issues," Marvin said. "How many embryos go back? Would they do an induced reduction? An amnio? If somebody won't do a fetal reduction, we would never put back more than three embryos." But doctors may incorrectly assume that patients who agreed earlier to reduce will later do so, if needed.

As Roxanne suggested, however, patients, desperate to have a child, may agree in advance without fully appreciating the risks of twins or multiples. Patients thus commonly consent to potential later reduction because doctors require it as a precondition for transfer. If reduction is ultimately needed, potential parents often reassess, given risks to the whole pregnancy, grappling with emotional, not simply statistical, issues and frequently discounting the risks of extra births.

Patient Wariness of Partial Abortions

Decisions about reduction can excruciate both doctors and patients. A patient's subsequent refusal can surprise physicians, who generally feel they must simply accept this decision. "Your likelihood of success is probably less than 5%," Jill told one patient. "I put back three. I didn't believe she'd ever get embryos again. She's now 10 weeks pregnant with *triplets*, and decided not to reduce, thank you very much. So, you don't know what's going to happen. You inform and educate patients, and hope you make the right recommendations."

Prospective parents can vary in the nature and strength of their preferences and reasons, and physicians must decide how much to coax. Jill's patient "wasn't completely adverse to reduction. She didn't say: 'I will *never* reduce. It's against my moral or religious beliefs.'" But ultimately, despite Jill's encouragement, the patient "didn't *like* the idea, and continued the pregnancy."

Even for pro-choice patients, reduction can provoke moral and religious qualms. Women who have been trying longer to get pregnant may be especially reluctant. Amanda, who works in a high school, and her husband, a clothing salesperson, "had to discuss those options and what to do with extra embryos; and those religious factors appeared. Luckily, we didn't have to face that choice. It would've been very difficult, after trying for so long [to get pregnant], to then reduce." Patients may refuse reduction because of medical risks to both themselves and the entire pregnancy. If reduction "would have 'ensured' that the other babies would be OK," Amanda concluded, "I'd have done it."

Patients must thus weigh multiple risks and uncertainties—potential complications for twins, if born, against possible loss of the entire pregnancy. After long, unsuccessful efforts to have children, abortion can feel very counterintuitive and scary. "In reducing, some women have unfortunately lost the entire pregnancy," Suzanne, an IVF clinic psychotherapist, observed.

The costs of reduction are also rarely reimbursed. "A lot of insurers don't cover it," Steve noted. Patients vary, too, based on whether they already have offspring: "Usually, patients who reduce already have one or two children and don't think they can handle twins."

Challenges for Providers

Physicians differ in how much they appreciate these struggles or believe women's prior agreements to reduce. "The doctors feel it's a pretty easy, rational decision," Suzanne felt. "It's going to be safer for you and the baby." Other clinicians recognize that long-stymied patients cannot fully anticipate these trade-offs. "Whenever patients say, 'I'm OK with reductions,'" Steve said, "I always think inside, 'No, you're not.' They have no perspective. You can't blame them."

Patients and providers appear more willing to reduce from three fetuses to two than from two to one, feeling less compelled by the data that twins encounter more risks than singletons. "Women reduce with three," Suzanne said. "However, usually with two, they don't feel the absolute necessity, because it's safe, though it's a tougher pregnancy." Yet twin pregnancies are not wholly "safe" but "safer." Reducing from two fetuses to one makes some physicians uneasy as well. "Proponents of more active reduction propose going from two to one, but we struggle with it," Steve said. "As a practice, it makes us uncomfortable: three to one or three to two doesn't seem to bother anybody, but reducing twins to one makes people uncomfortable. Most twins are OK."

Providers commonly let twins continue to term but are bothered by the elevated risks, seeing tensions between the goals of establishing pregnancies versus terminating fetuses. "We're about helping people get pregnant," Steve commented, "not: get pregnant and then have a termination."

Who Should Decide and How?

Physicians wrestle with who, ultimately, should make these decisions and how. Many providers feel that ultimately, these decisions are the prospective parents'. "To go from two to one is usually a *patient* decision," Steve explained. "We don't do reductions here. We send those to a major academic medical center." Infertility specialists may find it easier to transfer additional embryos since the actual abortions are performed not by them but by other physicians.

Providers may disagree with a patient's decision yet feel that their role is limited—that they should make recommendation but in the end let the patients choose. Jill treated a gestational surrogate, who "got pregnant with quadruplets," and "ultimately, elected not to reduce. I think I have a responsibility, as unpleasant as it may be, to strongly recommend things, if there is a reason to do so," Jill said, however she let the birth of the quadruplets proceed. "The birth was a little premature, but the kids are doing OK." Still, Jill may have had some responsibility to transfer fewer embryos.

Regarding another patient's refusal to reduce any of three fetuses, she pondered how far exactly clinicians' responsibilities go. "Even though she got pregnant with triplets, I don't know that I made the *wrong* decision," Jill commented. "The outcome just isn't what *I* wanted. If somebody said, 'she's 26 years old, you can only put one embryo back in,' I'm not sure that would've been the right decision, because it's important to allow people that level of *autonomy*." In simply following patients' autonomy, Jill sees potential problems but nevertheless leaves to patients the decision of how many embryos to transfer.

Yet views of these decisions as wholly the patients' clash with countervailing ethical principles of social justice and non-maleficence (or avoiding harm). Most triplets encounter medical problems. Especially in under-resourced healthcare systems, severe complications will consume many resources. "We wrestle," Henry said, "with the competing ethical areas of patient autonomy versus the physician's social responsibility to do what's right for society."

In sum, current professional guidelines contain ambiguities and lack enforcement. Not only limited insurance coverage but other non-economic social, psychological, cognitive, medical, logistical, and professional factors heighten rates of twins and triplets. Even in countries that offer insurance for IVF,[52] many patients prefer twins. Both patients and providers commonly minimize the resultant risks.

These decisions can be highly dyadic and dynamic. Doctors may discuss with patients the numbers of embryos to transfer,[53] but these discussions vary widely in quality. Providers may discuss these critical issues only for five minutes on the day of the transfer, rather than for longer or before.[54] When providers minimize the risks, patients may do so, too.

Professional recommendations in the US state that transferring additional embryos is justified if patients are informed about the risks, but patients commonly misunderstand or deny these potential harms. Patients may feel they can disregard these dangers since any one patient faces relatively good odds of success with twins—about 53% of twins will be healthy, and the chance of each risk occurring is less than 50% (i.e., 46.7% of twins are born prematurely and 38.8% have normal birth weight).[55] Since, on average, most embryos will not lead to a live birth, patients may think that transferring two will be more likely to produce a "take-home baby." Prospective parents may also rationalize that since they are healthy and exercise, these risks do not apply to them. Yet many patients then have twins, of whom over 40% have complications, and other prospective parents remain childless.

While past studies found that most clinics deviate from guidelines due to patient requests,[56] patients often fail to grasp the risks. Hence, doctors should not always simply follow a patient's desire but instead ascertain how informed the patient is and how strong his or her preferences are.

Guidelines should push providers to discuss carefully with patients all relevant details, including potential dangers. Patients also benefit most when told neither absolute nor relative rates alone but rather both together, along with proportions—"one out of ten," rather than "10%" of patients.[57,58]

Fertility specialists often do not consider the high expense to hospitals and the healthcare system as a whole, partly because they do not observe or pay these costs. Rather, pediatricians are the ones who care for the children born with complications. Generally, in assessing risks and benefits, people have cognitive biases, drawing conclusions based on their own *personal* experiences of bad events.[59] Infertility specialists often lack these personal observations.

In focusing on patient autonomy, physicians also downplay their own responsibilities to the future children and to society as a whole. At a certain point, wider social and public health costs and risks to the mother and future offspring may outweigh the individual patients' autonomy. Clinicians should at least seriously consider these other concerns, especially given their potential conflicts of interests and patients' possible misunderstandings.

Statistics regarding the relative pros and cons of extra embryos and of remaining childless are hard to weigh. A single embryo can yield either one or no baby, while extra embryos can yield two, one, or no offspring. Additional embryos may thus seem a better bet for having a child. Yet patients may be unsure or miscomprehend how to compare these multiple odds. Increased insurance coverage may help, but other approaches, such as improving provider–patient education and communication about these issues and attention to clinics' rates of twins, can assist, too.

Professional organizations and/or governments should further encourage providers more to decrease the numbers of embryos transferred. The ASRM, for instance, currently states that for patients under 35 "providers should only transfer a single embryo, and not more than two embryos,"[60] which sends mixed messages, sanctioning transfer of two embryos, and could be changed, to state that physicians should *"discourage"* DET in healthy young women.

Better clinic reporting requirements could also help. In the US, current reporting regulations motivate providers to transfer extra embryos. The CDC should thus require that all clinics report all initiated cycles and outcomes, based on numbers of embryos transferred, along with how the infants fared.[61]

Some doctors will surely resist divulging such information.[62] But such enhanced reporting could improve transparency and perhaps "name and shame" non-adherent clinics. Inclusion of aggregated, anonymized outcome statistics from each medical practice, segregated by patients' sociodemographics and types of prior obstetrical or medical history, can also aid potential parents in comparing clinics and making decisions.

New technologies thus let patients and doctors shape the number and health of future children. Additional dilemmas subsequently arise, however, concerning into which wombs to place these embryos.

9 | Choosing Wombs

Andrew and Charles, a gay couple, wanted to have a child but worked for nonprofits and had limited funds. Their Midwest state does not recognize surrogacy contracts. They could hire a "gestational surrogate" in California, where such contracts are recognized; but doing so would cost over $100,000. They explored their options and decided to use so-called traditional surrogacy—they would collect their sperm and send it on dry ice to a woman who would impregnate herself and carry their child. The cost was far less, around $35,000. On the Internet, they found Nicole, who agreed to provide this service if Andrew and Charles would pay her directly in cash. But at any point, if she changed her mind and decided to keep the baby, it would be hers.

The gamble was big. But the men felt they had no choice if they wanted a child biologically related to them. They traveled to meet Nicole and her family. She agreed but at the last minute, prior to insemination, got nervous and backed out. Andrew and Charles again searched online and several months later found another woman, Evelyn, who consented. They drove to meet her. She promised to avoid alcohol and recreational drugs and eat only healthy foods during the pregnancy. The men met her husband and three kids—who would be the new child's half-siblings. Through ties of blood, their families would be forever bound.

Andrew and Charles FedEx'd their sperm to her. She got pregnant.

Then they all waited. The gay men wondered what would happen if the pregnancy went awry. She had her own health insurance, but other costs could accrue. What if she ended up using drugs or drinking? Every six or seven weeks, they visited her. The pregnancy continued.

Nine months later, she was ready to deliver. Andrew and Charles drove to her bedside. She gave birth to a girl.

On the second day of the child's life, Evelyn could have legally maintained full custody of the child, but she relinquished her rights. Andrew and Charles got their baby.

Other gestational surrogates have, however, reneged on these agreements at the last minute. Hence, countless prospective parents find this approach far too risky and prefer more legal protection, though it costs far more.

Many would-be parents face dilemmas in choosing not only eggs, sperm, and embryos but wombs—whether to "rent" one, to hire a woman to carry their baby in her womb, and, if so, whose and under what conditions. For medical reasons, many prospective mothers cannot carry a fetus themselves. Single and gay men who want a biological child have no choice but to use a surrogate. Celebrity actresses such as Sarah Jessica Parker have also hired a surrogate to avoid undergoing pregnancy themselves. The demand for such surrogates grows enormously, but debates about it rage.

In *traditional* surrogacy, which Andrew and Charles used, a woman agrees to be inseminated by a man and carry the fetus, while in *gestational* surrogacy, prospective parents provide the embryo, and the surrogate only lends, or "rents," her womb and does not provide the egg. Prospective parents can try to find a family member or friend to be a surrogate for free, or they can arrange to hire one.

In the 1980s, commercial surrogacy gained wide media attention in the US through the Baby M. case. William and Elizabeth Stern wanted to have a child, but Elizabeth had multiple sclerosis and was concerned that pregnancy could harm her. An infertility clinic placed a newspaper ad looking for a surrogate. Mary Beth Whitehead, a high school dropout, responded, signed a contract, and was inseminated with William's sperm. Mary Beth then gave birth to a baby girl but decided to keep the infant. The Sterns sued. In 1987, the New Jersey Superior Court awarded them custody of Baby M. In 1988, Mary Beth appealed. The New Jersey Supreme Court gave custody to the Sterns but also visitation rights to Ms. Whitehead and invalidated all future surrogacy contracts in the state.[1] In 2009, a New Jersey court decided that the ruling bans gestational surrogacy as well.[2] Subsequently, several states banned all paid surrogacy.

A few states, however, most notably California, sanction paid gestational surrogacy, upholding legal contracts that prohibit the birth mother from keeping the baby. Gestational surrogacy has become a big business in California and in developing countries such as India, Thailand, and Nepal, where the costs are far less than in the US (around only $5,000, rather than $100,000). Multimillion-dollar industries have spawned in several developing countries, along with ethical, legal, social, and psychological concerns about the potential exploitation of women. Couples from industrialized countries who arranged for such surrogacy elsewhere have also divorced before the child's birth, leaving the infant, unwanted by any of the parties, marooned in an orphanage. Additionally, in many countries, the surrogate can legally change her mind and decide to keep the baby. The surrogate's health, which may be suboptimal and not necessarily well monitored, can also adversely affect the child.

US states range considerably in whether they permit, prohibit, or limit surrogacy and in how they enforce whatever laws they have.[3] As of 2011, traditional surrogacy is allowed (since it is not explicitly prohibited) in 16 states; permitted by statute without much detail in five; permitted by statute with restrictions in two; permitted only if uncompensated in four; permitted but with unenforceable contracts in nine; practiced, though contracts are prohibited, in two; not practiced because contracts are prohibited in four; and unpredictable in nine.[4] Gestational surrogacy is allowed by law in three states; allowed (since it is not explicitly outlawed) in 22; allowed by statute without much detail in seven; permitted with restrictions in six; allowed with unenforceable contracts in one; supported but with no law in six; practiced, though contracts are prohibited in five; and not practiced since contracts are prohibited in Washington, DC.[5] States differ in other ways too, such as how much surrogates can be compensated (e.g., whether more than basic expenses can be paid); whether surrogates can change their minds and, if so, in what time frame; whether court approval and state residency are required; whether at least one of the intended parents must provide the sperm and egg; and whether the surrogate's eggs can be used.[6]

Like all pregnant women, surrogates face potential shorter- and longer-term risks, including high blood pressure, anemia, diabetes, preeclampsia (or toxemia), and depression. Factors such as obesity raise risks of complications.[7] As a result, doctors usually choose gestational surrogates who have previously been pregnant and given birth and are thus aware of what to expect.

When exactly a woman is too old or unable to use her own womb is not always clear. Many women face uncertainties. "We were considering surrogacy and/or egg donation because I wasn't having success," said Roxanne, the Michigan marketer who had a son through in vitro fertilization (IVF) and was now seeking a second child. "We were leaning more towards surrogacy, so the child would at least have been made from my egg and my husband's sperm." But she remained unsure and queried her doctor. "I asked, 'Would you have *your* daughter do this?' He said yes. It made me feel, 'OK, this is what I'm going to do.' Yet it was very expensive. Hence, to save money, a good friend kindly offered her services. I don't know how realistic she was. She said she 'really' wanted us to have a child, and would be willing to carry it for us."

Roxanne and her husband remained puzzled whether to have this friend carry their future child. "We talked to my regular OB/GYN about it," Roxanne explained. "Unfortunately, our friend has epilepsy and takes anti-seizure medication, which would not make her an ideal surrogate, even though she already had children of her own. My OB/GYN wasn't comfortable with that," Roxanne continued. "So, we said we'll do one more round of IVF, and if that doesn't work, look at adoption. But I got pregnant and had my son." Friends or relatives may thus offer to be unpaid surrogates, though obstacles and unpredictable complications can nevertheless arise, and many prospective parents remain wary or uncomfortable about it.

While California laws clarify relationships between gestational surrogates and prospective parents, challenges remain elsewhere, as in the Baby M. case. The surrogate, even if related to the future parents, could decide to try to keep the child. Jill reported one such instance: "A gay male couple used an anonymous egg donor, and one of the partner's sisters carried the pregnancy. The sister was counseled, and signed her consent, but did not meet with the psychologist, and has now sued for custody, and has temporary joint custody. She knew what she was doing, and agreed to it. She's not biologically related to these children – her brother's partner provided the sperm. Why should she be able to parent them as the intended parents would?"

Tensions can also emerge concerning the surrogate's behavior while pregnant. "The intended parent may freak out," Jill observed. " 'Oh my God, she's carrying my baby. . . . I can't believe she wants to go to the amusement park!' " A gestational carrier may want to drink wine, eat certain foods, exercise, or travel. Prospective parents and surrogates can argue. "Typically, to use a gestational carrier, patients need to have significant financial means," Jill continued. "Without making rash generalizations, such patients are used to pushing people around."

Legal disputes are relatively rare but occur—especially if the intended parents and the gestational carrier live in states that do not legally recognize these contracts. "The arrangements are usually successful," Jill added. "The biggest challenges are managing the relationship between the intended parents and the gestational carrier."

Careful counseling and informed consent can help prevent surrogates from changing their minds, yet challenges linger. "We haven't had the problem of surrogates wanting to keep the baby," Marvin said about his clinic. "But we spend a lot of time getting consent and counseling our patients. We have very detailed legal documents. They protect all parties. We require mandatory psychological counseling for all parties. They've got to wait three months. That gives them time to set everything up, get adjusted to it. But some potential parents may simply drop out and go somewhere else."

State laws that forbid paid surrogacy may not reflect most citizens' views but nonetheless be hard to change. "Somewhere along the way, someone said that surrogacy is terrible. I don't know exactly why, how, or when," Jill reflected. "Yet nobody's bothering to fight it because change requires time, energy, and money." In fact, advocates have tried altering state laws in New York and elsewhere, without success. Opponents tend to be either right-wing Christians (who often oppose all assisted reproductive technologies, seeing them as immoral), or feminists (who argue that women who agree to be surrogates will mostly be poor and thus be taken advantage of).[8] Yet no data exist on who these women in fact are and why they agree.

Questions arise of why women choose to be surrogates. They may do so for both financial and personal reasons. "They're doing it for the money, but that is only part of it," Jill continued. "Usually, they're extraordinarily motivated. Some are lovely. Others are controlling: they think they know everything. A colleague feels that they are somewhat narcissistic and like feeling highly valued. Other areas in their lives might not give them that same kind of appreciation and respect." Each of these motives may be necessary, but not sufficient in itself. Anecdotally, many such women who receive $100,000 for nine months of pregnancy are middle class, fully understand the risks and benefits involved, having almost always previously given birth to their own children, and feel that the rewards are worth it.

These complex relationships can pose difficulties for providers: "I'm now the primary doctor for both an intended parent couple *and* the carrier," Jill said. "Is that in everybody's best interests? I'm not sure. The surrogate has an infection that was really of no consequence. It probably shouldn't have been tested in the first place. But she is all in a tizzy about what the genetic parents are going to think. Then her estrogen level was a little low. Again, she was all in a tizzy about what the parents were going to feel. It's heartbreaking because all she wanted was to do well."

Altruistic and humanitarian reasons motivate women, too. "Some of them are really doing it out of the goodness of their heart," Jill continued. "They want to help somebody. Some really liked the way they felt being pregnant. For some, it is the ability to do something important and well and receive that level of recognition. Few would do it without the money, but that's reasonable, if you think what they're going through and the risks. Yet it's not *all* about the money. Maybe 40%–50%."

Prospective parents employing gestational surrogates in the developing world confront added challenges. In 2007, for instance, a Japanese couple,

Ikufumi and Yuki Yamada, arranged for a gestational surrogate in India, using Ikufumi's sperm and an anonymous donor's egg. One month before the child's birth, however, the Yamadas divorced. Ikufumi then wanted to raise the child alone as the father. But when the child was born, Yuki refused, and the Japanese embassy thus rejected the child's passport application, due to Japanese birth citizenship requirements. Indian law did not, however, recognize Ikufumi as a single father. Hence, baby Manji, as she became known, remained in an Indian orphanage, stateless. Only after many subsequent and protracted legal actions did Ikufumi prevail.[9]

Prospective parents who use gestational surrogates abroad to save money face not only medical but legal risks, involved in bringing the children home. At times, local governmental authorities have prevented couples from transporting newborns across national borders, arguing that "the mother" is the woman who gave birth and remains within the country. Concerns also arise about potential exploitation of these gestational surrogates. In addition, the Yamadas and some other couples from industrialized countries have separated or changed their minds, no longer wanting the child. Baby Manji and other children have consequently been abandoned in foreign orphanages.

In Nepal and Thailand, surrogates have carried fetuses, unaware initially that these were for gay couples, igniting controversy due to homophobia. Moreover, research is lacking on how parents, surrogates, and children fare medically, psychologically, or socially over time. "A few patients go abroad for cost reasons, but I don't get follow-up on many of them," Brenda explained. "I hear about the problems afterwards, getting the children home. It's not quite so simple as the parents think it's going to be."

Prospective parents interested in renting wombs thus face medical, ethical, legal, social, psychological, and financial challenges, exacerbated by growing surrogacy markets in developing countries. Arrangements for paid gestational surrogacy require significant legal efforts to protect the rights of all these parties—the biological parents, birth mother, and future offspring. When done appropriately, with full informed consent and legal protection for all parties involved and without coercion or undue influence, surrogacy enables prospective parents who lack or can't use their own womb to have children. Many jurisdictions, including New York, prohibit it but arguably should consider allowing it *if* it is very carefully done and closely monitored. More education of patients, providers, and the public at large about these issues is also key to addressing these novel challenges, which all parties may not appreciate beforehand.

10 | Choosing Children
WHETHER TO ADOPT

"This is my last IVF, because it's all out-of-pocket," said Wendy, the 37-year-old Irish Catholic secretary who has failed four cycles. "If this fails, I'm going to adopt. I have an adoption agency ready to go, once I decide. It would be somebody else's child, but I will treat it as my own. I will love it as my own.

"A woman in my mind/body workshop adopted, and was not disappointed. She felt, 'I had to try IVF first; but it all ended up working out the right way.' I thought, 'Whatever's going to happen is the right thing.' I knew I would be a mother—that was in my control. I wouldn't have to relinquish that desire."

At some point, Wendy and many other prospective parents who fail to get pregnant using these technologies consider adoption as an alternative, but can then encounter obstacles, too. "Adoption is not unreasonable, but is a lot of money," Wendy continued. "At one place, you do a home study and pay $10,000, and wait in the lobby. The nurses try to convince 16- to 17-year-old girls having a baby to terminate their rights, and give their child up for adoption. The nurses come downstairs with the child. But if the mothers want their child back within 30 days, you're out 10 grand. With closed adoptions, a lot of people scam you. A lawyer charges you $12,000 to $18,000, and finds you a birth mother. Once the baby is born, he walks away. Yet the mothers just gave up their child; their hormones are raging. They don't know what to do, and contact the lawyer: 'I want my baby back.' Once you pay, if it doesn't work or the mother changes her mind and takes the child back, you don't forfeit all that money: they work on getting you another child.

"In open adoption, the child's parents stay involved with the child until it's 18 years old. The child may see the mother once a year. They can Skype or send pictures. Within the first year, most birth parents actually drop off. It's too painful for them. But some actually form a family bond and stick with the child through its 18th year. The agency has counseling for birth parents and doesn't just drop them, but helps them transition, having just given away their child. They have a 90% placement. But, the wait at an open adoption agency is six to 18 months. If you do it out of the country, you only have to go there two to three months. Yet, that is too expensive for us.

"We also looked into foster parenting. But it didn't work out. The state requires too much information on you—family and friend references. They make you attend 10 weeks of classes. The sessions were boring. Awful. They basically want to know how much money you have and plan on spending; and whether you can afford to support somebody else's child for six to eight weeks before you get reimbursed. The questionnaire to friends asked: 'How are they with money? How much money do they have? Are they frugal or going to 'use' the system?' My friends didn't know how to answer: 'I don't know how much money you have. I know that you have a beautiful house and beautiful cars. So, I have to assume you are good with money.'

"I'd know nothing about the kid coming into my house. If the mother's a drug addict, or not doing the necessary paperwork, I'd have to wait 15 months for that child to become legally mine. Frankly, I cannot take a little baby from a drug-addicted mother who won't do anything to get herself better to get her child back. It wouldn't have worked for me. You want all this information about *me*? What about the people coming into my house?"

Many infertility patients investigate adoption as a backup plan, but then face quandaries of how long to keep trying IVF before pursuing this alternative or accepting childlessness—whether having a biologically unrelated child is better than not having a child at all. They observe adoptees and try to draw conclusions, either for or against this option.

Since antiquity, adults have adopted children. Pharaoh's daughter adopted Moses; Julius Caesar, Octavius. But in recent years, adoption has become harder. Adopted children have been getting fewer and older. In 2014, 69,350 children were adopted in the US by unrelated parents. Only 26.4% were infants, down from 47.9% in 1992.[1] Most adoptions were from foster care, where, in 2015, the mean age of children was 8.6 years and only 7% were infants.

Adoption from other countries has also become tougher.[2,3] From 2005 to 2015, children adopted into the US from abroad dropped about 72%, and the children's ages rose from mostly being younger than two years of age to mostly being older. Older children were adopted far more than infants.[4] The children's countries of origin have shifted as well. In 2015, around 42% of adoptions to the US came from China, 16% from sub-Saharan Africa, 14% from South Korea, and 13% from eastern Europe.[5] Costs ranged up to around

$64,000 in 2015. Domestic adoptions have also evolved from mostly "closed" (with the birth and adoptive families not knowing or communicating with each other) to "open."[6]

Wendy and many other prospective parents consider adopting, especially if they are older and have repeatedly failed IVF using their own or another woman's eggs.[7] They want to raise a child, and needy children lack homes.

"I always had adoption in the back of my mind," Karen, the physician who underwent fertility treatment, said. "I expected the first IVF to work. When it failed, it was so disappointing. I became really depressed. I didn't want to do that again. I convinced my husband we should adopt." They then did so.

Women who have successfully used assisted reproductive technologies (ARTs) in the past and want to have an additional child may also adopt. "I lost two pregnancies. Then, they gave me extra progesterone, and I conceived my oldest son," Ginger, the IVF clinic psychotherapist, said. "A year later, we were thinking of trying again, but I had a very high FSH [folic]le-stimulating hormone] level. So, we turned to adoption."

Yet other would-be parents instead pursue ART as much as they can. Friends and family repeatedly asked Sally, the pregnant Maryland website manager who had terminated two pregnancies due to cystic fibrosis, "Why don't you just adopt?" How important could it be to have a biological child?" These conversations perplexed her: "I don't know that it can even be reasonably explained, because it's nothing to do with the reason. It's more visceral."

Past personal experiences and perspectives mold these perspectives. "I have a personal bias against adoption," Sally explained. "My dad remarried and adopted two sons with his new wife. These kids have had so many problems! I know it's not only because they're adopted; but it has left a bad impression with me. My sisters and I, conceived and born through our parents, didn't have those issues. So, to me, adoption is a last resort.

"I probably would use a donor egg before I adopted," she continued. "Because of the visceral need to give birth, and the fear of inheriting someone else's problems—as defective as part of my own genes may be. What you eat and do in pregnancy, and how the mother bonds and acts after birth all affect the child. I also wanted the experience of being pregnant. My husband didn't feel as compelled as I did. He probably would have rather adopted. *I* was definitely pushing us forward. Obviously, you're going to love any kid you have. An adopted child can be wonderful, but I have a lot of fears." Adoption can be hard, too, because it means relinquishing desires for one's own biological child—for creating a physical embodiment and union of oneself and one's spouse. "I think about my connection to my grandmother," Sally elaborated. "I really want the child to be somehow physically connected to me like that."

Sally and many others remain wary of adoption, seeing it only as a last-resort final option, "second best."[8] Prospective parents may be leery about the health of the adopted child's biological mother, and adopted children consequently having more problems, perhaps genetically exacerbated.

For patients who have offspring, adoption may be less difficult. "Since I already had one child, it was easier," Ginger, who adopted her second child, observed. "Going from fertility treatment to adoption, you have to give up your dream of having a child who is a blend of you and your spouse's genes, and give up the whole birth process. You can breastfeed, but it's not quite the same." Many prospective parents see a biologically-related child as uniquely integrating the love of two people—a literal melding of themselves. Still, institutional obstacles exist. "The world is getting more adoption-friendly," Ginger added, "but that's not the focus in IVF clinics."

Many prospective parents consider adoption but ultimately prefer child-lessness if they can't use their own eggs or sperm. "My best friend had infertility and adopted," Yvonne, the Jewish Philadelphia psychologist, said. "For some people that's really comfortable and lovely. But my husband and I both knew it wasn't right for us. We would either get pregnant or have to come to terms with failure to do so. Other people adopt. We would not."

Certain couples, if in relatively new relationships, emphatically reject the possibility of raising another couple's child. Nancy, a 44-year-old IVF clinic nurse, has sought infertility treatment herself and "wouldn't want to adopt." She is now in her second marriage to "the love of my life. We each have four children already, but want to have children together, if it's at all possible."

Improved ARTs also often lower prospective parents' interest in adoption. "Many patients are just scared of adoption," Helen, the Wisconsin psychotherapist, observed. "Everyone tells them, 'You could always adopt.' But it's hard for people to look at adoption. Some patients ask, 'Am I supposed to accept second best?' People are more likely these days to move on to do a third-party reproduction, like egg donation, if they can afford it. In the past, they would have moved on to adoption, and not spent as long in treatment. Somebody would have said to them, 'We've done all we could.' IVF has changed that."

Patients who may have previously adopted frequently now try donor gametes instead. "It's almost *impossible* for people to stop IVF," Helen added. "Some patients decide, 'if we are going to spend all this money on IVF, and it might not work, let's move on to adoption.' But typically, people do IVF. Still, questions of how many cycles to do and whether to stop—when is enough enough?—are terribly difficult." Many patients go to extraordinary lengths, pursuing additional treatment to have their own biological child, rather than raise another couple's child—potentially reflecting deep evolutionary drives to reproduce and pass on one's genes, as well as wariness of, and obstacles to, adoption.

These varied financial, administrative, and other logistical obstacles to adoption can prove daunting. Due to costs, many individuals consider but finally reject this possibility. "Adoption is not a bad idea, but is not cheap," John, the Texan mechanic who had had the vasectomy reversal, said. "We could spend more on adoption than on having our own kid." High prices can thus drive these decisions but lead patients to question whether to spend more on IVF.

As Wendy suggested, other prospective parents consider and have to balance the relative pros and cons of various types of adoption—open and closed, domestic and foreign.

Many Americans seek to adopt children from abroad but then face further dilemmas. Adoption from overseas costs somewhat less but still remains relatively expensive and can add barriers and uncertainties. "When we adopted, I wasn't able to work for two months, and had to go stay in South America," Karen, the physician, said. "It was pretty expensive and time-consuming, harder than I thought, stuck in a desolate town in a foreign country. The region used to be rainforest. They chopped down all the trees and it became a desert. The town had no postcards for itself—nobody cared. We were isolated and didn't speak the language. I didn't know what to do with a newborn infant, how to give him a bath. We were so far away from our families. I was afraid the local people would see how terrific he was, and take him away."

Overseas adoptions also raise broader social, cultural, and political questions. "I wonder why parents are happy to adopt a child from Ethiopia, but not a black child here," Ginger, who had adopted a child, said. "People in the Netherlands adopt a lot of our black babies. So, why are *we* running to Ethiopia? Because it feels like a rescue fantasy—going all the way across the world? Or are we afraid of the birth mother showing up if we adopt in the US, but not if we adopt from Ethiopia?

"As international adoptions have decreased, more children are being adopted from foster care," Ginger continued. "So, if more people look at foster care, we may have done a good thing." Yet adoption of older children through foster care, rather than of babies through other means, can pose challenges, too. Foster children may be easier than infants to adopt, but many patients, such as Wendy, remain cautious. "Some people adopt five kids out of foster care. But I'm not one of those people," Ginger stated. "I couldn't do it—the challenges are huge. I grew up in a family with five kids. It was chaotic. So, I don't like the idea of bringing in five grown foster kids. I know someone who's got 12 kids from foster care. But you need to be very organized and get professional help from somebody who knows about adopting older kids. Some parents are able to turn these kids' lives around. It's amazing! I admire that. But you need support, counseling, and money."

Finances, time, attitudes, and patients' ages can shape decisions of whether to adopt infants or foster children. Adoption plans can collapse. "My older sister was adopted. So, I never really felt I needed to give birth to a child for it to be 'my' child," said Amanda, the Delaware high school administrator. After six unsuccessful intrauterine inseminations, she now planned to pursue adoption if the one last cycle of IVF she could afford failed. "What scared me most about adoption was that I could get right to the point of getting a child, and it could fall through. The biological parents could change their mind. That terrified me. Adoption through the foster-care system would be the most cost-effective. But that meant that the parent had six months to complete a case plan. If they finished it, they got their child back, after you've been living with the child for half a year. Emotionally, I didn't feel strong enough for that."

Patients struggle with how to decide about these emotional, medical, and statistical complications and often draw on inner gut feelings, rather than empirical data alone, feeling they would intuitively "know" when they had had enough ART and would thus adopt. "Women who had children through IVF and others who were unsuccessful and adopted, all said that I needed *to know my limits*," Amanda reported, "that I would know when I was truly done with ART. Somewhere inside of me, I knew I hadn't gotten there yet."

Given the complexities and meanings involved, spouses can also disagree about adoption. Helen felt, "more often, the husbands oppose it." Francine saw a more mixed picture, based on contrasting desires and views of potential costs and benefits: "Husbands say, 'I don't want us to spend any more money on IVF. We need to adopt, it's guaranteed,' even though adoption is *not* guaranteed. Or maybe the husband is more interested in continuing on the infertility journey than the wife, who has been through a lot, and is tired and ready to move on."

Prospective parents generally receive input on these issues, both pro and con, from an array of people. These decisions rapidly become social phenomena, with family members and friends strongly pushing for or against in ways that patients may oppose or accept.

Clinicians vary in how much they try to shape these decisions. "Maybe the doctor is saying, 'We can do IVF *this* way next time,'" Helen said. "It's really hard when nobody is stopping you from getting more treatment instead." Yet other doctors help facilitate adoption. "At my clinic, if the doctor hears some interest in adoption, he's very quick to say that his staff can tell patients about it and go through the journey with them," Ginger explained. "That's about as good as it gets."

Still, in forming a family, adoption can be just one of many ongoing steps. "After we adopted my son," Karen said, "it was easier to undergo IVF again. Then, at least I was a mother and had a child."

In sum, when efforts to use their own eggs fail, many prospective parents debate whether to seek more treatment, other women's eggs, or adoption—which may seem easier since the baby is already alive. Yet, as IVF has developed and grown, the world of adoption has also evolved. Adopting babies from both the US and abroad has become harder. Nonetheless, given that most IVF cycles fail, especially for older women, many prospective parents consider or pursue this option. Would-be parents must reckon exactly how far they are willing to go—exactly how much they want to pass on their own DNA. Potential parents must determine whether the genes they would transmit would be better or worse than a stranger's.

III | Choosing Adults

11 | Choosing Doctors

"I'm such a better doctor than when I started," Karen, the physician with infertility problems herself, confessed. "I've learned: instead of saying to patients, 'You're just being a pain in the butt,' or, 'You should stop talking,' or 'I have to get out of here,' I need to just shut up and listen, and believe my patients. They're not all lying to me! When I was having my first IVF, lying there with my feet up in the stirrups, and they were sticking this giant needle through my vagina into my ovaries, and sucking out eggs, I would say, every time they stuck the needle in, 'That hurts a lot!' The IVF specialist said: 'Oh, the ovary doesn't feel the pain, that's your perineum.' He didn't believe me. [As] a doctor, pregnant women gaining weight would tell me they weren't eating too much. I didn't believe them—until I was a patient, too. When patients said they had 'weird' reactions to drugs, I didn't believe them. But since then, I've had very strange reactions, too. No matter what the science says, every patient is an individual. Give them the benefit of the doubt!

"The bedside manner of the office and the practice helps: how the receptionist talks to you on the phone when you first call, how they treat you when you walk into the office. Everything!"

Yet clinicians confront competing pressures. "One receptionist was really stressed out, having an emergency situation going on," Karen recalled. "So, she cut me off short. But as soon as she said, 'We have an emergency. I need to call you back,' I totally understood." Explanations of *why* good communication can be difficult can therefore help.

Human factors shape use of these assisted reproductive technologies (ARTs) in dynamic ways. Providers' individual traits and experiences can influence responses to these challenges, though not always based solely on simple predictable characteristics such as gender. "It's personality-dependent," Karen believed. "Some men have very closely watched what their wives have gone through, and can be compassionate and non-judgmental. In general, I think women are a little bit more patient and less judgmental. Every woman has also had a period, and knows what that feels like. Yet that sometimes makes them *more* judgmental, because they tolerate theirs just fine, and just don't believe women in agonizing pain."

Prospective parents choose not only eggs, sperm, embryos, wombs, and children but doctors. Decisions about which technologies get used and by whom are not made by patients alone but rather with clinicians. Patients are not passive recipients but heavily involved, often increasingly so as they move on their individual journeys. Yet high costs, treatment failures, and taboos talking about sex and reproduction create barriers. How doctors recommend and discuss these technologies powerfully shapes whether and how patients proceed.

How then do patients choose their doctors? Patients select infertility specialists based on published success rates and quality of service,[1] but how do they assess and weigh those factors? In general, patients seek good communication and "patient-centered care" that respects their values and preferences and includes spouses. But patients often feel disappointed,[2] perceiving inadequate communication regarding long-term treatment consequences (59%), reimbursement (50%), and anxiety and depression (40%).[3]

Physicians value such patient-centeredness less than patients do, underestimating how much patients want such communication rather than "continuity of care."[4,5] Physicians commonly focus on scientific data, while patients value more their own individual experiences, preferences, and needs.[6] These differences are important since about one-quarter of infertility patients do not follow treatment recommendations.[7] Many questions thus emerge concerning why these tensions exist and how they can be alleviated.

Patients, I soon saw, confront obstacles in not only getting referred to infertility specialists but in then interacting with them, wanting good interpersonal skills in doctors but encountering difficulties evaluating and balancing these traits against technical expertise—often "doctor-shopping" to find "the best" physician.

How Patients Choose Fertility Doctors

Patients range in how readily they find a clinician with whom they feel comfortable. Though patients frequently get referred to an infertility specialist through a prior doctor in another field, they often end up seeing additional ART specialists as well since most treatment cycles fail.

Occasionally, patients like the first infertility specialist they consult. "Choosing a provider and clinic was pretty easy," Isabelle, the pregnant Connecticut office manager who underwent preimplantation genetic diagnosis (PGD) for Emanuel syndrome, said. "We didn't shop around. We got a recommendation from our doctor . . . We were very happy with them . . . very *comfortable*. They were local."

Still, feelings about providers can shift over time, prompting patients to debate whether or not to give providers a second chance. "I didn't like one doctor, and complained," Isabelle continued. "But he actually did the embryo transfer, and I ended up liking him. Maybe it was a misunderstanding on my part. Ultimately, he did a great job. I was happy I gave him a second chance." Over time, patients may alter their perceptions of physicians, either positively or negatively. Early antipathies can prove ungrounded.

Ultimately, many patients consult multiple infertility specialists. Gaining a personal sense of a clinician's human qualities is not always easy. "I'd heard about the practice from different people, and did some research," Roxanne, the Michigan marketer who had been undergoing ARTs for eight years, said. "But I wanted to meet the person first, and see if I like them, and have a good feeling about them. You need confidence, a comfort level, because it's such a personal thing. You lose a lot of privacy. It might take a couple of doctors."

Roxanne "went for a second opinion after our third miscarriage. People said, 'Have you gotten a second opinion?' My doctor was very open to getting it: 'I'd expect you to. If you didn't, I wouldn't be doing my job. If I miss something, I want to know.' So, I was very comfortable with him.

"A friend went to a doctor she couldn't stand. He was 'terrible,' a 'jerk.' But his pregnancy rates were so good, and the practice itself so well known that she endured his ill-temper. He wasn't nice, but she endured it to get pregnant. I don't need that. She told me I should see him. I didn't. A doctor can have a good bedside manner *and* be a good doctor.

"My second doctor was surprised that my first doctor had gone to IVF so quickly. I felt it wasn't so quick. But this second doctor said: 'I'd go back to IUIs [intrauterine inseminations] first. It's less expensive, less invasive. You have been pregnant yourself without help.' I said, 'We were trying to expedite things, because we've been going through this for so long.' I never felt overly comfortable with his solution."

Doctors can differ in both what treatment they recommend and how they communicate about it, offering conflicting advice that patients must then gauge.

Physicians vary, too, in their receptiveness to patients seeking second opinions. Some patients felt that doctors should encourage these external consultations more and that patients should unhesitatingly consider these if dissatisfied with care. Physicians' openness and communication regarding such possible outside professional views can in turn heighten patients' comfort.

Unfortunately, insurance coverage may restrict which other doctors patients can visit. Nonetheless, even within one clinic, patients may have choices and opportunities to select physicians. Francine, pursuing treatment in Mexico, said her first clinic "was covered by my insurance. We went there mostly for that reason. But I researched the different doctors there, and picked the one I thought would best suit my case. I've been treated at two different clinics, but consulted *six*. I wanted more answers and information, to make sure we had the best possible chance! Is the problem the lab, the doctor or me?"

After her second failed cycle, she stepped back. "We only had so much coverage. How were we best going to use it? If we ran out, we'd have to wait, and pay out-of-pocket. So, we did more research and consultations. Nobody could say definitively, '*This* is a better course of treatment.' So, we stayed with one doctor because we were comfortable, and liked the staff. But after *that* cycle failed, we put things on hold. We did our fourth cycle at a *new* clinic because I got different insurance. The insurance covered *both* clinics, but we decided to change."

Especially when treatment languishes, patients commonly reassess, prioritizing not just a doctor's overall success rate but their own personal failures and resources, pondering whether to switch. Patients may change clinics to understand *why* treatment failed and to ensure that they leave no stone unturned.

"The person who knows me and my body best is *me*," Francine continued. "Initially, I just did whatever they told me. Then I started doing my own research, and realized I didn't exactly fall 'inside the box.' My doctor didn't want to look at immunological factors: 'I don't feel it's in your best interest. These tests are extreme, and costly, and could use up your coverage.' But I wanted to try. So, I went outside, and got testing done.

"Doctors will probably look at an obscure study and think, 'if this was reproducible, or had any benefit, I'd have heard about it,'" Francine remarked. "But, if *I'm* the patient it works for—and got pregnant because of it—I wouldn't care. *I don't care that it's anecdotal—just that it works!* As long as it doesn't harm me, trying something new is not bad." Desperate for a child, patients may feel less swayed by limited scientific evidence, overestimating the benefits and underestimating the dangers.

Especially concerning use of donor eggs, patients may disagree with providers. Older women who have floundered using their own eggs may succeed with another woman's eggs but resist this option. "Some clinics say 'Our goal is to have you have a family. Not leave you perpetually in treatment,'" Francine explained. Such clinics may therefore more strongly shift to donor eggs. "If you know that, walking in, and that fits more with your mindset, you'll probably be more open to what they're going to say."

Ultimately, Francine and many other patients find a clinician they feel is both skilled and emotionally attuned. "My doctor now is very supportive and concerned about my mental, not just my physical, well-being," she added. "He says, 'I just want to make sure you're OK. How are you doing? What are you doing to take care of yourself? Are you going to therapy? Support groups? Talking to friends?'"

Changing Doctors Because of Medical Factors

"I'm now on my fifth and last clinic," Wendy, the Irish Catholic secretary, said. Prospective parents face questions of *how many* different doctors to consult, when to change, and how to decide. "One friend went through seven attempts before having a child, and changed facilities several times," Wendy continued. "My new doctor reviewed my records, and said the embryos 'were too immature.' I said, 'So, you're telling me I just wasted $40,000, and gained 40 pounds, for nothing?' It made me mad. Each doctor has an excuse for the last one. I went on to a fourth facility, but they wanted to do IUI—though the earlier IUI didn't work! I said, 'That's absurd! I just want IVF.' So, I moved on to another facility." Over time, as treatments fail, she and many others come to feel more empowered.

"Friends who have had children through IVF love and recommend their doctor; but unfortunately, I've been to most of those doctors and *had bad experiences*," Wendy added. "One friend did one IVF and had a child. It depends on the experience they've had." Whether patients end up with a "take-home baby" profoundly affects their views of doctors, outweighing all other considerations. Since the field is new and rapidly evolving, the specific causes and optimal infertility treatments for a patient are often unclear. Providers can differ about various biomedical as well as holistic approaches and philosophies, offering explanations for other doctors' failures. Yet these interprofessional disagreements can frustrate patients.

Physicians may diverge particularly in how much they will use a patient's own eggs before switching to donors, how they communicate about these issues, and how much they pursue less guaranteed procedures. "Some doctors who work on their own are doing more experimental stuff, or work with patients whom no major medical centers will see," Valerie, the psychotherapist and single-mother-by-choice, explained. "If a patient gets insufficient numbers or qualities of eggs after three rounds, most major medical centers would say, 'don't waste your money.' Those patients then go to the *last-chance places* that provide the least education, and take the most steps."

After treatment failures, patients may also seek more holistic, patient-centered approaches, finding comfort in these, though providers range considerably in their attitudes toward these approaches—from advocating them to remaining circumspect. Ginger, the psychotherapist who underwent ART and adopted her second child, works at "an 'East-meets-West' clinic. They do cutting-edge Western medicine. Next door is a wellness center with acupuncture, Chinese herbs, fertility massage, and nutrition. They often combine these for a patient, and are very relaxed and accepting. It's not like an 'institution.' Many of my patients swear by acupuncture. Especially before and after transfers. Apparently, it raises the pregnancy rate." Patients may see such holistic approaches as psychologically and medically helpful, even if not all doctors do so.

Though these alternative therapies may not work effectively, many patients at least want to know about them and feel disappointed if providers do not mention them. "I went through six rounds of IUI, and no one told me about acupuncture!" Amanda said. "I found out about it on my own! Someone said, 'Did you consider acupuncture?' Why wouldn't you tell me that to begin with?"

Yet while numerous patients seek such unproven, alternative approaches, physicians may view these skeptically, as delaying effective interventions. "Patients go to amazing lengths, hearing about alternative treatments," Helen, the psychotherapist, reported. "'I didn't have enough acupuncture or herbs. Maybe I should go see this person in Chinatown.' Patients who become very informed, using reason and being strategic, sort of throw that out the window. Patients jump from doctor to doctor. There's always another treatment, another doctor. They'll go to anybody who gives them hope. They'll go for a second opinion, who will usually say, 'I'd have done *this*.' The patient then does that. If it doesn't work, they'll seek a *third* opinion. Another doctor says, 'I'd have done a touch of *this*, added a bit of *that*.' Patients go across the country, have immune stuff done, go to mental health retreats, go to Mexico, London, hang upside down."

Since success rates may be relatively low, particularly for older women, patients commonly seek not only good communication but optimism.

Unfortunately, clinicians may hype procedures, fueling overexpectations and false hopes.

To pursue treatments that doctors view as unpromising, patients may even prevaricate. "I've seen patients who've lied to various clinics about how many IVFs they've done," Suzanne said. "I've seen patients working with *two* clinics at the same time."

Differing Perceptions of Bedside Manner

Though doctors generally try to communicate well about these technologies, patients may see these interactions as ranging widely from good to bad. Medical and emotional complexities and uncertainties as well as characteristics of particular providers and clinics can hamper providers' communication and relationships.

Providers can end up feeling closely connected to patients and find the work very rewarding, personally, as they together endeavor to create new life. Clinicians can become highly invested in the results. "It's fun to watch those children grow," Joe, a reproductive endocrinology and infertility specialist (REI), reflected. "Many patients send me pictures every Christmas of their families as they grow. Those have been some of the joys—silent accolades, but they mean the most. I savor them."

Doctors can become emotionally involved in prospective parents' failures as well. "Patients bring us up or down," Steve confessed. Even when treatments flounder, patients frequently appreciate these physicians' efforts, "The most humbling letter we get," Steve added, "is 'I didn't get pregnant, but thank you for taking care of me.'"

Nurses, too, play vital roles and can become very engaged in patients' care, both professionally and emotionally. Paulette has "lost sleep over patients I've grown strongly attached to. I really want them to be pregnant. During those 12 days after the transfer and before their pregnancy test, I'm sweating it out as much as they are."

Especially if they end up with a baby, patients tend to reflect back positively on their doctor, admiring the physicians' compassion, dedication, and skill. But, particularly when treatments repeatedly falter, would-be parents may feel dissatisfied with both outcomes and conversations. "I want a doctor who has a high success rate but *also* understands what I'm going through: that I'm miserable, not having children," Amanda, the Delaware high school administrator, said. "That's lacking! Doctors do this for a living. But for me, it's my life! It becomes routine for them. It's understandable: it's their job, they do this every day. But for the patient, it's anything but. At a lot of places, you feel you're in a baby factory. A huge emotional component is lost."

Patients can feel lack of empathy from not only doctors but also nurses and other staff. Sterile, mechanical procedures can incite intense emotions. "I had an old-fashioned nurse—all business," Amanda continued. " 'Here are some needles. Go do this.' No softness. I felt she tried to give me false hope with the IUI. With the miscarriage, I started bleeding. She said, 'It happens. Just don't worry about it. Come in for an appointment.' Another nurse, the first time I met her, spoke about all the nuts and bolts, but said, 'I just want you to know that I'm really sorry that this has happened to you—that you have to go through this.' I thought, 'Thank God, *somebody* understands.'

"I went to an office with five fertility doctors. The doctor I was referred to, whom I wanted to see, wasn't always my doctor. He came up with the plan, but then the doctor on call that day decided the next step. That was very difficult. I didn't feel any one person was in charge of my treatment." The structure of clinics can thus impede continuity of care—physicians commonly rotate on-call schedules. Patients may understand these constraints but still feel frustrated. "At a more private office, that wouldn't have been the case," Amanda continued. "But this office is supposedly one of the best in the state, and is five miles from my house. So, it was more realistic for me, but difficult. Still, I was very pleased with them. They were compassionate and skilled. One doctor in particular would sit for two hours if you needed it to go over every question you had." Even within one clinic, providers can range significantly.

Despite the frustrations she encountered, she also ended up giving birth to twins and reflected back favorably on her experiences.

Given the stresses they face, patients often feel that physicians could present and disclose possible barriers more carefully. Clinicians may make casual, offhand comments that inadvertently leave patients distraught. "Doctors are sincere," Sally, the pregnant Maryland website manager, admitted, "but maybe not properly trained on how emotionally fragile women are going through this. Doctors should never say, 'Maybe something's wrong with your eggs.' That could be true, but patients don't always need to hear the worst-case scenario. I may not be able to handle it right now. Sometimes doctors are giving us too much credit: 'here are the options' "–wanting patients to decide for themselves.

She and many other patients recognize that they are needy but still feel that providers could respond better. "The nurses are impatient with women in our situation," Sally added, "because we are high-maintenance, often calling a lot: 'This is happening. Is this normal?' Pregnant patients are neurotic: 'I have this. I feel a cramp here.' Nurses need some education about why it's important to be sensitive, and have understanding and compassion for women going through this. I've had to push back and ask to talk to the doctor. I know they're busy, but it's important to have someone who knows and cares about who you are."

Individual providers themselves can range widely in specific behaviors and aspects of communication that may seem small, but affect patients, such as remembering details of past interactions. "Most facilities just herd us through. At some facilities I felt like cattle," Wendy said. "I go in for my blood work and sonogram and leave. The doctor reviews it after I'm gone. And they call me in the afternoon. Until the transfer and retrieval, I had no doctor–patient contact." Wendy does "not like when a doctor sits down and looks at my chart, trying to figure out who I am. Read my chart *before* you come into the room! I have had doctors who are really good at this: they have hundreds of patients, and probably don't remember me, but they seem to. That's comforting.

"I called to tell them to say I had a miscarriage," Wendy continued, "and had no contact after that. They did not follow up with me. When I called about problems with the medications, they said, 'Oh that's normal, don't worry about it.' They didn't look into it. They just left me." Patients' criticisms of care may not always be wholly objective or accurate, but nonetheless reflect their strong feelings.

Dynamic Tensions in Communication

Dynamic strains and challenges—medical, physical, emotional, cognitive, logistical, and financial—can fuel patients' perceptions of lack of caring. Failed interventions can trigger mutual disappointments. Patients can feel angry and blame doctors, who can themselves feel frustrated, helpless, and/or guilty and thus withdraw or distance themselves. Doctors may have trouble discussing setbacks with patients, creating vicious cycles.

Some patients understand and accept these failures. "A great reward," Joe said, "is those patients who in the end say, 'We know you tried your best.'"

Yet when treatment fails, patients, who have usually paid large amounts of money, endured physical burdens, and had high hopes dashed, feel anger and stress. "Women feel like pariahs if the cycle hasn't worked, or if they have a miscarriage," Isabelle, the pregnant Connecticut office manager who underwent PGD, explained. "Nobody really talks to them. Nurses are beginning to do that more, but doctors should reach out to the patient. It goes a long way to hear from the doctor—that the doctor is very busy, but actually cares. These women feel like failures, and that the doctor isn't going to be interested in them because they failed, and haven't contributed positively to the doctor's success rate."

Physicians distance themselves from patients whose treatment does not work, worsening tensions. "The doctors may not want to reach out and then be blamed," Suzanne, the IVF clinic psychotherapist, said. "Doctors can feel guilty that they failed, too. After pregnancy losses, a lot of the doctors feel very sad and guilty, even if there was no negligence or malpractice. And we're a litigious society. A lot of patients can't accept that bad things just happen. They blame the doctors. Sometimes it *is* the doctor's fault, but not always." Worries about possible litigation can further impede communication. "Physicians may fear potential lawsuits," Suzanne added. "But if they show a humane approach, I think they're less likely to be sued."

As Karen, the physician who herself underwent fertility treatment, suggested, unless becoming patients themselves, many clinicians may dismiss complaints and not fully appreciate these difficulties.

Providers generally receive little, if any, training on how best to address complex and difficult emotions. Factors ranging from bodily pain to desires and shame can exacerbate these tensions. After all, patients, not providers, experience the physical burdens of treatments and may feel that clinicians ignore clinically important complaints. "The doctors are basically surgeons," Valerie, the psychotherapist and single-mother-by-choice, observed. "A lot don't have great bedside manner. In the sixth week of my pregnancy, I had a sonogram, and two hours later had bad cramps. I went back to the doctor, who made it seem like I was being histrionic. In fact, I ended up with an ovarian torsion and needed emergency surgery. I personally liked my doctor, but he handled that really poorly."

To patients, having a child provides vital and unique personal meaning and purpose, while to physicians, these treatments can become "routine." "There's a certain kind of day-to-day grind to it," Steve, the Virginia REI, confessed. "I can certainly get stuck in the tasks that need to get done: I have three embryo transfers that represent incredibly stressful days for those three couples. But it's 'yeah [nonchalantly], I have three embryo transfers.'"

Complex statistics can also be hard to grasp, convey, and apply for any one particular patient. "A lot of what I try to do is manage expectations," Jennifer, the Northwest medical center doctor, admitted. "I just try to give patients realistic expectations, so they're not disappointed. A lot of providers use adjectives: high, low, moderate. . . . Some IVF providers say: 'This procedure is very likely to be successful'—in their own universe, compared to

47-year-old women who have had ten miscarriages. But that may translate into only a 25% rate per cycle. It is likely to be successful for a woman with no infertility issues." Providers struggle with how best to communicate statistics, given such varying adjectives and numbers.

"I gave a reality check," Jennifer elaborated, "to a 43-year-old woman who hasn't yet found Mr. Right and was trying to decide whether to use a sperm donor: 'You're 43. Even though you've never before tried to get pregnant, these are your odds: just being scientific about this; with one cycle of IVF and PGD, without any reproductive or medical issues, 25% is only a success rate. If you're 43, those odds go down by an order of magnitude.'" As in other areas of medicine, providers face tensions between being either too definite or too vague, too optimistic or too grim. Patients seek hope and may see sobering odds as overly harsh, even if these are realistic.

Generally, patients seek not merely statistics but ways of interpreting and making sense of these numbers and can struggle to apply to themselves the numerical averages physicians offer. "Doctors have to use statistics, but sometimes I just want *an honest opinion*," Roxanne explained. "I finally had to say, 'What would *you* do?'" Patients may commonly see statistics alone as not straightforward, wanting instead clear answers on how to proceed.

Profound uncertainties in treatment outcomes can be tough for providers to convey and for patients to grasp. Doctors' predictions may seem too definitive and later end up wrong. "My doctor said, 'This is an art and a science. It's not just one or the other,'" Sally, the Maryland website manager and cystic fibrosis carrier, reported. "It's a little bit of hocus-pocus: he wanted to trigger the cycle that led to me being pregnant with my daughter. I pushed back: 'I think I should go another day. Maybe the eggs [aren't] mature enough.'" The doctor said ok. "I said, '*Should* I wait?' He said, 'If you don't go another day, and it doesn't work, you're going to think that's why.'" Doctors thus strive to balance optimism and realism, likelihoods of failure or success.

"They told my husband to redo the sperm test, but he hates those tests," Sally explained. "So, I said, 'Just wait to the last minute.' The next month, we got pregnant on our own!"

Given competing pressures, physicians may have insufficient time, diminishing communication. "It would help if doctors gave a little bit more time to their patients, rather than just coming in, doing a sonogram, and going out," Suzanne said, "But that's hard, because doctors have to see a lot of patients. But these patients need some emotional understanding." Providers may try to convey treatment limitations, while patients have trouble acknowledging and processing the information. "We hear part of it," Yvonne said, "but we're so numb."

As Amanda suggested regarding on-call schedules, clinics' structural and logistical characteristics can affect these patient experiences. Clinics vary widely in size, personalities, and other characteristics. "Multi-doctor practices are not *there* for the patient," Wendy, the Irish Catholic secretary now on her fifth infertility clinic, felt. "Too many hands are in the pot. Too many people saying, 'I see it *this* way.' With one-doctor offices, the doctor, even though you don't get to see him all the time, makes himself available to you, and is more there for the patient: 'It's breaking my heart to see what you've gone through, and what you've spent, when you should have just come to me from the beginning.'" Nonetheless, the doctor at this last clinic, though expressing empathy, may also be somewhat biased in suggesting that his treatment would have fared better than her prior physicians'.

Smaller practices may seem more personable and caring than larger, multi-doctor ones. Ginger, the fertility service counselor who conceived one child through ART and adopted her second, works "in a very small clinic: we give people individualized care. I've had patients who were initially in a larger clinic, and felt like a number."

In the end, patients are often left with hard trade-offs and conflicting impressions, information, and approaches. Especially within larger clinics, patients' experiences can also be very mixed, both good and bad. Many patients stick with their initial physician, even after obtaining second opinions, but may remain uncertain. "My first doctor was very very smart," Nancy, the IVF clinic nurse who herself underwent IVF treatment, said. "My insurance used her, and she works across the street from my job. But her bedside manner was horrible. She knew what she was doing, but was very abrupt and made me cry. My husband didn't like her. The women I know who saw her wouldn't return to her. I totally loved the second doctor—though I had to drive much further, and it would have been much better if they gave me the injectable form of the medication, and [if] I produced six eggs not one."

Traveling Elsewhere for Treatment

"We're going to Mexico for treatment," Francine, who underwent four failed IVF cycles, said. "A reproductive immunologist found problems with my body, and recommended various treatments. One approach, lymphocyte immunization therapy, requires the trip. They take my husband's white blood cells and inject them under the skin of my forearm. It's supposed to help my leukocytes. I have a hard time explaining it, but my body is attacking the embryo. The treatment is not available in the US. I feel very hopeful, because this doctor has very high success rates. People [who] have had 10 or 11 miscarriages [are] able to carry a pregnancy to term."

As infertility treatment has become global, patients now routinely travel to and from other states and countries for care. Developing countries, with looser regulations, can proffer certain treatments, luring desperate patients, undeterred by potential dangers and lack of scientific evidence. Many doctors, Francine felt, "don't like us going to Mexico, but it's not *their* decision."

Travel to other cities or states can result partly from differences in physician expertise in particular procedures. "Nobody in our area deals with PGD directly," Cathy, from Alabama, said. "They sent us to doctors out of state."

Countries vary widely, too, in available interventions and costs. Nations' laws range in what specific procedures they prohibit or allow. Several Western European countries are more restrictive than the United States. "People go to the US, to India, and elsewhere for certain procedures," Valerie, a psychotherapist, said.

The US, generally, is an outlier. In various countries, certain medications may also be unavailable. "We have a lot more available," Sally said. "When I lived in France, I tried to get progesterone suppositories, and they didn't have them. A friend brought them to Ireland, and I met her there."

Many infertility patients thus travel to the US for procedures unavailable abroad, though perhaps less so as services expand elsewhere. "Fewer patients come here than 20 years ago when they couldn't get these services in their own country," Jill said. Yet, changing costs and foreign exchange rates also affect these trends. "When the economy downturned, many more international patients came here, because it was actually a bargain for them," Jennifer said.

High costs and limitations on experimental treatments also compel US patients to get care abroad. Foreign companies in fact target US patients in various ways. "I see flyers: use frozen eggs for egg donation in Spain," Calvin, the southern California REI, said. "It costs about the same as in the US, but you get a free vacation." Many procedures, in fact, cost much less in certain other countries.

Desperation propels Francine and others to travel elsewhere for treatments that are experimental or unavailable in their own country, yet such foreign interventions may entail lower quality and more risks. Jill described a patient who "went to Mexico for her tubal reversal. They flew to Texas where somebody met them and brought them to a place in Mexico. For $2,000 she had her tubal reversal—not so well, as it turns out. She got pregnant anyway. That happens here, too. There's a big difference in fees. Here, it would cost $16,000–30,000."

Additional problems arise, depending on the particular procedure, particularly in using foreign egg or sperm donors. Offspring may never be able to learn about their genetic origins. Using donors from abroad concerns Brenda, "since their screening standards are different, and not always as stringent as ours. I don't like obfuscating somebody's genetic origins—which happens when you go abroad for donors."

Many patients thus consider treatment in the developing world, to save money, but ultimately fear and reject such options. "You can go to India and get it done for a third of the cost, but that doesn't sound safe," said John, who undid his vasectomy and has a low sperm count. "Travel there would be expensive. I think about some guy sitting in a back room in a little hut somewhere with a dirty knife. You're in *their* country under *their* rules and regulations, with no protection or safety that we have here."

In sum, patients struggle with how to pick doctors—how to gauge and weigh communication against technical skills, success rates, and financial resources and how to respond when clinicians seem to fall short. Patients frequently struggle with whether to shop around and, if so, how much and how to decide. They often expect success and want definitive answers, technical skill, and patient-centered care but face difficulties assessing and weighing these qualities against one another. Disappointed, many patients consult more than one physician[8] but frequently still grapple with how to balance interpersonal comfort against technical skill,[9] finances, and hope.

In other areas of medicine, about half of all patients "doctor-shop"[10,11]; and poor doctor–patient relationships affect patient satisfaction and decisions.[12,13] But with infertility, these issues take particular forms since patients accept only one outcome, a "take-home baby," and view partial success—a pregnancy that culminates in miscarriage—as failure. Unfortunately, most IVF cycles fail. Many women pursue childbirth "at all costs." Especially if they delayed childbirth for a career, women now often "race against the biological clock" but have lower success and consequently struggle to be proactive, readily switching providers. Since ARTs are relatively new and evolving, providers differ in recommendations, fueled by interclinic competition. Providers may criticize each other's approaches and are busy and not necessarily well trained to address patients' elevated stress and expectations.

Clinicians also face quandaries regarding how exactly to address the inherent uncertainties and emotions involved—how to frame and manage expectations, convey ambiguities, and prepare for and present bad news. Clinicians can easily appear either overly optimistic (to attract patients) or overly harsh (in giving "reality checks"). Yet both situations can generate tensions. Providers need to communicate the relatively low success rate of IVF, to avoid promulgating overly high expectations of a "take-home baby," along with the possible psychiatric and medical side effects of fertility medications and the emotional difficulties of miscarriage. But patients may not want to hear and have difficulty accepting this information.

Patients' complaints of lack of compassion may not all be wholly accurate but reflect their feelings and are thus critical to note. Whether and how much these physicians are remiss is unclear. But patients' perceptions of deficiency can impede whether and how they pursue treatment.

These processes are dynamic. Physicians may not communicate well concerning would-be parents' emotional difficulties and expectations, and patients may then hold unrealistic expectations and switch doctors, losing valuable time and resources. Providers can thus often help patients more.

Yet while patients are assessing doctors, doctors are also evaluating these prospective parents.

12 | "Will They Be Good Enough Parents?"
CHOOSING PATIENTS

"Who is good enough to be a parent?" Steve, a Virginia infertility doctor, asked. "I find that question difficult. I find very odd people whom I don't relate to. Or I am concerned about whether they're going to be good parents. If they didn't have infertility, they'd *already* be parents—I wouldn't get to judge them. But they have infertility, and want me to help them. So, what am I going to do? One woman had an elective abortion, and six months later, 'wanted to be pregnant immediately' with the same partner. Basically they just weren't getting along six months ago. I'd be bringing a child into a world—a relationship—that I already know is not the most stable."

Just as I had wondered how well my friend Abby, who asked me to donate sperm, would raise our child, providers grapple with patients' parenting abilities in general and in specific, unusual combinations of prospective parents. Many adults make poor parents. Consequently, Steve and other providers wrestle with how much responsibility they have, in using these technologies, to uphold the rights and well-being of the future child, whether to treat all such patients and, if not, whom to reject and how to decide—what to do, for instance, if the future parents' relationship seems unstable.

More and more adults, including novel combinations of prospective parents, such as related kin, are pursuing these technologies. Physicians see not only "traditional" arrangements of married heterosexual couples but "nontraditional" families, including gay and lesbian parents, single-parents-by-choice[1,2] and more "unconventional" types of reproductive arrangements, such as gamete donations between sisters and brothers as well as mothers and daughters serving as surrogates for each other.[3] As these technologies expand but remain costly, such unconventional combinations will no doubt increase further. Clinicians generally want to assist patients but sometimes wonder about the best interests of the future child.

Questions About Parenting Abilities

Eventual child-raising abilities can be tricky to predict. Individuals with disabilities, for instance, may face certain challenges in parenting but clearly have rights to reproduce. Nonetheless, in the past, these rights have unfortunately been disputed. Most egregiously, the US government supported efforts to sterilize mentally ill individuals. In 1927, the Supreme Court decided, in *Buck v. Bell*,[4] to forbid adults with disabilities from reproducing.[5] In that ruling, now recognized as seriously misguided, Chief Justice Oliver Wendell Holmes wrote, "Three generations of imbeciles are enough."[6]

Today, however, adults with disabilities can freely have children. Clinicians thus need to avoid unfairly discriminating against potential parents, based simply on the existence of medical or psychiatric conditions. But whether doctors should draw any boundaries in individual cases and, if so, where, is unclear.

The American Society for Reproductive Medicine (ASRM) has stated that clinics "may withhold services from prospective patients on the basis of well-substantiated judgments that those patients will be unable to provide minimally adequate or safe care for offspring."[7] Clinics "should develop written policies and procedures for making determinations to withhold services on the basis of concerns about the child-rearing capacities of prospective patients." But "persons with disabilities should not be denied fertility services solely on the basis of disability."[8]

Still, it remains murky what, practically, constitutes such "well-substantiated judgments," how such judgments should be made (based on what criteria), with what frequency doctors withhold treatment on these grounds, and whether they have written policies and procedures.

The ASRM states that "marital instability" constitutes criteria for rejecting egg, sperm, or embryo donors.[9] But mental health providers (MHPs) who work with in vitro fertilization (IVF) clinics then face confusion about whether their role is to educate or evaluate patients as gatekeepers.[10] The ASRM guidelines are aspirational, open to interpretation, and unenforced, creating a gap between guidelines and practice. MHP roles and expectations may vary across clinics, and how guidelines are implemented is frequently unclear.[11]

How do clinicians view these issues and proceed? One of the only studies on this topic found that Australian and New Zealand MHPs were concerned about the future child and sought to follow laws and guidelines but struggled, "adamant that their role in counseling was one of support rather than screening and gatekeeping."[12] They "emphatically denied a role in screening patients for parenting fitness," but "they were indeed" doing so.[13]

Most clinics lack a formal policy for screening patients for potential parenting ability but feel they have a right and responsibility to do so and refuse about 4% of patients each year.[14] The majority of doctors doctors would refuse HIV-infected women and single men and women, and around half would refuse a gay couple. Less than 3% of clinics have treated HIV-infected patients, partly due to cost and fears of cross-contamination—though such infection has not occurred.[15,16] Attitudes may be shifting somewhat but not entirely.[17] In 2008, the California Supreme Court ruled, in *Benitez v. North Coast*, that providers could not, based on their religious beliefs, refuse to treat an unmarried lesbian patient.[18] Yet questions persist about other groups, and states vary widely. Among obstetricians and gynecologists (OB/GYNs), 14.2% would discourage and 9.6% would refuse to treat unmarried patients, and 16.5% would discourage and 11.0% would refuse to treat single patients.[19]

In seeking to determine child custody, courts regularly ask psychiatrists to assess parenting abilities. Yet forecasting future parenting ability, including both abuse and minimally competent parenting, is fraught.[20,21,22] Certain past behaviors, especially criminality, violence, and prior history of child abuse, correlate with future parental maltreatment (as opposed to minimal parenting competence) but not enough to predict whether a specific individual will mistreat a child.[23] Psychiatric diagnoses may play roles, due to the impact on future parenting, not to the mere presence of a diagnosis per se.[24] Chronic unemployment and social isolation can also impair childrearing. Yet serious maltreatment itself is relatively rare[25] and probably even scarcer among patients consulting IVF providers. Standards are also lacking for minimal parenting competence, as opposed to more severe abuse.[26] Methods of measuring minimal parenting competence remain limited. Parents' self-reported abilities, though useful guides, can be biased. John Robertson, the legal scholar, suggested that even when a future child's social environment may be suboptimal, the parents' decision to have the child "does not ordinarily harm the child who has no other way to be born, and will usually fall within the procreative liberty of parents."[27] Carl Coleman, a law professor, argues that for patients with disabilities, other reproductive and parenting options, such as adoption, may be available[28] but that these disabilities would have to impede parenting abilities in "truly extraordinary" ways to justify a clinician withholding treatment since childlessness would pose "significant burden to the patients without any benefit" to the child. In custody disputes in divorces, the law tends to defer to MHPs' judgments of whether each parent is adequate.[29]

Yet questions remain about how infertility providers in fact view and make these decisions—whether, how, and when they confront these issues and what they decide to do.

Providers Wondering About Parenting Ability

"All sorts of people come in and want to have a baby," Helen, a psychotherapist, reported. "A lot of people are making babies that were not making babies in the past." Clinicians then struggle with whether to try to assess and predict patients' potential parenting and, if so, how and what to do if questions linger. Adults' rights to reproduce can collide with the eventual child's rights to be born into a stable and loving home. Providers face quandaries about their responsibilities to the potential parents versus to the future child, and often look to psychotherapists, who may also feel constrained.

Infertility clinics face a range of ethical challenges concerning parents' suitability. Clinicians generally want to assist patients but may doubt parents' abilities and have to weigh these qualms against both desires not to discriminate and the fact that fertile parents can reproduce without a clinician's approval.

Moreover, over time, social and professional attitudes and practices are loosening, creating challenges. "Guys in their 70s are now ready to become parents. We don't want to discriminate, so what do we do?" Helen asked. "There has to be a policy. It used to be somewhat common for doctors not to want to treat lesbians. Most [physicians] do these days—a lot of lesbians, single women, single men. Gay guys go to certain states." Helen sees, "a variety of people," about whom she wonders "what kind of parents they would make, and if they can ever obtain successful parenting skills." Yet balancing these conflicting rights and concerns can be hard.

Finances, too, can affect patients' abilities to care for themselves and their prospective child. Helen described a "kind-of homeless couple," who came for free IVF. The clinic helped the couple, "who were doing a really bad job of looking after themselves, create a baby. The clinic didn't have anything that would give them reason to turn someone away."

Why Providers Are Concerned

Several types of potential medical, psychiatric, and social uncertainties can worry clinicians. Patients may have serious medical problems that can hamper the pregnancy or the baby's health. Doctors must then decide how to proceed and face predicaments regarding, for instance, patients with mental health problems. Yet while some such cases are clear, others are gray—whether to prioritize the future child's well-being over the parent's autonomy. "When do you treat someone who has schizophrenia, in and out of mental hospitals, completely dependent on their parents?" Helen asked. "They want fertility treatment because they want to have a baby. It's very tricky."

"Rarely do doctors say 'No,' but it does happen," Peter, a West Coast reproductive endocrinology and infertility specialist (REI), explained. "Patients have significant medical problems . . . enough to interfere with their lives, and ability to hold on a job, and now they want to have a baby; but the pregnancy is going to interact with their chronic medical problem. High-blood pressure in women over 50 is an exclusion criteria for us. So, we will do an embryo transfer to a 50-year-old, but not if she has high blood pressure. We just say this person should have a gestational surrogate." Pregnancy, he feels, will worsen the patient's medical problem and "make the pregnancy more risky for them and for the baby. But they say, 'I don't care. My doctor thinks it will be OK for me to carry the pregnancy.' I say, 'But you're on disability. Every piece of evidence here says it's not OK.' I've turned those people away." Such patients may push back or consult other infertility clinics.

Clinicians also confront questions about patients' broader coping, intelligence, and functioning. "Without seeming unkind, some people might not be of entirely normal intelligence," Jill said. "But is it my place to decide that they shouldn't be a parent?"

Providers grapple here to define their roles. Ultimately, they tend to feel that making and acting on such determinations lies beyond the scope of their responsibilities. Jill has therefore treated patients "with significant mental illness."

Clinicians face quandaries, too, about potentially suboptimal but not severely harmful environments, such as having "neurotic" parents. Many providers feel uncomfortable with these situations, but none thought it was, in the end, their job to make these judgments definitively. "Some 'normal' couples cannot feed their child," Jill elaborated. "I'm not sure if being highstrung or neurotic is a diagnosis, but I'm sure that I wouldn't want to be their child. Yet *it's not my place* to decide. If you release a child into an unsafe home, that would be a separate issue. But that doesn't really come up."

Clinicians thus wrestle with what standards to use. Patients may be "good enough," yet not ideal. Still, even defining *that* threshold is hard. While "unsafe" might appear relatively clear, "normal" is highly contested, with blurry boundaries, covering a broad spectrum. Providers confront uncertainties and may help patients despite doubts. "I treat a couple who seemingly don't have two cents to rub together," Jill said. "But is it *my* place to decide that they shouldn't try and get pregnant?"

Social Factors

Relationship Instability

Suzanne sees couples who are "not in a very good place with each other. They don't have really good coping or communication skills with each other to deal with difficulties in life, or good ways to confront and resolve conflicts together. They don't seem to like each other very much right now. There's a lot of disdain."

Parenting skills range widely, from ideal to potentially harmful, but are inherently subjective and hard to predict. Couples together for several years may still seem to cope or communicate poorly. Clinicians thus wrestle with whether to be concerned about not only clearly abusive parenting but suboptimal parenting that may be less severe than abuse per se. OB/GYNs see unstable relationships as well but have less moral responsibility—ensuring the health of the already established pregnancy, not creating the pregnancy itself.

As Helen suggested earlier, regarding a "kind-of homeless couple," very limited finances can also worry clinicians, but how and to what degree to assess these financial, not only emotional, states are unclear. "Octodoc has gotten a lot of grief about how [his patient could] support eight children," Henry, the Midwest REI, said. "We don't want to be the ones to say, 'you have to be making X amount of dollars before we'll treat you.'"

Single-Parents-by-Choice

Until recently, many clinicians resisted treating single-parents-by-choice but are now generally shifting. Henry's office was "one of the long-time holdouts. Most providers would treat a single mom or a gay or lesbian couple. Before, we didn't treat single women and lesbian couples. But now we do." His clinic, however, still does not treat gay men.

Doctors may try to justify their reluctance, which can reflect fears or rationalizations. Henry perceived, "ethical issues, and problems with nursing. A lot of the nursing and ancillary staff are a little uncomfortable with it, at least at the outset. They feel we ought to be helping couples where there's a high likelihood that the child is going to be born into a supportive and loving family. So, for instance, if we had a couple where we knew one of the partners had molested children—which we've come across—we won't treat them. Or, if there's substance abuse." Yet the example he provides—of molesting children—would presumably not preclude the vast majority of gay couples, whom he also does not treat.

Over time, shifts in wider societal norms and logical reflection and re-appraisal have transformed many providers' attitudes and practices. Henry has colleagues who "said 'data suggest that children born into two parent families are more stable, better adjusted, etc., and so shouldn't that be our ideal?' But we've come to realize over time that being married at the time of IVF doesn't guarantee that that will continue in the future. So, we've lifted those restrictions."

Gay and Lesbian Parents

"Gay couples are so routine for us that we don't even pay attention to it," Marvin, in Massachusetts, said. But elsewhere, obstacles prevail. More gays and lesbians, singly or in couples, are also having children but nonetheless face challenges. Social attitudes have been changing, but difficulties persist. In large urban areas, gays and lesbians, if they can afford it, generally en-counter few problems obtaining assisted reproductive technology (ART). "Gay and lesbian couples don't face problems *here*, but I'm sure they *would* in Mississippi," Suzanne said.

Gay men in fact succeed in IVF more than heterosexuals, who seek treat-ment because of medical infertility problems. "We do ten gay couples a year," Steve explained. "They're the most successful cycles in our practice, because they've got a gestational carrier, an egg donor, and sperm that *work!*"

Given regional differences and attitudes, many gays and lesbians travel to clinics in relatively more tolerant states. Demand by gays and lesbians thus varies by geographic region. "Half the clientele of colleagues in California are gay male couples," Henry observed. "That would never happen *here*. There's not that large a population."

Still, Henry's practice is "establishing a surrogacy program. . . . We have talked to the ethics committee about what we will do and not do. I think we will be treating gay male couples in the future."

Over time, numerous providers have also shifted their own personal views. "If somebody had said to me before that I would be doing it, I'd say, No, I'd be terribly uncomfortable," Calvin confessed. "But they have put a lot more thought into having a family than young couples who just get pregnant and have kids because it's 'the thing to do.'"

Yet in society and among some REIs, prejudices persist. "Certain physicians have problems with same-sex parenting," Calvin continued. "I heard of a case in the Midwest of two doctors in practice together. One has no problem with inseminating a same-sex couple. But he happened to be out of town, and his partner refused to perform the procedure. The couple sued. There are still doctors who don't do it. But the vast majority is willing to."

Other doctors take a while to feel comfortable and remain uneasy. "I would now say we've gotten our head around it; I think we are OK with it," Peter said, though suggesting some lingering hesitation.

Additional legal challenges emerge since four parties may be participating. Gay male couples, for instance, must have an egg donor and/or surrogate. "It gets very complicated when multiple people are involved," Roger, a New York REI, elaborated. "We've done two gay men, a donor and a surrogate several times, but very subtle things turn out to be very big things. What's the kid's last name going to be? Are we going to use only one man's sperm to fertilize the eggs, or half and half? Are we going to transfer all of the embryos from one man's sperm? It's way more complicated than just saying 'do half and half.' First, they all have to *agree!* Lawyers are involved. Then, often because they are not married, *two* lawyers are involved, who of course never agree with each other. Then the surrogate says, 'What if we are going to transfer three embryos? Should it be two of one, and one of the other?' "

Moreover, despite rapidly evolving attitudes, gays and lesbians who use ARTs may still face stigma in raising their children. "The world is becoming more open to gays having kids," Ginger said. "But a few years ago, one partnered gay man with a child told me how sad he was that he had to raise his kid in a state where he and his partner couldn't get married! They always still have to explain themselves. Their kids have to come to terms with the fact that they are a different kind of family. I don't know what it would be like to be a gay couple in a red state."

Transgender Parents

Roger was "very torn" about a case: "Two girls were around 18-years-old. One was under psychiatric care, and transitioning from female to male. She came to me because she wanted to get her eggs before she became a male. They were going to fertilize those eggs with a donor sperm and transfer them, as embryos, to the other girl. So, there were several issues: should we do it? I was hung up on: technically, she's 18, and can do what she wants. If she just wanted her eggs removed, I would do it. But I'm now going to make them into embryos! She had a well-written letter from her psychiatrist who agreed, saying she's not a crackpot, and has clearly thought this out. Yet a part of me said: 18 is extraordinarily young. Then, part of me said: I'm being very paternalistic. They are clearly in love. The shrink said it's cool. *But* they're not married. I didn't want them to make any mistakes. I wanted them to be able to look back in five or ten years and feel comfortable with these decisions—because it's now no longer just about the two of them, but about a *third* person."

Despite enormous progress by gay men and lesbians, individuals who have transitioned from male to female or vice versa still face prejudice. Various combinations of transgendered men and women may require ARTs to have children. Before undergoing gender-altering surgery, they can store their gametes[30] but may not always do so or be allowed to. Several providers said that they and their colleagues felt awkward treating these patients.

Many providers have not worked closely with transgendered patients and initially hesitate. Complex challenges may be involved, including obtaining valid consent and weighing patient autonomy against provider paternalism.

Unsure what to do, Roger consulted his hospital's ethics committee. "It raised a lot of eyebrows. A sanctimonious philosophy professor said, 'You are out of your mind! Who's the quack psychiatrist who said this was normal?' Another member said the exact opposite: 'Women have the right to do anything!' So, at the end of the day, I didn't get an answer." Several issues— lesbianism, transgenderism, and youth—split the ethics committee. Roger could remove and freeze her eggs, but the broader context troubled him.

Evolving sexual mores can confuse providers, who wrestle with their attitudes and society's more broadly and may feel conflicted about how paternalistic to be. "My own paternalism bothered me," Roger continued. He thought, "Listen to yourself: You're basically saying you don't feel comfortable with it, so she can't do it. That's obnoxious. Then I said, 'No, I'm not supposed to do things just because I *can*. I'm not an auto mechanic. This has serious consequences, for which I'm going to have to answer—if not to other people, at least in my own heart. Maybe I'm not strong enough. I should get other opinions.'"

He spoke to clergy, who were helpful, which surprised him. "The stereotypes about what church people are going to say are largely wrong. The Jesuits, for example, were extraordinarily insightful. They didn't have an answer, of course. But they looked at her overall health: 'You are saying that her mental health is somehow less important than her physical health. If a nice, cute couple—an 18-year-old boy and girl—came to you, and wanted to have a child, and nothing else was wrong with them, you wouldn't have any qualms. So, you are clearly making a value judgment here. You have to decide if you feel comfortable with that.' I thought, *Oh my God!*"

Roger has also grappled with the patient's age: "18-year-olds can be very impulsive, but these were extraordinarily sophisticated 18-year-olds. How many 18-year-olds would show up at a clinic like mine? They are awfully young, but not idiots. They have clearly done their homework. They know a lot more than some of the 40-year-olds who come into my clinic who don't even know when they can get pregnant. So, I was torn. We ended up doing it. But I felt better knowing that I wasn't the only one conflicted."

HIV Infection

Doctors have struggled, too, with whether to treat HIV-infected patients. Thomas's East Coast academic hospital has discussed HIV "half a dozen times over ten years. Our policy evolved, vetted many times over many years. In the mid- to late 80s, we said we won't treat HIV-positive patients. We've turned 180 degrees, and feel it's unethical *not* to do it."

HIV-infected men raise dilemmas but may do so less now, with the development of improved sperm-washing procedures. "Initially, HIV-positive men posed ethical issues," Joe recalled. "But with more effective therapy, and better isolation and washing of sperm, that was less of an issue." The advent of successful treatment reduces the potential risks. Still, doctors may question the infected parent's future health and the effectiveness of sperm washing. Demand for and willingness to perform this procedure also differ by region. "One big city center does a lot of HIV discordant couples," Henry said. "There's not a big market for that *here*."

Other Unconventional Combinations of Parents

"Unusual, nonconforming social relationships are one of the most challenging issues," Marvin said. Physicians see a wide range of unconventional combinations of possible parents. For many women, sister-to-sister egg donation, for instance, can provide an important option.[31] An infertile female patient using a sister, rather than an anonymous third party, as an egg donor or gestational surrogate can still have a genetic connection with the child and avoid paying high costs to strangers to fill these roles. But these arrangements can strain families.[32] Other familial arrangements can raise additional problems related to real or perceived biological consanguinity or confused roles for eventual offspring or others.[33,34]

Clinics vary in which such unusual combinations, if any, they treat. In 1998, 90% of US ART clinics said they would accept eggs from family members and 80% from friends, and 60% would use family members' sperm.[35] Clinic directors were generally more restrictive than their organizations' policies— 14.8% of directors would personally restrict use of family members' eggs, despite their clinics' official policies otherwise,[36] possibly due to fear of litigation.[37] Among these directors, 67% would allow a male patient's brother without children to donate sperm, but only 29% would allow a male patient's father to donate sperm (since the future child's father would then also be his or her half-sibling, causing potential cross-generational confusion) and 18% would allow a female patient's mother to donate eggs.[38] Seventeen percent of US OB/GYNs would discourage single parents and 14% would dissuade unmarried or lesbian patients. Male OB/GYNs were twice as likely to discourage these groups, and religious OB/GYNs were three times more likely.[39]

The ASRM has stated that, "The use of intrafamilial gamete donors and surrogates is generally ethically acceptable when all participants are fully informed and counseled,"[40] but the organization recommends prohibition of two related individuals each providing gametes. A brother donating sperm to fertilize his sister's egg, for instance, would be incestuous.[41,42] Moreover, "Child-to-parent arrangements are generally unacceptable, and parent-to-child arrangements are acceptable in limited situations."[43] The European Society of Human Reproduction and Embryology has published similar guidelines.[44] Other arrangements may appear to outsiders to be incestuous when that is not in fact the case—such as a mother carrying a fetus for her daughter or a brother donating sperm to his sister who is, along with her husband, infertile (which would not be incest since the embryo would not be formed from these related individuals).

With intrafamilial donations and surrogacy, some critics have argued that truly voluntary decisions are impossible, due to emotional and financial pressures in families: close kin may feel coerced or unduly influenced to agree.[45] Moreover, once the child is born, familial roles and relationships can become complicated, especially if the child encounters medical, psychological, or developmental problems, which parents may then blame on these arrangements.[46] Dilemmas also emerge regarding what to tell the offspring about these relationships. Hence, the ASRM emphasizes the need to avoid coercion or undue influence within families and to proceed cautiously, especially with relationships that may make others wonder about possible incest.[47]

But how do clinicians in fact view and make these decisions?

Providers turn out to face a range of challenges and dilemmas concerning both the content and the process of deciding about such unconventional combinations and differ in what they do and how.

Types of Unconventional Combinations

Clinicians grapple with not only intrafamilial donations but other complex relationships that may be interfamilial and intergenerational. Marvin described, for instance, "a couple that is not divorced, but he's living and sexually intimate with another woman, not his wife. He wants to have a kid with this girlfriend. His wife, however, will carry the baby. His current girlfriend couldn't or didn't want to carry the baby, because she's a model. The ex-wife will do it because he's paying her a ton of money. You couldn't make that up!" Clinicians thus encounter arrangements they never anticipated, reflecting complex *social* as well as biological connections, with murky roles and responsibilities. Unusual social arrangements are now logistically possible but can entail psychological and ethical uncertainties. Requests for these arrangements can range from mere possibilities to firm priorities.

In these cases, REIs usually ask for careful legal contracts. Marvin and his colleagues require such patients "to see lawyers, and go to counseling, and we require documentation." Yet these decisions can remain tough.

Such arrangements can also shift over time. A patient might begin with a relatively conventional arrangement that then becomes unconventional. Spouses may become uncertain whether to stay together or alter their reproductive plans. Marvin described women who, "come in, and one cycle they're sexually intimate with somebody, and the next cycle, they're not sure they still want to be a parent with that individual. One month, they do, and the next month, they'll want the person just as the sperm donor."

Unusual intrafamilial combinations create problems regarding not only creating but raising the eventual child. "A 65-year-old professor who is retiring wants us to help him pick an egg donor, and is going to use a surrogate, but who's going to raise this baby?" Diane related. "He's got it all planned out: he's going to retire to where his sister lives, and *she's* going to raise it for him. She is 63, and knows nothing about this. We say: 'No.'"

Informed Consent Challenges

In these cases, informed consent can be knotty—as in other medical situations, such as organ transplants, that involve multiple individuals. Family members may agree, due not to explicit coercion but to complex long-standing psychological relationships, perceived familial obligations, or guilt. Intergenerational donation can raise particular questions. "A mother gets remarried, wants another baby, and wants her daughter to be a donor," Helen reported. *"The daughter says 'yes' because she feels she can't say 'no' to mom!* Doctors shouldn't do 'upward' generational donation. It's an informed consent and relationship issue. Plenty of mothers will do things for their daughters, but that's a more natural, motherly act. Japanese grandmothers carry babies for their daughters since that country does not approve of egg donation except in those kinds of situations." These practices and norms can vary significantly across cultures, related to varying laws and moral perspectives.

Still, the voluntariness and autonomy of the individuals involved can be hard to gauge completely. Gamete providers may feel torn but willing to proceed. "A sister is going to donate for her sister. You meet the donor sister, and she tells you that she doesn't want to do it, but just can't say no to her sister," Helen continued. "Or she *tells* you she wants to do it, but she is *acting* like she doesn't." Donors may say yes but seem reluctant or unsure. When one of the parties may have reservations, clinicians should assess motives to ensure that participation is wholly voluntary.

In other situations, family members may all fully agree to a plan, yet providers may refuse. Informed consent can be necessary but not sufficient since providers may be concerned about the future child's best interests. Peter "turned down" a couple because "the husband was 75 and the wife was 60, and they wanted to use a gestational surrogate and select for a male because he did not have a male offspring. He said, 'My daughter will take care of the baby if something should happen to us.' I said, 'This is not *if*, but *when*.' He was a wonderful guy. At the age of 75, some neuron fired in his head and said, 'You don't have a male offspring.' So, he wanted to get a donor egg and a gestational surrogate so he could sire this male child. His daughter agreed that she would take care of the baby. So, in effect I'd be producing a child for the 40-year-old daughter that's going to be sired by his sperm. It was too odd for me, so we backed out." Family members may all consent to an arrangement that nonetheless poses ethical problems. Moreover, providers may refuse to treat patients due to concerns about not only conceiving but *caring for* the child. Clinicians may have difficulty, however, articulating explicit ethical principles and simply feel patients are "too odd." Many providers thus draw on implicit intuitive moral thresholds but have difficulty defining or articulating these a priori and recognize these limits only when hitting them.

Initial moral intuitions and the "yuck" factor may prevail. "That was a tough case," Peter said about the 75-year-old man. "Because if everybody really agreed to it, it was not endangering anyone's life. This baby would have an unusual start, but presumably have a stable home and everything else would be OK." But, in the end, Peter refused because the request still didn't feel instinctively "right": "It was a non-verbal, personal kind of decision. I said, 'No. It's our policy not to work with old couples,' but that was a cop-out—although ultimately that was the problem. Producing babies for a couple that was *that* old, with no realistic expectation that either one of them would be around, was not right. We discussed it, and all decided we were uncomfortable."

Personal "comfort," rather than explicit analyses per se, may thus sway clinicians. Doctors may refuse to offer treatment because of an initial non-verbal feeling and then seek reasons that may not precisely fit the case but seem legitimate excuses. Providers may also consult colleagues.

Physicians could simply follow the patient's wishes, but the proposed arrangement could be difficult for the offspring, blurring familial roles in ways too far outside perceived norms. Peter felt it could still disturb the child, "who would be raised by the grandparents for as long as they could do it. Then the child would go back to the de facto mother—the daughter. It didn't make any sense to me," Peter said, drawing on his perceptions of social mores. "I thought it stepped outside the norm."

Cases may also feel unacceptable due not to one aspect but to several together—parental age, sex selection, and children being raised, at least partly, by other family members.

How to Decide About Parenting

Doctors must determine not only *whether* to offer treatment in these situations but *who* should decide and *how*. Physicians' personal, professional, and political characteristics; institutional structures; and financial motives can all shape whether and how much they support such requests. Clinicians may initially oppose an arrangement because of intuitive moral revulsion but then later carefully and logically reconsider, explicitly articulating and weighing competing ethical principles and coming later. "In some cases, the 'yuck factor' will get into play before they think about it," Diane said. "So, they just say, 'No.'" She described a man who wanted to donate his sperm to his twin sister and her husband, who needed both an egg and a sperm donor. This arrangement may appear incestuous but, after close analysis, is not. "The patient is in her 40s and single, and needs donor eggs to conceive. Can she use her twin brother as a sperm donor? Everybody initially went, 'Ewwww.' Then we went to the ethics committee and hashed it out. We all said, 'Well, why not?' It was her gene pool. It wasn't incest.' She was using donor eggs, and her brother and his wife were very accommodating. She gave birth to a baby, fathered by her brother. It was lovely."

Many clinicians try to address these issues logically and systematically, using clear, rigorous, evaluative steps, but ultimately still rely partly on instinct and "the smell test." "First, we ask if this is medically reasonable,"

Roger said. "Then, is it in the best interest of each patient? Thirdly, is it in the best interest of the child? Fourthly, does it pass a *smell test* in our profession? Fifthly, how does it relate to practice and ethical guidelines? Sixth, does it make sense for us as a practice to do this, or should we send them somewhere else? Then, what kinds of medical, psychological, financial, and legal management are we going to need? Any ethics problems? So, we go through all those aspects, and try to determine what to do. Eight out of ten times, we tell people, 'If you do these things, then we'll be happy to provide service. But if you don't want to do these things, we can't.'"

Yet, physicians' own financial motives may also play roles. "In some cases, it's all about the money," Diane said. "They'll do pretty much anything as long as somebody pays for it. Those are the practices that make the big, newsworthy mistakes." Yet clinics range considerably. "Another practice does all kinds of alternative family building pathways," Steve said, "but *we're* a pretty buttoned-up practice."

Providers' Strategies

Clinicians respond to these quandaries in several ways, looking for countervailing considerations, providing mental health input, discussing concerns with patients, or hoping certain patients drop out. Yet these strategies can all trigger complex challenges. Troubled by particular requests, providers may seek offsetting justifications, such as amounts of social support. Jill, for instance, "had a patient who was blind and single. Should she get care? Is she able to take care of a child? Does she have adequate support? She actually did. Ultimately, she dropped out, and met some guy and is very happy."

Concerned about single-parents-by-choice, doctors may simply require that such patients bring in a designated "co-parent." "I made up my own rule," Peter explained. "The patient needs to come in with a co-parent, and demonstrate that she has support systems—to bring another human being who will say, 'Yes, I will help her take care of the child.' Couples come in, and I don't know if they are going to be together next month; but at least the husband came in at the time of the treatment and indicated his support of the pregnancy—at least at the moment of conception. *That's as much reassurance as I have with heterosexual or homosexual couples.* So why not ask someone who is going to be a single parent to come in with a person who is willing to be their support system?"

All couples are potentially unstable. About 55% of married heterosexual couples eventually get divorced.[48] Many providers thus seek to ensure that single patients have strong support, even if not a "significant other."

Based on "gut feelings" rather than objective facts, providers may feel uneasy but want to avoid unfair discrimination. Consequently, clinicians may decline, giving patients reasons that may be relevant but not reflect the providers' main concern. Regarding the "kind-of homeless couple" mentioned earlier, Helen added, "Most infertility clinics see people with physical disabilities. But if there's a problem, and you feel the person is not going to be able to care for themselves, let alone a child, most clinics probably don't come right out and say it, but use more *avoidant* tactics: 'We don't think you're ready right now. You need to go get some therapy. Come back when you've resolved whatever it is.' They just go to another clinic, because they can. *Somebody* will treat them."

Given the difficulty of resolving these conflicts between moral responsibility and discomfort as gatekeepers, providers are often glad when worrisome patients fail to follow up. "I had a schizophrenic woman who was fine," Jill explained. "I could talk to her, but had concerns. Ultimately, I gave her a slip to get blood work done, and she didn't do it. So, the population often *self-selects*. The treatment is demanding. . . . I'm not disappointed that patients self-select. That makes it a little easier." When such patients end treatment, providers avoid having to make tough decisions and possibly oppose these patients' autonomy.

Mental Health Providers' Roles

Steve usually asks patients such as the woman who underwent an abortion six months before and now wanted a child with the same man, "to take a step back. I'm pretty blunt. . . . I ask them to see our therapist." Physicians, not only MHPs, may be concerned about and try to address these issues in some way.

MHPs may have treated a patient over time or be affiliated with a clinic. "If we have evidence of active substance use, we ask them to get a statement from a counselor, psychologist, or psychiatrist that they're free of substance abuse," Henry said. "We will do testing as well. If we obtain a positive result, we'll cancel the cycle." "A couple of times a year, a psych session leads us to not going forward," Diane explained. Yet clinics vary in how and to what extent they utilize MHPs. *Who* decides about psychological factors may affect *what* is decided. "It depends on the clinic," Diane continued, "and who's running it and what their personal views are."

MHPs not only evaluate patients but at times provide psychotherapeutic input. These professionals can help patients become "good enough" parents, aiding them, for instance, in acquiring additional social support. Many psychologists in private practice, unaffiliated with ART clinics, see women struggle with these issues even before visiting an ART clinic. "I see people who don't really seem to have their life in order," Suzanne said. "I worry about how they're going to become parents. I see a lot of single women considering parenthood. Most have their life in order. But I worry about a single woman being able to bond and attach with a baby if she has no support system or adequate financial support to take time off, or has never in her life had a significant relationship. So, I talk to them about the importance of having a support system: 'I don't think you can do it alone.' I try not to be judgmental about it—'Why are you in this situation by yourself?'—but rather, 'There are organizations for single-mothers-by-choice, ways of connecting with a community, so you have some support.'"

Providers wrestle with unpredictable future parenting abilities and relationship stability, larger ethical questions about providers' roles and responsibilities to the potential parents versus the unborn child, and troubling ambiguities. "How good are we at predicting people's behavior?" Brenda wondered. "If I look at somebody and think 'Wow, I don't know what kind of parent you're going to be,' am I even going to be right? I struggle with that. One woman had twins at 62 with her husband through a gestational carrier. Is that right? Not right? Should that be *my* decision? Those questions keep me up at night." Providers are left facing a series of troubling conundrums, especially whether making these decisions is indeed within the scope of their responsibilities.

In helping to create a new human being—a daunting feat—many clinicians feel heightened responsibilities and tensions regarding their duties to the future children. "Personally, the most difficult ethical issue is 'What is my role? *Am I a gatekeeper?*'" Brenda stated. "Just because somebody has fertility problems does not mean they have to answer to a different standard than somebody who does not. Is that fair?" Providers struggle with what criteria to use in deciding.

To make these determinations, physicians commonly look to psychotherapists, who often, however, also see these decisions as beyond their purview. Screening potential parents seeking to adopt children poses similar but also different issues. "Professionals have a different role in adopting," Helen explained, "where it's very clear about the child's best interests, and the fact that we're going to evaluate the parents. The evaluations are actually not a big deal, but everybody thinks they are. But *we've got kids there*, and want to watch out for them. A lot of people think that the infertility world should watch out more for the kids—*not* the adults' needs to have a child."

With adoption, the child already exists, while with infertility, a child remains only a possibility. Yet that distinction may not wholly justify infertility providers paying less attention to parenting abilities. The fact that clinicians are required in creating a human being, and consequently are complicit, arguably elevates their moral duties.

Psychotherapists often feel that serving as the judge simply lies beyond their proper role. "Many mental health providers do not feel their job is to be the gatekeeper," Helen added. "They are procreative libertarians: 'I don't know how good a mom she's going to be, but in the end, that's not my responsibility!'"

Nonetheless, many MHPs can offer therapeutic insights and/or intervene, enhancing mental health treatment and support. They often work to improve patients' parenting abilities but do not want to be the final arbiters. Suzanne, for instance, can "discuss deeper attachment issues, show my concerns, suggest that therapy or outside resources may be useful. I will discuss learning how to resolve conflict with each other as they move forward in their life in general. But I'm not in the position to tell people whether they should have children or not. I am not a gatekeeper. If somebody has characterological problems, they're probably still going to be able to go forth and have a baby on their own. They could do it if they didn't need any intervention." Clinicians feel uncomfortable approving or blocking treatment and that their role is limited since fertile individuals who will be suboptimal parents can have had children on their own. Yet, here, providers are actively involved in creating the child.

MHPs' roles also vary, depending on whether they work within a clinic or independently—are "embedded in a clinic or just come in to consult," Helen said. Within clinics, physicians pay MHPs, who may be more motivated to support treatment, though creating tensions. "It's tricky," Helen added, "because staff want things to work out, and get very swept up in the process."

How to Decide About Parents

Infertility providers vary in whether they use formal or informal decision-making processes and/or consult others. Many clinicians rely on their "gut feelings," rather than empirical evidence (which is generally lacking or ambiguous) or explicit formal ethical analysis. Yet, discussion of the ethical issues in formal external ethics committees or informal consultations with staff can help.

Especially when affiliated with hospitals and medical centers, providers may develop formal policies or mechanisms. "Instead of doing it on an individual basis, we try to make a decision about the whole group of patients," Henry said. "These are difficult areas. For a long time, we didn't treat single women. Then, after reviewing the data and meeting with the ethics committee, we decided it's no longer an issue." Yet clear policies, though offering benefits, can be difficult to establish. "The hardest issue is establishing the policy," Henry explained. "Once it's set, we don't feel it's our place to decide on an individual basis." But given countless unanticipated clinical scenarios, developing a "one size fits all" policy is hard.

Given these ambiguities, providers and regional cultural attitudes can shift over time. Most physicians rarely, if ever, decline to treat patients because of parenting concerns but vary, related to personal and social views. In certain offices and areas of the country, single-mothers-by-choice still face resistance. Though professional guidelines "don't support turning somebody away based on their marital status or sexual orientation," Brenda pointed out, "parts of the country and individual physicians still have their individual beliefs."

Though acceptance is gradually increasing nationwide, single-mothers- and single-fathers-by-choice and gay and lesbian couples can face much more resistance outside large cities. "In a major metropolitan area," Valerie, the psychotherapist and single-mother-by-choice, said, "nobody looks twice if you want to have a baby on your own."

Providers' attitudes toward single mothers are also evolving. "Years ago, I thought it was crazy," Calvin said. But when he sits "across the table from them," he feels "much differently" than "in the abstract."

Ultimately, providers appear rarely to refuse potential parents. "Maybe twice I've recommended that this is really not a good idea," Helen said. Egg donors are generally placing themselves at medical risk in return for compensation, with no direct medical benefit for themselves, and may not follow through with the extensive required medications and procedures. Providers may also feel more responsibility for infertile patients paying for treatment for a medically related problem than for healthy women selling their eggs, essentially as a paid job.

In short, unconventional combinations of potential parents request these technologies in ways that fall across a wide spectrum from acceptable to problematic, from siblings and other family members to friends. These arrangements differ, based on patients' respective roles (from donating gametes to carrying the fetus and/or raising the child), relationships (from biologically related to unrelated family members to friends), and perceived norms. Providers face questions about blurry and confusing roles, future children's welfare, and competing medical, moral, personal, scientific, and social considerations.

Providers do not always know in advance how to evaluate or weigh these tensions and may shift from initial "gut reactions" to more careful contemplation of ethical principles. By definition, the nontraditional and unfamiliar combinations that these relatively new technologies make possible are not the "norm" and can thus make clinicians morally uneasy. Yet certain phenomena, though not the norm, may still be ethical. In the past, many providers, due to their gut feelings, felt justified in declining gay, lesbian, transgender, or single patients but later altered their views. Mores and ethos (what is generally *done*) can differ from morals and ethics (what one *should do*). A clinician may at first feel that such an arrangement does not "smell right" and only subsequently, if ever, more carefully and explicitly analyze the ethical principles entailed.

Faced with uncertainty, individuals in general often make fast, instinctive, and emotional decisions, rather than slow, logical, and conscious ones.[49] But some providers may simply follow their emotional impulses without reflection, withholding treatments from patients who should receive care. Mores and feelings of discomfort will surely evolve further, underscoring the need for clinicians to think carefully through each of these types of combinations.

Given competing ethical, financial, and other concerns, physicians can change from "fast" to "slow" thinking about complex moral issues, overcoming feelings of disgust. Emotions of disgust might coexist, lessen, disappear, or be overshadowed by other considerations or be wholly reversed.

While some providers have unfairly discriminated against lesbian, gay, or single patients, uncertainties increasingly arise, not regarding patients in these relatively well-defined sociodemographic categories but related to far murkier concerns about potential future parenting abilities among married heterosexual as well as other patients. Clinicians confront decisions about other characteristics such as possible future mental health symptoms that are less distinctly defined or ascertainable than relatively objective sociodemographic categories (such as marital status or sexual orientation) and involve balancing the future child's well-being against the prospective parents' rights. Questions surface based not on clear present harm but rather on unclear possible future risks.

While the mere existence of a medical diagnosis or disability alone does not warrant refusal to treat, dilemmas arise regarding whether *any* limits would be discriminatory and what risks to the unborn child might outweigh patients' rights to care. Even if clinicians could assess parenting ability, what minimal threshold they should use is hazy. Married heterosexuals' and others' family environments may be suboptimal and stress children, but not constitute clear physical harm or abuse per se. Laws tend to grant parents relatively wide discretion in how they raise their children, generally permitting anything short of physical or sexual abuse.[50] Laws also tend to give clinicians relatively wide room to make their own decisions, based on their own judgment.[51] Infertility providers struggle because laws fail to resolve quandaries of whether married, straight, or other parents will be "unsafe" or whether parents may be less than ideal ("I wouldn't want to be their child"). Though children have a right to an open future,[52] what that means precisely is often unclear when helping to create a child—*how much "less than ideal" parenting is OK?*

The ASRM states that "Child-to-parent arrangements are generally unacceptable, and parent-to-child arrangements are acceptable in limited situations"[53]; but questions emerge about whether such arrangements should in fact "always," rather than "generally," be prohibited and, if not, what exceptions are permissible and how these should be determined, especially given concerns about informed consent. Despite ASRM guidelines, how physicians do or should assess parenting and exactly how stable relationships need to be are unclear.[54,55]

Infertility providers have to assess and weigh competing, difficult-to-reconcile ethical principles that involve ambiguities and the rights and well-being of *different* individuals. Clinicians worry about not only patients' mental health but financial and other concerns.

To justify treating infertile patients who may be suboptimal parents, providers frequently rationalize that fertile patients often make poor parents but are nonetheless allowed to procreate and that infertile patients, if they had been fertile, could have had children on their own, even if they would not be ideal parents. But this logic is questionable since these patients are in fact infertile and consequently, providers need to be actively involved and possess a degree of moral agency.

Adults have rights to procreate but not necessarily to receive unlimited ART. Political theorists have distinguished between negative rights (to be free from state interference) and positive rights (to receive certain services).[56] Positive rights could include receiving essentials for a healthy life, yet problems surface in defining what is "essential," especially given the financial costs of providing services on a wide scale. Parents may thus be suboptimal but nonetheless have rights to reproduce, though not necessarily to receive ART.

Providers attempt to address these moral strains and uncertainties[57] by referring these decisions to other professionals, hoping patients drop out, assessing social support, arranging for possible co-parents, assisting psychotherapeutically, or discussing these concerns with patients. Ultimately, mental health and other providers often remain *gatekeepers*, wrestling uncomfortably with how much to play this role and what threshold to use, given the moral and empirical uncertainties.

Clinicians should try to acknowledge and address these strains directly and explicitly with prospective parents, rather than just hoping that these patients drop out. Physicians should appreciate as much as possible the challenges these MHPs face.

The ASRM should clarify its standard that judgments of problems with parenting be "well substantiated." Guidelines could address, too, combinations that involve friends and how carefully providers should consider child raising. Though the ASRM has recommended that clinics develop policies and procedures, questions remain regarding whether and how exactly providers do and should proceed. Anecdotally, many clinics have informal, implicit customs rather than formal explicit guidelines.

The Society for Assisted Reproductive Technologies and the CDC should collect data on rates of LGBT patients and single-mothers- and single-fathers-by-choice, to illuminate current practices. Longitudinal research is needed on what other factors affect parenting, how interventions might improve parenting skills, and how providers should define "good enough parents." These providers and patients thus encounter multiple choices about these technologies and a series of financial, social, and psychological stresses, and must decide how to respond.

IV | Confronting Stresses

13 | "How Much Is a Child Worth?"

CHOOSING BUDGETS

"How can I save up the money to pay for my IVF?" Nancy, the IVF clinic nurse undergoing infertility treatment, asked. "I'm already working every single day of the week, working double shifts at both hospitals, 18-hour days, sleeping for three hours, then going to work at the next hospital, earning as much money as I can. I think I can scrounge up enough . . . to do an IUI [intrauterine insemination], but not enough for the medicine. I'm banking on an inheritance my uncle left me, which I'm hoping comes by August. I'm then going to use the ten grand for IVF. My cousin, who's handling the affairs, says it should be coming. But I can't bank on that. I'm freaking out because I just turned 44. My biological time clock is ticking.

"Because my anatomy is so good, and I had a pregnancy at 42, my doctor said 'I don't think you'll have any problems at all. I think you'll need just a little bit of treatment.' But he turned out to be wrong. I probably wouldn't have utilized so much of my prescription coverage the first time. I maxed out so much—$2,200 for prescriptions. I had 15 follicles" —a relatively large number. "So, the second time, we used half the medicine, and I had eight follicles. I really didn't need to blow up my ovaries; and would have had more medicine to do more IUIs.

"Had I known in the beginning, I probably would have figured out a way to make my health benefits go further . . . gotten three rounds of injections, instead of two. But it's too late now. You learn as you go."

Nancy and many other patients also learn only afterward that doctors' fees can vary. "The hospital where I work drills in our head: 'Anything that you can do at our hospital, has to get done *here*. Otherwise we don't cover it. They charge $1,450 for a hysterosalpingogram [an X-ray to examine the uterus, fallopian tubes, and surrounding anatomy]. But one doctor said, 'I could have sent you somewhere else for only $350!' . . . I could have used that thousand dollars for a third try."

Doctors may want to help financially but be limited. "The doctor had a free online coupon. She said she would try to keep stashing samples of drugs, but I don't know how many samples she can get."

Clinics advertise "all-inclusive" plans online that may lure patients but contain hidden costs. "I saw a physician who offered a bundle package," she continued. "$985, no matter how many times you had to go in. That was nice. But *the medicines were not included!*

"A lot of clinics offer free phone consultations. *That* was really helpful: a free 20-minute, and then another 30-minute consultation. There's a program that helps pay for the medicines, but unfortunately has income qualifications: you have to be poor.

"A website on how to obtain financial assistance would be totally awesome—help with the medicine, subsidy programs." Such a website how-ever, listing legitimate free resources and services, does not yet exist.

Dilemmas of how much to pay for these technologies shape patients' decisions. Prospective parents commonly draw on present or future savings or income that they might otherwise dedicate to graduate school or other long-term goals, yet uncertainties can persist about how long treatment will take. Medical, financial, and legal stresses mount regarding both the total cost and the eventual extents of insurance coverage. Physicians cannot wholly foresee the kinds or amounts of necessary treatment, frustrating patients.

A patient's age and health can affect outcomes but not predictably. Physicians' forecasts can thus prove incorrect. Responses to any particular medication or procedure are also unknowable, making it difficult to tell, in advance, how best to allocate limited resources.

Countries range widely in their laws and regulations concerning insur-ance coverage but generally have restraints. In the United Kingdom, despite some coverage through the National Health Service, 78% of patients pay for treatment themselves or at least partly.[1] About half of patients plan to use sav-ings or retirement funds, a third go into debt, 12% anticipate refinancing their home, and 37% said cost influenced their treatment.[2] Financial constraints are the major reason British patients stop treatment, and 10% regretted not stopping treatment earlier.[3] In Brazil, cost is the biggest source of anxiety for IVF patients, affecting 82.6%.[4] In 2002, partly due to financial constraints, only 38.5% of US infertile women used infertility services,[5] with wealthier and college-educated patients using more.[6,7]

Some observers have argued that insurance coverage for IVF should in-crease in order to reduce the incidence of twins,[8,9] but in the United States at least, this proposal confronts several obstacles. Currently, only 16 states require that insurance companies provide any coverage for ART, and only nine states cover IVF, with the specific amounts and requirements ranging widely—whether all insurers must provider coverage, whether coverage includes IVF, and whether the infertility needs to be due to strictly "medical" causes.[10,11], Partly due to state mandates, some insurance companies have begun to cover limited amounts of ART, but many insurers still cover little or none. The Affordable Care Act does not mandate infertility coverage as an "essential benefit."[12,13] The American Society for Reproductive Medicine (ASRM) advocates that insurance coverage be increased since many poorer patients cannot afford care,[14] yet providers could also potentially lower costs. The ASRM has cautioned that clinics should also fully inform patients about all relevant details—including full expenses, disadvantages, and realistic odds.[15]

Crucial questions emerge, too, of who should profit from these technologies and how much. Critics have argued that many clinics are too "entrepreneurial."[16,17] Clinics with more patients paying out-of-pocket transfer more embryos per cycle[18]; possibly because of more competition from other clinics. Excess costs have also been criticized.[19] Elsewhere in medicine, 73% of patients and 61% of others see high costs as very serious problems.[20] Approximately $650 billion per year, one-third of US healthcare costs, are wasted, partly due to expensive procedures, unproven to outperform cheaper alternatives.[21]

Insurance Uncertainties

"What's this going to cost?" John, infertile despite a vasectomy reversal, asked his physician about IVF. "Can you tell us?" His doctor replied, "I can't guarantee anything, that's with God—He does that. We just do the work, and let Him handle the rest."

"It's an awful lot of money for us to spend and not have anything," John grumbled. "The doctor talked about $15,000 for just one cycle. It didn't actually include getting the sperm out of me. All they do is fertilize the egg. We considered IVF three years ago, but couldn't afford it. So, we checked into reversing my vasectomy." Online, John read questions of whether it would work. "A urologist said, 'I've done many, even older than yours, and they work.' . . . It was only supposed to cost 10 grand, but ended up over 13. But here I am. Still nothing!" Costs can mount more than expected or planned, and patients' finances can shift.

Doctors may denigrate colleagues' past treatment decisions that failed. John found a doctor who does "two cycles for $13,500. He said we 'wasted our time and money doing the reversal after 15 years. It sounds like the doctor just wanted to make some money off of you. You could have spent the same money on IVF and would already have had a kid.'" Doctors may seem overly optimistic that their treatment will succeed. When costly procedures flop, despite the providers' prior assurance, patients can feel misled, deceived, and angry. John felt he got "false hope."

Only too late do many patients grasp the implications of insurance limitations—restrictions on lifetime total coverage and hidden costs. Insurance companies may cover some expenses but not others—paying for certain procedures but not the necessary medications or vice versa. Insurers may also communicate unclearly about what they will reimburse. Eligibility may depend on differing and ambiguous definitions. "You had to match certain criteria," Isabelle, the pregnant Connecticut office manager, explained. "They told us that even for IVF, they would never cover 100%. So, we were taking a risk and might have to pay the whole thing on our own, because we were in that gray area. We weren't considered infertile, because we had gotten pregnant on our own the year before. The trial-and-error part was very taxing emotionally and financially. Technically, they *should* cover us, but [we] never really knew. Everybody on the phone said, 'I can't imagine it won't go through; but I can't say that for sure.'"

Dealing with insurance companies is "a hard, long process," Isabelle complained. "The crappiest. Keeping really good notes of who you talk to, and what they said, helped." But it "could have been a lot easier and a lot shorter . . . not 6-8 months to cover a claim." To her surprise, "Insurance ended up covering everything."

Insurance stresses can be among the most arduous aspects of patients' journeys. Though insurers may cover certain causes of "medical" infertility, the reasons for a patient's problems may be murky. Insurers may cover certain laboratory tests but not others. Joe, the physician who specializes in male infertility, has "operated on men with similar issues who were *not* trying to get their spouses pregnant, to preserve their fertility and testicular function. Most insurance companies look favorably upon *that*—but rarely cover vasectomy reversal or sperm retrieval, which can be very expensive."

Insurers may simply want to avoid committing to coverage in advance, instead preferring to review each claim case by case. "Insurers may not want to make such promises up front or in writing," Tammy, a graduate student, told me. "We were told they'll never send a letter saying that they will definitely cover it. If your husband gets laid off or quits his job, and you have such a letter, they have to cover it—whether he's still employed there and paying his insurance or not.

"They said, 'The policy was written in another state, so we don't have to cover it,'" Tammy explained. She had started graduate school and then moved, but her insurance policy, from her prior job, remained in her prior state. Hence, "I'm going to have to pay cash. I printed out all the stuff that I found on the insurance commissioner's board and gave it to the lady who does the insurance," Tammy added. "I guess she's looking into it. I don't know who else to [talk to]. I guess I can call the insurance commissioner and ask them some questions about how we go around this."

She and her husband are now using their life savings for IVF, rather than for other important purposes, such as future education—sacrificing other opportunities for IVF. "I could probably be using this money to pay off student loans," Tammy confided, "so I wouldn't have so much debt later. . . . Graduate school is going to be over $100,000, and I've still got undergraduate loans! We didn't realize [treatment] was going to cost us a couple of thousand dollars. We waited a little while and tried to save the money.

"My husband thought of looking for a different job—just to get better insurance. But it's not easy to look for a job based on the insurance coverage, because they don't give you all the details in the new employee handbook." Tammy's husband in fact changed jobs and then "had better insurance, but it *still* wouldn't cover the treatment." Changing employers to obtain better insurance can be arduous and unfruitful.

Other patients feel stuck in jobs and locations that they don't like but that provide a modicum of IVF coverage. Only afterward, in retrospect, will patients know if their decisions of how to allocate these resources worked.

Seeking answers, Tammy and other patients try to communicate with large corporate bureaucracies but frequently end up frustrated. Insurance companies may erect logistical barriers with vexing fine print, altering processes and policies over time and between states. Cathy, the Alabama customer service representative, whose infant daughter died of trisomy-22, said that her new insurer, "told me to send a pre-authorization. So, I'm not sure if they're going to cover it or not. I've talked to other moms. Some have gotten their insurers to cover costs. Some haven't." Patients must balance expenses, resources, and likelihoods of success and may postpone or forgo treatment or exhaust their savings.

Patients like Nancy and Tammy, who remain childless with depleted insurance coverage or finances, commonly rue their earlier choices, feeling they could better have selected their treatments and allocated funds.

To obtain treatment, many patients not only change jobs, like Tammy's husband, but move to jurisdictions that mandate IVF coverage. Yet relocating due to IVF generates tough trade-offs and stresses. "I didn't really want to move," said Sally, the pregnant Maryland website manager, who had switched states to get better insurance coverage. "I liked our apartment, our friends. We were closer to my family. But it was going to cost $20,000 a year per cycle here. If we moved there, insurance would cover it. So, we moved."

Yet, even after migrating to regions with mandated coverage, patients may still fail to have a child, which can be crushing. "When the two cycles didn't work, we were devastated," Sally added, "because we had moved there, and didn't really like it. We had shifted our whole life for a baby, and it wasn't even working. I gave everything up, and still had nothing. We were so stressed, I had to wear a mouth guard at night because I was grinding my teeth and my jaw was starting to hurt badly. When you try naturally, it's very disappointing every month when it doesn't work. But when you're going through IVF/PGD, you never 'feel yourself' or can tune it out at all.

"It doesn't seem fair. If you have a life-threatening disease, it's covered. If you have infertility, it may or may not be." Eventually, she moved overseas, and only then became pregnant. "We had always talked about living in Europe for a while. My husband was able to get some consulting work there." She then got pregnant "naturally."

Searches for Free and Discounted Treatments

Clinics' "all-inclusive" packages, including money-back guarantees, can prove limited costly "gambles." "A pre-pay IVF company would charge me $45,000 up front for three IVF cycles, based on my factors," said Amanda, the Delaware high school administrator who ultimately decided she could afford only one IVF cycle. "If it didn't work, I would get a portion of that money back. But I was not willing to make *that kind* of investment." One clinic, for instance, said, "If we can't get you pregnant after three cycles, here's 80% of your money back," John, the Texan who had had the vasectomy reversal, reported. But not all pregnancies end in births.

Online ads for free IVF clinics and services generally prove misleading. "You get a lot of empty promises out there," John continued. "One site says, 'free IVF,' but when you get in and start looking into it, it doesn't mention anything about it being *free*. *A lot of bogus stuff is out there*. It's hard to find the stuff that isn't making a dollar off you. All they're looking for is 'put your credit card number here. Pay us $49.95, and we'll show you how to get the money through your family, or going abroad.' There's always someone trying to separate you from your dollar."

Some sites "want two, three years of your tax returns, how much money you have in your checking account, just to get a $10,000 grant," John reported. "That's too much information. It looks like a legit site, but they want a donation of $25 or $250 a month. So, I don't see why they call it free. There's always a loophole."

At times, providers and patients have also sold patients' leftover medications to aid other prospective patients, but problems can ensue. "People donate drugs back and discreetly give them to help these women," Diane, the IVF clinic mental health professional, explained. Online 'underground' markets buy and sell unused infertility medicines at a discount. Nancy described "a black-market website where a lot of women sell leftovers at a much cheaper price. It seems OK. Women post their leftovers, usually with a little story: 'We've stopped trying now, and are adopting.' That'll probably end up being my next route. I just hope they're not putting baby powder in those bottles!"

Clinicians can play key roles but range widely from reducing to inadvertently increasing these tensions. When a patient's insurance coverage ends, a few providers have continued treatment for a period at lower fees. "Doctors try to help out their patients and give them discounts, if it's been a few cycles," Brenda, a psychotherapist, reported. But overall, that approach seemed rare.

Occasionally, physicians give patients free drug samples that may not be the most effective. Patients may also ask a physician to dissemble on their behalf to insurers—to list other, reimbursable indications for procedures. Doctors generally refuse. As Nancy reported, "the billing department said that if I could have the hysterosalpingogram resubmitted under a different code, like endometriosis [a condition in which tissues that usually cover the uterus grow in the fallopian tubes, ovaries, or elsewhere] rather than infertility, they would be glad to resubmit everything; and reimbursement would be a piece of cake. But my doctor's not willing to do that. Malpractice scares her."

How Much Is a Child Worth?: The Stresses of These Decisions

"How much money is it worth to have a child?" Amanda, the Delaware high school administrator, asked. "I never aspired to be a doctor, lawyer, or teacher. My only dream was motherhood. People say: 'go for it at all costs!' Unfortunately, that's not reality. People who haven't gone through it don't understand. Each round of a procedure represents a gamble. Yet the odds involved are hard to assess. . . . Do we want to spend $16,000 on a 40% chance [or] $25,000 on the chance of adoption? I didn't know where I should invest the money. . . . Then, there are all the options in between: donor eggs, surrogates. It came down to a *financial investment*—your cost analysis.

"With IVF, I knew all the emotional components: I'd be prepared. If we put this investment into adoption, I didn't know what to expect. By the time we got to IVF, I was emotionally drained. I thought, 'At least I'm familiar with this emotional set.'

"Everybody around me said that we should start with IVF. I said, 'You're crazy if you think we're going to spend $15,000 for a 50/50 chance!' So, we started with IUIs." In retrospect, she wishes she had done IVF first. "If your insurance covers it, why not go where your odds are the highest?

"I drew lines, erased them, drew them again, erased them, and drew them yet again. If you asked me at the start how many rounds of IUIs I would do, I would have said one or two. I wound up doing six! I wasn't willing to go into complete debt, to take a second mortgage on my home, to get so far behind that by the time we had a child, we had nothing for that child. We joke that when the kids ask for a bicycle, we'll say, 'Too bad. You were born!' I also took advantage of a 0% interest credit card."

Prospective parents thus confront terrible dilemmas—*having to put a price tag on creating a child*. At multiple points over time, many patients re-evaluate and readjust how much they are willing to spend, facing harsh trade-offs of whether to continue or stop—deciding what sacrifices to make to achieve this goal. Individuals commonly draw sharp lines, beyond which the price seems too high, but vary significantly, balancing money, statistical odds, and desires. The pros and cons of more IVF versus adoption versus ongoing childlessness are tough to compare. Deep psychological, economic, and existential factors shape patients' responses. Humans are generally poor at "affective forecasting"—correctly anticipating their emotional reactions to future events.[22] Given these uncertainties, patients frequently receive input from friends and family, but it can counter their own preferences.

"If somebody had said to me, 'This might kill you, but might work in getting you pregnant,' I would have done it!" Sally mused. "Having a child was more valuable than my life!" Offspring can provide deep meaning to parents, who may see bringing one's own child into the world as priceless, more treasured even than their own lives, fulfilling lifelong dreams. Many outsiders fail to appreciate the degrees in which expensive failures can decimate hopes.

Couples who succeed tend to feel that these costs were justified. John, who reversed his vasectomy, knew a couple who "spent $30,000, and had a baby girl. They thought it was the greatest thing. They couldn't be happier with anything in the world, but are now burdened by debt. But to them, *every penny was worth it.*"

Spouses can disagree about how much to spend—how much a child is worth. "The number one thing that will tear up a marriage," John stated, "is finances." Joe, a physician, has observed patients who "will do *anything* and not stop – take out loans, do strange things. One man's wife was putting a lot of pressure on him to sell his Harley-Davidson motorcycle. He thought long and hard: 'This isn't worth it! *She* isn't worth it!' I thought he was going to break down in my office."

Changing Policy?

Increased insurance coverage would aid patients financially and reduce incentives for twin or higher multiple births but faces significant political and economic obstacles. Most patients and providers feel that ART coverage should be mandated. But to cover more infertility treatment, governments or private insurers must take the funds from elsewhere. "I wish the insurance companies would cover it, but someone would have to pay for that too," John admitted. Yet he, in a conservative Midwestern state, and several others argued that government resources in other areas could be better used on ART. "A lot of money is wasted on other programs." Politicians and special interest groups fiercely debate such government allocations.

Controversies erupt concerning whether any limits should be drawn and, if so, where and by whom; who should be eligible for coverage; and how these decisions should be made. Diane, the nurse, felt that the number of children "should be limited. Two per family is reasonable. . . . *That* could be federally mandated."

Political resistance may hinder increasing coverage anytime soon. Opponents may see infertility treatment as elective, not a priority. These treatments do not eliminate life-threatening disease. "It's like cosmetic surgery," Valerie, the psychotherapist and single-mother-by-choice, admitted. "Preventative care for pregnant women, and treating all the diabetics and related diseases ends up on top of the list," Bill, the New England REI, added. "The fact that somebody can't have a baby is lower down."

Concerns About Entrepreneurialism

"IVF brings out the very best and the very worst of medicine," Edward, a physician, said. "Wonderful techniques to help couples have children. But doctors get into this *entrepreneurial* spirit, and are doing procedures just to generate cash flow, without regard to outcomes." While western Europe closely monitors and controls these technologies, the US imposes few constraints, spawning a highly commercial, largely unregulated market, the "Wild West." Several providers felt that colleagues helped patients but that at times, mixed, financial motives prevailed.

Crowded markets spur such behaviors. "Unfortunately, competition breeds certain kinds of stuff," Nicholas, another REI, added. "There's a lot of entrepreneurship. You draw lines, but every year they get blurry."

A few providers argued that competition among clinics enhances clinical quality. "There are some outliers, but most private places are doing very well, and take very good care of patients," Calvin, the southern California hospital REI, said. "Their success rates are probably as good or better because they *have* to produce, or won't have a business. A lot of institutions do IVF so they can say they do it, and don't yet have a really good program." He and many others argue that private clinics do not pose significant problems and in fact provide good care.

But many patients believe that profit motives may overly sway doctors' decisions. "Patients feel that some doctors take advantage of women," Francine, the Mexico-bound legal assistant, stated, "ordering tests, or having women go through cycles with low odds of success."

Concerns About Specific Procedures

Even clinicians wonder whether certain colleagues may at times order unnecessary tests or procedures because of profit. As Calvin suggested earlier, concerning the possibility that some physicians may be overhyping egg freezing, providers may observe and question certain colleagues' decisions. "Every patient we see who comes from one doctor always has a laparoscopy, whether the patient needs one or not," Diane, the nurse, noted. "These women don't need one! The doctor is only doing it for the money; and the hospital makes tons of money."

Doctors may overly order certain procedures that lack demonstrated effectiveness. "ICSI [intracytoplasmic sperm injection] is grossly overused," Roger, a New York REI, said. "It's a big deal at some institutions, but is really best for male factor infertility, and patients with prior failed fertilization. But it's used *over the top*. Insurance companies are now asking doctors why. If doctors have to justify it on paper, the number of cases decreases. So, it's income-driven. Doctors get higher fertilization rates: if you naturally fertilize eggs, you can get 60% to 70%. With ICSI, you can get 90% or 100%. But in the end, you don't get more babies! Still, patients will do anything. They'll say, 'That sounds logical: if I have more embryos, there's a better chance I'll have a baby.' Unfortunately, the data just don't support that."

For several years, physicians also overused preimplantation genetic screening (PGS), assessing numbers of chromosomes to increase live birth rates and reduce miscarriage (rather than merely to avoid disease in the off-spring), which turned out to have limited efficacy and cause harm, including decreasing chances of the patient having a child.[23] Steve described one prac-tice "routinely doing PGS on three embryos, charging patients $15,000, which I think is criminal."

Concerns about profit motives arise strongly regarding egg dona-tion, which insurers do not cover, forcing all patients to pay on their own. A crowded infertility market can also lead physicians to let older women use their own eggs when the odds of success are nil. "In a *competitive* environ-ment," Steve thought. "It's usually a no-go to refuse 43- to 45-year-old women wanting to use their own eggs, because they might eventually want to do third-party egg donation."

Critics argue that doctors pay egg sellers too little and profit too much. "Doctors make money with donor eggs, which bothers me." Paulette, an IVF psychotherapist, said. "Insurance is not going to cover anything done to the donor: All the medications, screening, anesthesia, are *cash*. So, doctors are setting their own fees, and can charge whatever they want—as long as the recipient's paying." Paulette doesn't think that egg donation is over-used, only that the profit motive is too unchecked.

Patients may not feel that doctors are overcharging but that these fees can accumulate without producing a child. Especially if attempts fail, women can feel that the industry is too profit-oriented, taking unfair advantage of them. "I haven't heard anyone say flat out that their doctor was doing things *just to make money*," Amanda, the high school administrator, said. "But small things happen. A clinic ordered one woman the wrong medication, which she could not then return, so she was stuck with a bill. They told her how much of another drug to order, but she only used a third of it, and had all this extra medication. Those little things add up to big bills. You think, '*Is this all one big money-making thing?*' And it sucks when you're spending money, and not getting anything in return. Almost anyone who has gone through failed attempts has said at some point, '*Someone's just making money here!*' We're probably scapegoating the doctors, but how can all this be *that* expensive?"

Many feel that medications, in particular, are too pricey. "For almost every prescription drug, the costs are absolutely ridiculous and insane," Amanda continued. "It's such a fragile, emotional human procedure, you think 'How can someone charge $800 for this little bottle?'" The industry "seems to feel justified charging that amount because they know you're willing to pay to have a baby. That's where you start to feel taken advantage of."

Factors Fueling Entrepreneurialism

"When you have the highest concentration of REIs, you have the most intense competition," Nicholas, a New Jersey REI, observed. Several underlying features of clinics and patients may affect these perceptions of profit motives and entrepreneurship. Especially in major urban centers, multiple clinics exist that can foster entrepreneurialism. Initially, providers may be cautious about new procedures that other clinics perform despite unproven efficacy. But competition may then prod the spread of these new methods. "One clinic was trying to sell egg freezing on their website several years ago when it was definitely still experimental," Nicholas added. "A few years ago, I wasn't doing that. Now, I'm going to start offering it."

Smaller infertility practices, rather than larger groups or hospital-affiliated clinics, may engage in these practices more. "Usually the people who do this kind of stuff are in small one- or two-doctor practices, not the group practices," Diane, the nurse, clarified about some doctors overperforming laparoscopies. Valerie, the psychotherapist and single-mother-by-choice, has seen "a lot of charlatanism in a lot of the smaller practices. A patient with repeated miscarriages saw one of those guys early on. A doctor at a major medical center later did a hysterosalpingogram and said, 'You have scar tissue all over your uterus; no wonder nothing will implant.' This new doctor scraped her out, instead of putting her on her fifth or sixth round of IVF, and said, 'Take a few of these shots, go to Bermuda with your husband, and have sex.' Hence: my godson!"

Patients may also complain about these issues when more costly treatments, for which they have paid relatively large amounts and have failed. Understandably, patients may then feel irritated, if not bitter, and see these doctors as "cowboys," unfairly taking advantage of women. "If it works, patients are happy," Suzanne, a psychotherapist, commented. "If they paid and it fails, they're angry at the profit motive—plunking down $15,000 for something that didn't work."

Possible Solutions

Providers and patients both pondered how to lower the effects of profit motives but felt that solutions may be limited. "I'm not sure how it gets regulated—whether anything can be done," Valerie admitted. Suspicions that providers are offering procedures largely or primarily because of profit are also difficult to assess or prove. "It's hard to judge motivation," Edward, a physician, said. In part, mechanisms for reporting or overseeing clinics are scant. "Whom do you report them to? Diane, the nurse, asked. "Who's going to blow the whistle on them? It gets tricky. The doctor may not even belong to ASRM, so what does he care? He's making money! You'd have to go to the hospital, but it's making money off this guy hand over fist." Patients could complain to the clinic itself or to the larger institution, if there is one, but it can have inherent conflicts of interest (COIs), making recourse hard. "There's the state medical board, but doctors are very unwilling to report [problems]. So, these doctors are out there."

Diane has "done nothing" regarding the unneeded laparoscopies she described earlier "because: *what is going to happen to me if I complain?* Are they going to believe me? But every time I talk to a patient who's seen this doctor, I say, 'Tell me about your laparoscopy.' The only patients who haven't had it are those who absolutely refused.

"ASRM isn't going to take that on. They've refused to take on issues that they should, because they are so afraid of being sued. The ASRM wants to be nice, and has no enforcement power. They could kick somebody out, but so what? A state-licensed board, not ASRM, has some clout, and is the body to do it." Professional organizations could do more to address these concerns, establishing guidelines and targeted education; but they have hesitated, may face challenges, and have potential COIs. State medical societies can exert more of a role here.

Monitoring by colleagues could also help: "A peer review process to make sure doctors are doing procedures for the right reasons," Edward, the physician, said. "A better auditing process for the data would be very good."

In short, insurers generate dilemmas because of not only *limited coverage* but ways they *communicate* about it—frequently deciding only afterward, case by case. Consequentially, patients must pay first and later struggle to get reimbursed, which can take months. Patients may move and/or seek treatment in other states or countries or feel "stuck" in jobs because of insurance coverage, pursue "free" treatment online or "free" medications through providers or the "black market," or make various financial sacrifices. For other medical conditions, such as chronic disease, patients are grateful for partial reduction, even if not elimination, of symptoms. But fertility patients find incomplete success unacceptable. seeking only one goal—a "take-home baby," not pregnancy alone.

Couples tend to make these decisions jointly but can disagree, having to negotiate how much to keep spending, making complex risk/benefit calculations about how much a child is worth and how much debt to incur. Desires for children and the ever-ticking biological clock intensify these pressures. Many doctors try to help patients battle with insurers; but other physicians hype their own procedures, and criticize colleagues' prior approaches. Still, not all insurers may fit these descriptions.

Patients must quantify the significance and worth of a future human life, translating these profoundly life-altering, personal quests for fulfillment into numeric, economic terms. Individuals seek sources of symbolic immortality—connections to provide a sense of ultimate meaning, often through offspring or religion.[24] Many individuals value having a child above all other goals—as providing immeasurable psychological, social, cultural, moral, and existential meaning and purpose.

Patients often worry whether physicians' profit motives may play too large a role, particularly regarding certain procedures such as egg donation, PGS for miscarriage, and ICSI. These perceptions may not be wholly accurate; but the fact that providers and patients repeatedly raise these concerns is noteworthy since these perspectives may make patients, family members and the wider public wary of treatment. Moreover, since prices for human eggs will doubtlessly increase, following the ASRM's Kamakahi settlement that eliminated suggested price caps on women's eggs, physicians may now gain even more financially.

Clinicians should avoid performing procedures driven by financial motives, rather than patient benefit. Practice guidelines can help reduce unnecessary expenses.[25] In other areas, reimbursement to providers based on outcomes, rather than fees for service, have been proposed but may have mixed effectiveness and be difficult or unfeasible.[26] Since fertility patients seek only a single outcome, payment based on that outcome may be impossible. Using pregnancy alone as an outcome would also be inadequate, given miscarriages. Professional organizations should urge providers and insurers to publish and clarify rates and reimbursement up front. Providers and insurers may resist doing so, but enhanced transparency can help.

These efforts can help patients as they confront strains due to finances and, as we will see, other sources.

14 | Emotional Roller Coasters
COUNTERING OTHER STRESSES

"It's hard on your body, your emotions, and your relationships," Wendy, the Catholic secretary, said after four unsuccessful in vitro fertilization (IVF) cycles. "My body hurts where it never hurt. Over the past five years, I've gained forty pounds from the hormone injections. You take more medications after the transfer, and as the pregnancy goes on. Your body never really bounces back. It's like when women go through menopause—they have the pouch in the front. You have to be ready for all these problems, because if you're not, it's going to hit you over the head like a brick.

"After the first IVF didn't work, we used frozen embryos," she continued. "The doctor said, 'Oh, *this* one's going to work, don't worry. It's going to be great.' It worked, but miscarried after nine-and-a-half weeks.

"The doctors hype you up so much that you're excited and believe it's definitely going to work. They should not hype it up or say, 'This is a sure thing. You're going to wind up with a child in the end.' Doctors have to be more honest with patients, and not give false hope that it's going to happen."

These life-creating technologies can cause not just physical and financial but also emotional, psychological, marital, social, and ethical stresses over extended periods of time.[1,2] Especially when treatments fail, these difficulties can accumulate, exacerbating one another.

Social contexts affect how doctors and patients use and experience these technologies. Physician hype can trigger not only doctor shopping but patients' past regrets. Desperate for hope, patients may also only listen for and hear positive scenarios. But in the competitive market, doctors may often be overly optimistic, igniting later patient disappointment and stress. The higher the expectations, the more devastated patients may feel.

Medications themselves can also cause powerful physical as well as emotional side effects. "I didn't like the whole procedure—the anesthesia, the shots, the medication," Roxanne, the Michigan marketer who has been undergoing assisted reproductive technologies (ARTs) for eight years, said. "When you're doing it month after month, it messes with your head. Who knows the effect of dosing yourself with all this medication?"

Ongoing delays intensify these strains. Francine "was supposed to get a result six days ago" revealing whether she had any viable embryos to transfer. "I got stressed out waiting another day. Unfortunately, it got pushed back and pushed back. Yesterday was the last straw. Until yesterday, I was able to manage my own expectations. Then, it was too much."

The high stakes involved intensify these stresses. "You're injecting yourself with shots of all these hormones, which are much, much, much higher doses than your body is hormonally adjusted to," added Sally, the pregnant Marylander who, because of cystic fibrosis, underwent three rounds of PGD to produce her daughter. "Physically, it's very challenging. You kind of get used to giving yourself injections. But it's just very emotional. There's so much riding on it: if this doesn't work, it's just money out the window."

"A friend," Sally continued, "said that women dealing with infertility have higher stress levels than cancer patients! Having gone through both, I believe it. For some reason, not being able to have a child is much scarier than even dying.

"At first, I thought, 'I'm young, I'm fertile, this is going to work! I got 25 eggs! This is great. Even if I have to do this a couple of times, no big deal.' But, the next day, they called to say that only three fertilized. I hit a wall. There was no explanation, except 'maybe something's wrong with your eggs, or with your husband's sperm. Or both.'

"When I couldn't get pregnant naturally for a long time, I started to feel horrible. 'Why me?'" Sally continued. "It was taking so long. I can't get pregnant naturally *or* through IVF/PGD. God, this is the worst-case scenario. I can't get pregnant, and when I do, I have IVF hanging over my head. It was the most awful feeling."

Failure can mean loss of purpose, identity, and dreams, deeply disturbing patients and precipitating existential quandaries. As high hopes and dreams collapse, many patients have to readjust their expectations, bitter and confused.

Couples may discuss or try to ignore these burdens—depending partly on how they have communicated and resolved conflicts in the past—but may end up clashing. These strains lead patients, even with insurance, to stop treatment.[3]

The Stresses of Failure

"The saddest part of my job is when I'm working with a lovely couple and treatment fails," Ginger, the psychotherapist and IVF patient, said. Those who have tried but not yet gotten pregnant frequently feel crushing defeat. Any one of multiple steps in the process can falter, generating distress.

Patients who quickly succeed generally feel pleased with their treatment and avoid certain tensions. "I feel lucky that the doctors knew what would work for me," Isabelle, now pregnant, said. She screened out embryos with the chromosomal abnormality that had impeded her earlier pregnancy. "We didn't have to try a bunch of times, or do trial and error. It didn't take us forever. A lot of people don't know what path they're going to take or what medications will work. That is more difficult." With a good outcome, she viewed the doctors and overall experience positively. If she had failed, she may well have felt otherwise.

The fact that having a child can provide deep sources of purpose and meaning in one's life can aggravate these failures. "I always pictured having kids in my life," Amanda, the Delaware high school administrator, said, "So, if it didn't happen, how would I define myself? What would my life be without children?"

These stresses are not just emotional but cognitive. Patients struggle to grasp new medical knowledge. Most other areas of medicine are more established and have more consensus. John, the infertile Texan, "never heard of half this stuff."

Limited knowledge of both the cause and outcome of infertility intensifies these frustrations. "The medical situation is usually unexplained," Suzanne, the psychotherapist, said. "Sometimes it's ovulation problems or 'male factor' or tubal issues. Women don't yet know that treatment is going to work."

Uncertainties heighten these difficulties. "It's nerve-racking," Cathy said. "You never know how things are going to turn out." These vagaries can strain relationships with clinicians. Patients have to figure out how much to follow recommendations and to stay with or change physicians.

Providers, too, can become stressed, searching for sources of failures, to understand how to proceed. "It's hard for both patients *and* doctors," Steve confided. "Recurrent pregnancy loss, for instance, is incredibly difficult to walk patients through. There's usually no answer."

Looking back, many patients feel they should have proceeded differently. "If I knew at 25 that I had a balanced translocation, I might have approached my whole adulthood differently," Yvonne, the Jewish Philadelphia psychologist, remarked. Instead, she got married at 28, went to graduate school, and worked for a year before trying to have a child. "Lo and behold, here we are—later than we wanted."

Earlier in their lives, some women, now infertile, had gotten pregnant but underwent an abortion that they now regret, given current failures to conceive. "Two years ago, I got pregnant," Karen said. "It was a casual relationship. We only had been going out for six weeks. I was all prepared to have the baby, but the guy turned into a horrible big creep. I couldn't take it, and had an abortion. I regret it. The only thing that helps me get through that is that the guy was no good."

Gender Differences Exacerbating Strains

"The burden of treatment is on the woman, not on the man," Brenda, the Vermont psychotherapist, said. "Going through this in your body 24/7. Some women never get time off from thinking about this, and feel powerless because they can't do more."

At numerous points, spouses can clash about how much they value parenthood—whether, for how long, and at what cost to do IVF, and whose eggs to use. In heterosexual relationships, women generally experience more of the physical and hence psychological onus than do men. "Women are the ones who give birth, getting shots, going through the physical part of it," Amanda, the Delaware high school administrator, stated. "My whole being is changing. To men, it's not real until they're holding that baby. It's just a concept. My husband is not as expressive as I am. He said, 'Be hopeful, but expect the worst.' I couldn't be hopeful without being disappointed. When I needed the most support, he didn't understand why I was so upset. A negative pregnancy test, or getting my period, were awful days for me.

"He wouldn't get mad at me, but he gave me no support: 'I don't know what you're so upset about.' . . . He felt, 'If things don't work, our life is still good.' I agreed, but always pictured *more*. . . . Yet at other moments, when I had my miscarriage, he was everything I needed."

Women, more than men, may discuss these travails. But couples can fluctuate, based on the specific stressors, agreeing on many, but not necessarily all, the decisions.

Men may experience these strains *differently*, however, rather than less. "Generally, women feel more hopeless," Suzanne, the psychotherapist, opined. "Men are more optimistic. That creates conflict. But I think they're just two different ways of coping with the same anxieties."

Couples can disagree, in particular, over how much time and money to spend—how to weigh medical odds against financial costs. Isabelle, with Emanuel syndrome, said, "My husband thought 'If IVF works and insurance doesn't cover it, we now have a baby, but huge debt.' For him, it was easier to try naturally. He still supported me, since I was the one physically going through the brunt of it." Ultimately, husbands, even if preferring to proceed differently, usually defer to their wives, undergoing the procedures.

Effects of Stress

"I understand why marriages break up due to infertility," Yvonne, the Jewish Philadelphia psychologist with a chromosomal translocation, said sadly. "It highlights all of the cracks and strengths in your relationship. There are astounding moments of strength, which surprised me. But the low moments are awful. I don't know if we could go through another miscarriage. During the last one, our marriage wasn't managing very well. We can barely agree if we're going to try one month or not, let alone do something more assisted. Our marriage isn't strong enough to manage another IUI [intrauterine insemination] or IVF."

Couples and individuals range in how and with what effectiveness they negotiate these unpredicted jolts. "It's put a strain on my husband and me," Wendy said, "though not to the point where it's killed the relationship. We talk everything out."

Social Stresses

Social support from family, friends, and others can be vital but varies and can feel inadequate. As Francine, now traveling to Mexico for treatment, said, "people told me that I need to stand upside-down on my head after I have sex, and just relax. Get drunk and have sex in the back of a car. My husband needs to wear boxers. They offer me their children or to be a sperm or egg donor. 'Oh, this is going to work for you—I know it!' I think: 'You *don't* know! You're not a doctor! You know nothing about my case. You feel like you're giving me hope. But *I* know my situation. We've been trying this for over three years now. It's not just going to happen. It's going to be more complicated.' I tell people this, and they get upset. I say, 'Why are you upset? I'm fine. I've come to terms with this. I'm glad I've got the support I do. You're just making it worse!' No one realizes: it's a very lonely path.

"Outsiders don't understand how it's sometimes upsetting to hear that somebody's pregnant, or that their children misbehaved. Parents complain about their kids, and say: 'You sure you want kids?' Or, 'Take him. You can have him!' I say, 'I don't want *your* baby—I want *my* baby!'" Parents who readily conceived children may not appreciate the difficulties of infertility—even hearing about others' children and successes.

External support ranges from negative to positive, and weak to strong. Family and friends may try to be supportive but fail to grasp patients' struggles. Beliefs about pregnancy and childbirth have deep cultural roots—the stuff of myth, folklore, and old wives' tales. Well-meaning outsiders may offer inaccurate input that patients experience as unhelpful and as dismissing physical and emotional tolls.

"Very simplistic suggestions are made to patients going through this," Suzanne, the psychotherapist, observed. "'Oh, he must be shooting blanks,' 'Why don't you just go on vacation,' 'Just adopt,' or 'Just use donor gametes. Then move on with your life and everything will be just fine.' These comments completely dismiss the emotional ramifications or . . . feel disrespectful of the couple's intelligence. If it were *that* easy, the couple would have done it. But it reflects people's discomfort with helplessness."

Patients may hide treatment specifics. "Most people knew we were struggling to have a child, but very few knew all the details," Roxanne, the Michigan marketer who has been undergoing ARTs for eight years, recalled. "Because their first response was to give advice: 'Did you try *this*?' Stupid things: 'Having intercourse this number of days? Doing it in this position?' I thought: 'You have two kids, and went through none of this! We've been going through this for a couple of years, and seeing a fertility specialist.'"

Friends or family may also question desires for particular treatments. "Patients doing egg donation and sperm donation are seen as selfish," Helen, the Wisconsin psychotherapist, explained. "'Why don't you just adopt?' People are very ill-informed and ignorant about adoption, and don't know all the issues involved. These patients are blamed for wanting what other people have."

Patients whose treatments have repeatedly failed may feel shame, stop informing friends and family, and become more isolated. "Initially, patients tell people because they want support. But then, too many people say, 'Did it work?' Patients have to explain themselves," Suzanne added, "and feel they're Debbie Downers, stigmatized, failures. Other people don't know what to say to them. Everyone else [seems to be] having a great life. Patients stop telling people. Sisters withdraw from sisters."

Friendships crack. Sally "ended up losing a couple of friends over it who already had kids, and I wanted kids so badly. It felt so unfair. Why would someone even bring up their kids to me? I hated seeing pregnant women on the street: it's everywhere, but *I* can't have it. It twisted me inside out."

Beliefs about the virtues of "natural" over "artificial" reproduction heighten stigma and prejudice, impeding communication. "Some people feel it's supposed to be a *natural* process. Since it wasn't, it's less important, less valid," Amanda reflected. "If a child's adopted, it's not *really* our child."

Work Strains

"The most difficult decision I made was lying to my boss and not telling him," Wendy, the Irish Catholic office manager, confessed. "I've told everybody else. I've been going through treatment for five years, and worked for this company for four years. At my prior job, I was pregnant and put on bed rest. I told two co-workers, but had a miscarriage and went back to work. My boss said, 'I don't know what to say.' I said, 'You don't have to say anything.' Everybody else hugged me, crying. After that, everybody just gave me those eyes: '*I'm sad for you.*' I'm not a leper! I don't have a deadly disease. Just look at me like you normally do!"

At her new job, she "runs the business" and has kept her infertility treatment secret. When she was hired, her new employer expressed "concern" about the possibility of her getting pregnant and going on maternity leave. " 'Don't worry!' " she assured them. " 'I will take care of all that, if and when it happens!' They didn't know that I'd already done two IVF cycles and been pregnant twice." She once overheard her boss saying, "I don't know why people go through IVF. Just get a dog. It's easier." When she needed a reference from him for potential foster parenting, he asked her "Who's going to watch the child? Are you still going to be here nine to five?"

She feels that her boss "wouldn't get mad at me," but she is nevertheless scared. "They'd be very happy for me. But I would leave them in a tizzy, because it's a one-girl office and they'd have to hire somebody for six to eight weeks or let me go."

She has to schedule doctor visits surreptitiously. Not all clinics are flexible. "I need to get blood work done at 6:30 a.m., so I can be at work on time. But some doctors' offices are open only at 9 a.m. The only thing that's made my treatment difficult is trying to hide it from my boss."

Employees face dilemmas of whether, what, and when to tell co-workers and employers. Infertility treatment can impede relationships with employers and co-workers, exacerbated by stigma, shame, and secrecy. Silence can feel dishonest. "I have no idea how people go through it when they're working full-time and have to be in an office every morning at 8," Amanda commented. "I was going to the doctor every day to get my levels checked, but had time off."

Patients may require prolonged respites from jobs, furthering strains. "A lot of women need to take time off of work, a prolonged maternity leave," Jennifer observed, "which does not always count as a disability."

In sum, stresses arise not only from infertility treatments themselves but from the psychological and social implications, including costs, uncertainties, failures, and loss of dreams. These strains, related to invasive procedures, self-injections, side effects, providers' hype, and ongoing costs, can aggravate one another. Prospective parents wander through novel terrains without clear maps, knowing their desired destination but finding themselves on new, bumpy, and muddy roads or dead ends.

Ticking biological time clocks and high stakes compel many patients to do everything they can, leaving no stone unturned, to avoid any retrospective regrets. The odds of success can be slim—less than 1%—but the payoff priceless.

These ordeals can hamper marriages, friendships, and jobs. Communication about these issues can help but prove difficult. Nevertheless, as we will see, countless individuals manage in various ways to cope and proceed forth.

15 | Choosing Supports

"My most meaningful relationships," Yvonne, the Jewish Philadelphia psychologist with Emanuel syndrome, said, "are now online with friends in other states. Five or six of us found each other when we were researching PGD on a message board. We've stayed in touch all these years. They're a huge support."

Many patients have learned to restrict insensitive relationships and seek more positive support from others. "We have dropped or deprioritized some relationships," Yvonne continued, "and given ourselves permission not to go to family events. My mom is always harsh on the phone. It's easier just to have quick calls, and not try to get more out of her. Learning *that* sooner would have helped. After a friend just kept talking about us being child-free, comparing herself to us, I told my husband, 'We're not getting together with her anymore! I just can't stand it.' Some people are aggressive, and bring things up to get a rise out of you. If you're going to go through an IUI or IVF, you don't want those other weights on your shoulders. But many women push themselves, going to Mother's Day every year, noticing there is no picture of them on the wall with the grandkids. I wasn't confidant enough to say, 'We're not doing that.' . . . Setting limits in how much you're willing to discuss your infertility with other people is key.

"People say to my husband, 'So when are you going to have another kid?' It's lighthearted. They don't know our history. But you have to come up with a way of answering. My husband will say something funny like, 'If we had kids, we'd *really* have our hands full,' just to get the topic away. Coming up with a way of containing those conversations, even the benign ones, is helpful."

"Women should ask themselves: What kind of support feels good to you?" Yvonne suggested. "Do you like it when somebody emails or calls you? How often would you like a particular parent or sibling to ask you about it? What could someone say as an icebreaker to check in about your infertility and loss? What's the next family event that you dread? How would you ideally solve that? What if you arrive an hour late?

"Books on miscarriages have chapters 'For women's families and friends on what they can do': explanations of what it feels like, and very clear dos and don'ts. Families and friends should find out how to check in. Good conversation starters—like, 'I know you may not want to talk about it right now, but I have been thinking about you a lot. I know it's a tough time'—allow the woman the choice whether to talk about it. Maybe women could give an info sheet to people on how to be supportive: ask the woman what feels good to her about checking in. Women differ. Say: 'Thank you for giving me the handout. I care about you. Do you want me to stay in touch with you about this? To ask or not?' Provide food—it's huge. A woman going through infertility and loss can use meals brought to her, just like anyone else with a medical condition: something concrete, so they have to think about one less thing. Be mindful about what family gatherings or other social situations might feel like. Obviously, you don't want to *not* invite somebody—that makes the person feel worse. But be mindful about what those traditions might feel like. Give the infertile person . . . latitude to come late or not attend. I found it helpful taking walks with a friend also undergoing IVF. I wouldn't have gotten any exercise otherwise. Self-care is also important because this is a marathon. Support her self-care and limits. It's about having a sixth sense, like when somebody's spouse dies. You never really know what they're going through, but you're prepared. A particular time of year might be easy or hard. They might be having a great day or fall apart. Love them no matter how irrational it might seem."

Patients thus struggle to cope and strategize in several ways—from intra- to interpersonal, from individual to highly social, from being proactive and trying to manage expectations to seeking various forms of formal to informal support and facilitating assistance from family and friends.

Psychological Responses

"I'm not the type of person to sit and cry about it," Wendy explained, "but to go out and get 'em. It hurt, but to cry about it isn't going to get me very far. Keep your head up, and hope and pray it works."

Given these challenges, many patients strive to alter their attitudes or perceptions, reframe their predicaments, and feel grateful for whatever steps they have achieved. After her second miscarriage, Sally, the Maryland website manager, "actually felt a little better. . . . Even though it didn't work out—I kind of knew it wouldn't—I felt, 'At least I got pregnant!' If I can stick with this, it will eventually work—naturally or through IVF. I'm more resilient than I thought. Women have been through even worse: they used to have eight kids, just so two would survive.

"I tell women at their low point: 'We can handle it. You might have short-term pessimism, but something will eventually work and you will be happy with the outcome.' Once I had my daughter, everything dissipated. All of a sudden I felt, 'It was four years of hell, but worth it.'"

Wendy and Sally eventually had healthy children and thus encourage hope more than do patients who have not yet succeeded—further highlighting how final outcomes can profoundly mold views of these journeys. Humans generally have "recall bias," with final outcomes determining retrospective remembrances of events.[1]

Nonetheless, as Yvonne suggested, patients who have not yet prevailed can still benefit from being proactive and taking as much responsibility as they can for these journeys. Patients commonly seek to educate themselves. "You kind of become doctors in the long run, but without all the schooling," Cathy, the Alabaman undergoing PGD, said. Patients must negotiate medical as well as logistic and bureaucratic aspects of treatment.

"You have to be your own advocate," Isabelle, the pregnant Connecticut insurance office manager, stated, "learning the process, the insurance, the code words and doctor terms, to know what is going on."

Such concrete steps can reduce helplessness and aid emotionally. "Having my injections to focus on, and sixteen things to do that month helped," Amanda said. "Injections every single day, something that was part of the process of conceiving a child. It was really hard when all of it stopped. When I did the last shot, I thought, 'I need something to do now.'

"I was 12 or 13 weeks pregnant before I let myself get excited. Initially, I was excited when I got the positive pregnancy test. That quickly turned to worry, because of my past miscarriage and chemical pregnancy." Patients try to manage their expectations and avoid becoming too optimistic, afraid of "jinxing" the pregnancy.

Understandably, many patients want definitive answers and directions, but these often prove elusive. "A lot of people are overwhelmed and fried. They just want someone to tell them what to do," Helen observed. "'Go here. Do this. It's all gonna be OK.'"

Back-up contingencies also assist—anticipating alternatives if treatment fails. "I try to help patients have a plan B," Suzanne said. "A lot of patients resist that. They feel they're giving up. But we reframe it for them. A backup plan can give them courage to continue."

Responding to Insufficient Prior Supports

Patients respond to inadequate support from family and friends in several ways—from trying to improve or limit these interactions to seeking new supports, whether from fellow patients or psychotherapists.

Prospective parents cannot always avoid unhelpful families, friends, and co-workers but can try to educate or distance them—though doing so can be tricky. Over time, these outsiders may come to understand fertility treatments better but not always.

"The only way we're going to overcome that ignorance is with education," Francine, the Miami legal secretary, said. "If patients aren't willing to share their stories, it's going to be very difficult to educate others. There's a stigma associated with [infertility]. It's a 'bad' thing."

"Outsiders should just say, 'I don't want to pry, so I'm not going to ask a lot of questions, but if you have some news, I would love to be the first to hear.' If they want to talk, you're there to listen." Fortunately, taboos about infertility may be lessening among many groups.

Gay and lesbian patients and single-parents-by-choice cannot hide the fact that they used IVF to create their biologically related children, but other patients face dilemmas about whom to tell. "Some women are incredibly open about their infertility treatment. People ask, 'Did you have your twins through IVF?'" Amanda noted. "But others are incredibly private." Numerous patients simply decide not to discuss their infertility and treatment with certain others but then face questions of exactly whom and what to tell—whether to reveal infertility, treatment, pregnancy, miscarriage, or tests for genetic mutations.

A few patients readily inform others, but they constitute the minority. "I'm proud of how hard I tried, and what I went through to have a child," Karen, the physician and patient, said. "I'm not ashamed and don't think there's a stigma."

But most patients, especially if they are not also healthcare providers, feel otherwise. Shame, stigma, and disappointment lead countless patients to mention their infertility and treatment to few, if any, family members; and consequently, derive little support from these individuals. "I haven't told my family about it," Tammy, the young graduate student, said.

Early disclosures can later backfire. "I told everybody!" Sally said about her first pregnancy. Later, she learned that the fetus had cystic fibrosis (CF), prompting an abortion. "It didn't occur to me that I should not spread this joyous news" about the pregnancy. Yet when she learned of the fetus' mutation, and chose an abortion, she faced "a horrible situation. A friend returned from vacation, and said, 'Oh, I thought you'd be a lot bigger by now!' Another family member sent a gift after we had already had the abortion. Then, I'd forget to tell someone. Not only was I personally grieving, but I had to deal with everyone else's responses! I saw this fear and misery in others' eyes, and was not prepared for it. They tried, but what could they do?"

Abortion due to a fetal mutation can also be difficult to reveal because outsiders may be pro-life. Yet silence can block the spread of vital information. Family members may not learn they are at risk for inherited problems. Cathy, whose daughter died due to a chromosomal abnormality, suggested to her cousin and his wife that they should ask the doctor about it, "as a precaution, so they don't have to go through it. I don't know if they did."

Secrecy about infertility, mutations, and miscarriages can hinder relatives from taking preventative measures. "My husband's cousin's embryos had tested for trisomy-22, and we got mad at her," Cathy said. "'You could have told us!'" But "People don't like to talk about it."

New Forms of Support

Perceived shortcomings of families and friends frequently lead patients to seek new social supports—from in-person to online, from formal to informal, from individual psychotherapy to support groups and large organizations.[2]

Psychotherapy

Psychotherapy can ameliorate these stresses,[3,4] but many patients who would benefit from it do not receive it.[5] Many assisted reproductive technology (ART) doctors routinely rely on mental health providers (MHPs), who frequently come to specialize in infertility. Yet stigma generally about mental health problems and treatment impedes many patients from receiving care. Only half do so.[6]

Infertility patients may see an MHP on their own—referred by various doctors. Some reproductive endocrinology and infertility specialists (REIs) require or strongly recommend that all IVF patients visit a psychotherapist. "We've got MHPs just for regular IVF because the process is stressful,"

Jennifer, a physician at a Northwest medical center, said. "A lot of IVF patients just go kerflooey with the process. . . . We try to help them think through this so they don't make bad decisions."

Physicians differ, however, in whether they require such referrals and, if so, when. "We use MHPs for all third-party reproduction," Calvin, the southern California REI, said. "At least, we offer it to everybody. We insist on it for some people—any surrogates or egg donors." Calvin, in fact, has a psychologist in his clinic: "We use her so much she rents space in our office!"

Patients may consult with their own past therapists, facilitating the process. Sally, for instance, "was fortunate to have a therapist from before, whom I was already comfortable with. That helped, because a lot of times counseling depends on whether you click with the therapist."

Patients may prefer to see an MHP *outside*, rather than inside, a clinic because of fears of repercussions if they complain about staff. "Patients who have access to a therapist within their clinic may see external mental health providers," Suzanne observed, "feeling weird, because of confidentially, talking about clinic procedures, about not getting calls returned or returned quickly enough, feeling like they're a number."

Patients may also not want to admit their apprehensions and trepidation to clinic staff. "They want to please the doctor, and not show they're worried," Helen said, "because otherwise they fear the doctor's going to think they shouldn't proceed, or the doctor is going to implant the embryo in the wrong place."

Functions of Mental Health Professionals

"I find out what patients are worried about," Diane said, "assessing pressures, unrealistic expectations, and beliefs that IVF and PGD are 'magic bullets.' I have to be the voice of reality, and say, 'They're not. They're good, but not perfect.' Younger patients, especially, believe they can have a 100% chance of getting pregnant with it. My role is dealing being the 'What if?' person: 'What if it doesn't work? If you don't have good embryos? If you don't make enough embryos? If they're all the wrong gender?' "

Psychotherapists can play vital roles, filling various functions, depending on the particular case,[7] helping patients emotionally and cognitively in anticipating and reframing experiences.

Therapists can serve as "reality checks," setting appropriate expectations. "The psychologist walked us through every step of the medical and emotional process and asked about my expectations," Amanda recalled. "What emotions I was feeling—the anger, inadequacy—what support I had. What if it doesn't work? How was I going to protect myself emotionally? Who am I going to lean on? She talked a lot about my marriage, and how we related to each other—points of disagreement and of incredible support—whether I was sharing the process with family and friends.

"The nurses tell you the procedures—how much medicine, how to do the injection—but the psychologist talked more about the emotional stuff. After the whole procedure, you start progesterone shots, and get this huge box of medicine with your own sharps disposal container. It's overwhelming. . . . Why do I have to keep giving myself shots?' The psychologist explained: 'It's supporting the pregnancy.' So, it's not just 'Here's what you have to do' but 'Here's *why*. This is what your life is going to be like for a while.' It was helpful. I just wish it had happened sooner." Patients are often very grateful for therapists anticipating obstacles and assisting in different ways than other staff.

Therapists can also help handle treatment setbacks that may trigger broader blows to self-esteem. "I felt, 'Why is this happening to me?'" Roxanne said. "'Why am I going through this? Why me? I'm a good person. I haven't really done anything terribly wrong.' I didn't have an answer. I was very doubtful. A counselor put those feelings into perspective: they're normal, but not helpful. Better to think, 'What am I going to do?' Have a game plan, think positively. If you want to have a child, one way or another, you will. Don't think it's not going to work, but 'I'm doing this because I believe it's going to work.'"

Still, altering unrealistic beliefs can be hard since patients want to believe that every intervention will succeed. "I walk a fine line," Suzanne, the psychotherapist, said, "because patients don't want to hear that it might not work. They can get very angry about that. They are seeking a tremendous amount of reassurance."

Ambivalence About Psychotherapy

Therapists evoke mixed feelings.[8] Patients may avoid or not recognize psychotherapy's potential benefits and, along with many physicians, remain wary. As Francine, the Mexico-bound legal assistant, commented, "More patients probably need therapy, but say 'Why do we need counseling? Our marriage isn't in trouble. We're just trying to have a baby.' Yet so many issues come into play that patients just don't know how to bring up and talk about. Infertility is a major life crisis—like a diagnosis of cancer. . . . If you were diagnosed with cancer, most people would tell you to go to therapy to work through issues. For *me*, it's been a lifesaver."

Insurance, though, often covers little, if any, psychotherapy—at times only if the mental health services are *not* due to infertility. Francine's insurance covers mental health issues "as long as they don't code it as infertility, but as depression or general anxiety."

Men, in particular, may be leery of psychotherapy. "A lot of people never cross the door of a mental health professional's office," Helen commented. "Once patients have crossed it . . . they're open. Patients who are not psychotherapy-friendly don't want to see a mental health professional, because *that's what crazy people do*." She and her colleagues try to normalize the experience. "I say, 'There are expectable stresses and psychological reactions to a situational problem. This is a normal experience. Not, 'You've got mental health issues' but 'You're going through a situational problem.' Though it can be both."

Prospective parents might benefit from mental health treatment but resent the cost. "A lot of patients want your help, but don't want to *pay* for it," Helen concluded. "Patients feel, 'This should be free. Doctors should provide this.' They want to have *the nurse* be their therapist." Yet even when it is free, patients may not seek it. "I got my REIs to pay for a couple of support groups," Helen added. "But patients wouldn't come, so the doctors stopped."

Physicians also vary in how much they encourage patients to pursue mental health interventions, and may instead focus only on strictly "medical" issues, seldom drawing on psychotherapists. "Some doctors are open and interested in using us more, but most are not," Helen observed. "They have a very different worldview—a medical model—things are quite straightforward. You go by the books. Doctors as a group are very conservative. It's big bucks. The mental health professional group has met incredible resistance. There are all sorts of political forces." Physicians dominate the infertility field and may insufficiently advocate for psychological services.

Nevertheless, physicians and their staff could potentially assist patients more in obtaining mental health assistance. "There should be some sort of support group or network recommended by every doctor's office," Amanda, the Delaware high school administrator, stated. Yet providers may not encourage or facilitate such groups, seeing these as beyond their purview. "I told my doctor they should have a group, and that I would be willing to coordinate it," Amanda reported. "But he said that people who get pregnant don't want to continue giving support afterwards."

Physicians, too, may resist paying for it. "Patients don't really want it, so most physicians do not want to support a full-time mental health professional and may resist psychotherapeutic input," Brenda, the psychotherapist, explained. "We think that patients need the support, but they don't. Or if they do, they don't want to pay for it. But I have to earn my keep. Most practices have an outside consultant who comes in, or they just refer patients out to the consultant's office."

Nonetheless physicians could easily assist, without incurring any costs, by better presenting and framing the benefits of mental health services. "Doctors could enhance a referral just by saying, 'many patients find it helpful, because treatment is stressful,'" Suzanne, the psychotherapist, suggested, "as opposed to implying that there's something *wrong* with you: 'You're neurotic and need to see a mental health professional.'"

Being Mental Health Professionals

Over recent decades, more psychotherapists address infertility issues. "There used to be only a handful of us around the country," Brenda said. "More and more are out there."

MHPs, in general, are seeing patients with fertility issues and may come to specialize in this area. "Some psychotherapists in the community recognize the limits of their understanding, and consult or refer to colleagues," Suzanne said. "Unfortunately, many therapists aren't very well versed in fertility issues, and don't understand why their patients are having such emotional difficulty with this. It gets glossed over.

"Some MHPs, particularly those of my generation who went through the women's liberation movement, may not fully understand the deep emotional repercussions of not being able to have children," Suzanne continued, "a deep need to continue their line that hits them now, later in life, more than earlier. A woman's career was supposed to be most important. Motherhood was a way to keep women in the kitchen, not in the workplace. Younger women don't necessarily think that way." Outside of infertility, psychotherapists, affected by their own prior views and informed by feminists, may emphasize the importance of women choosing careers over motherhood.

Therapists who have used ART themselves occasionally share their own experiences if they think these might benefit a patient. "It's very helpful when patients know they're talking to someone who has experienced it," Ginger said. "I become much more credible. I don't go on and on but, depending on what's happening with the patient, say, 'I had two kids: one through reproductive technology and one through adoption.' Later, they might ask, 'Can you tell me more about adoption?'"

MHPs find these efforts very rewarding, personally. "I love working in this area, and am glad I'm a part of it," Helen declared. It "feels more important to them than anything. Infertility is awful to go through. For years, people are stretched way beyond their limit. It gives me an incredible source of meaning in my life. I feel quite privileged helping people through it, a great sense of satisfaction helping people come to terms with limits, with the absence of guarantees. It's like AA's Serenity Prayer: when do you stop trying to change something, and start accepting? Life is so often like that, figuring out what's in your control, and what's not. Sometimes patients get mad at me, because *somebody* has to be blamed. The work can be very hard. But mostly, it's very rewarding."

Support Groups

Not only individual one-on-one psychotherapy, but support groups, too, can assist. Whether formal or informal, face-to-face or online, such groups can aid patients but also pose challenges. As Francine, who has undergone four unsuccessful IVF cycles, said, "I'd recommend that women find friends also undergoing it, through support groups, churches, educational meetings, or community or other organizations, who understand what you're going through, speak your language, and know what you're talking about, without you having to explain all the acronyms and tests. *It's its own little language!*"

Many prospective parents now join such groups for the first time in their lives. "I've never been a 'support group' kind of person," Roxanne, the Jewish Michigan marketer, said. "But I learned about an organization from a pamphlet in my doctor's office."

Nonetheless, other patients receive little, if any, information about such services. "In a perfect world," Helen stated, "there'd be support groups to face the stresses, risks, and lack of guarantees and help patients not feel alone."

Face-to-face and online groups each offer pros and cons. The Internet can facilitate such small networks but has its downsides. "When I started, I only had one friend from another part of the country who did IVF," Wendy recalled. "She was the only person I spoke to, and she was trying her best to help me. Through the years, I've found other friends, thanks to Facebook—a whole bunch of people who have gone through it. When you 'like' an infertility patient group, people then know. Some people are very open with what they post on their Facebook page, putting it on their status. Others are very quiet."

Yet she and others feel that local in-person groups offer certain advantages and that physicians' offices should establish more of them. "Each facility could get a group together of past or current patients, so that patients at least know what they're going to go through," Wendy added. Despite her online friends, "when you don't have somebody local, and have questions that the doctors aren't answering, a support group would help. *Online is not the same!* You can type whatever you want online, but you read and hear differently. When people talk, you can hear the *pain* in their voice, the sarcasm, the happiness. Online, you are reading it in *your* voice."

Internet support offers certain advantages, creating formal and informal subgroups of individuals confronting similar situations, especially about relatively rare disorders, and allowing anonymity, abetting candor. "With women you don't see, you could be even more open," Sally said. "I use my real name on the discussion forum, but a lot of women do not. We don't know each other's husbands, friends, or families. So, you can get support without feeling you're compromising yourself. In a way, it is better than talking to a friend, because the group can understand where I'm coming from.

"It's easier for someone who's had no context for this to be judgmental" Sally continued. Patients, who are otherwise strangers, can in certain respects help more than outsiders. "My husband says, 'Why aren't you over this already? It's no big deal. Let's just try again.' But months later, I still feel totally devastated. Other patients completely understand. That validation is really helpful." Fellow patients can empathize about feelings of loss and failure that even spouses may not fully grasp.

Specific niches emerge. "Separate boards exist," Sally said, "for women trying naturally after repeated miscarriages, and women in their 20s, 30s, and 40s or trying IVF."

Through websites, women contact Sally about specific embryo testing issues she has faced. "They message me: 'I'm trying to figure out what to do.' I'm always glad when a woman contacts me who's dealing with CF and wants to talk, because that helped *me* so much! Once you've been through it, you always have the *heart* for it, which someone else might not have. They have to make the decision themselves, but I can help them understand the pros and cons. They say, 'My mom and sister don't understand why I keep going back and forth on these options.' I know what it's like to terminate two pregnancies because of a diagnosis. People are worried: 'What if it happens again? Will I survive?' I can provide some comfort: 'You will survive it! *I* did.'"

Still, over time, individuals and groups and their respective needs can shift. A patient's particular needs and commitments can change. Sally "was very involved with one site for people doing PGD. But fewer people are on that site now. I still keep in touch with five women, all interested in PGD. It's been six years! We now all have kids. Pretty remarkable!"

More informal patient groups, lacking professional involvement or fees, have mushroomed both in person and online but can pose difficulties, too. "Free peer groups attract a lot of people," Helen said, "but I hear a lot of complaints. Things feel very out of control, or somebody dominates because she's in terrible shape, and the group is unable to manage that. Or people get vicious: 'You don't belong on this chat board!' Patients feel abandoned when others get pregnant and leave. It doesn't feel supportive."

Helen highlighted the specific skills offered by colleagues who "know what they're doing: how to create a sense of safety, a supportive atmosphere. The people who run non-professional groups are not trained to do that. Some patients love going—it's free, they meet people, and find their own way." MHPs are schooled to handle inter- and intrapersonal conflicts in ways that infertility patients may not always appreciate. Still, she thought that "most" patients "aren't going to anything."

Online groups can especially aid patients confronting rare genetic mutations and diseases, who may not know anyone else affected by the particular mutation and for whom interactions with fellow patients can provide unique benefits. Cathy, from Alabama, had "the only case in my state. So, we don't have that connection with anybody. But we share our issues online."

Particularly for a rare disease, such information can be crucial since physicians may have little, if any, information about it. Patients often come to know more than their doctors do. "The daughter of the mom who runs our online site was the first case in her hospital," Cathy continued. "She got all the information herself. There isn't a lot out there. Doctors could use more." Patient self-education can thus be pivotal.

Even after the children are born, these new friendships can endure. The Internet can help sustain these interactions, allowing initial anonymity that can then shift over time. "It's funny: you wind up knowing intimate details about someone's life, but might not even know their last name," Suzanne, the psychotherapist, mused. Yet these connections can be robust. "When one of us went in for her second round of IVF, we met her for coffee right across from our doctor's office, and waited for her until she got out. It was nice."

Prospective parents frequently form close bonds with one another, constructing their own informal, rather than formal, social systems. "I first met my best friend in the doctor's waiting room," Amanda said. "We had seen each other there twice. She said, 'This two-week wait sucks, doesn't it?' I said, 'Yeah.' We've been fast friends ever since. We formed our own support group. Three of us going through IVF became friends: 'Here's what I'm going through. I don't know what you've been through, but I'm guessing maybe something similar. Any advice?' Most people are pretty open about it."

Fellow patients can best understand and normalize the frustration, anger, and jealousy. "They know we're not hateful people—it's just part of the process," Amanda related. In talking to patients, "you realize it's pretty common, which is comforting."

Yet patients' needs can shift. Prospective parents can alter their associations to help them most at any one point, but doing so impedes continuity. "Two of us now have twins, but a third person was pregnant with twins and lost them," Amanda said. "*It was too painful for her to stay in touch with us.* I understood, though some people get offended."

Over time, support groups and networks can be hard to sustain because patients may fail to become pregnant and/or stop treatment. "The number of support groups goes up and down," Francine reflected, "as people resolve their infertility, and decide to be a support group leader, or to move on. Instead of championing the cause, as with breast cancer, a lot of people would rather forget it ever happened. If you end up with a baby, you say, 'I'm done. . . . Let's move on.' People who decide to live without a child are not part of an organization focused on having children. It's not like breast cancer: if your mother dies, you're forever going to walk on the Breast Cancer Cure day."

Prospective parents may thus self-select into different subgatherings, entering or quitting. As with psychotherapy, patients may also not want to pay for membership in a group. Free treatment can enhance accessibility. "If the group wasn't covered by insurance, I'd have said 'No,'" Amanda admitted.

In general, individuals also vary in their motivation to participate in groups. "Many women are not interested in being involved in a community," Francine observed. As Helen said, "People have busy lives, and fear groups and organizations."

Men, in particular, may feel awkward discussing certain topics not only with individual psychotherapists but with fellow patients. "It's worse for men because they don't speak to other men about these issues," Joe commented, "unless they have complete trust. Men don't have the types of relationships that women have. It's much more difficult to express what they're doing and feeling, what type of exam they had."

Other Kinds of Support

Many patients pursue alternative, non-biomedical approaches—from herbs and acupuncture to yoga. Sally's clinic sponsors a "mind/body workshop, which is essential for any woman going through IVF—learning deep breathing, meditation, and relaxation techniques." Such balms often help. "Almost everyone would benefit from stress reduction," Valerie, both a therapist and a patient, agreed. "I saw a hypnotist to find ways to stay calm and unstressed. It doesn't help get you pregnant, but helps preserve your sanity."

For Yvonne, "acupuncture was a nice way of having a person in your life, always kind to you. It's relaxing. Whether it actually improves your eggs almost doesn't matter. . . . Everything can't be about infertility."

Patient Advocacy Organizations

Formal patient advocacy organizations have also spread, not only offering information but enhancing political, public, and research awareness of these issues. Still, these entities can face tensions concerning whether to focus on education, research, service, or individual patients and can evolve over time.

Particular religious or other communities can form their own support organizations and even contribute to paying for services. Sam, for instance, works for an organization that aids Jewish patients. "We act as facilitators between the doctor and the patient, referring couples to genetic counselors or geneticists to explain options. If the patient doesn't understand something, we can help, regarding prenatal testing, IVF, or PGD."

Community organizations can also raise money and help members afford treatment but may face difficult trade-offs, given limited resources. "We cover PGD and CVS [chorionic villus sampling], if they are not covered by insurance," Sam explained, "if the test is not covered, or the physician doesn't participate in the HMO [health maintenance organization]. We have community fundraising, dinners, auctions, mailings."

With patient advocacy organizations, prospective parents have to decide whether and in what ways to be involved. When Francine got diagnosed with infertility, she "volunteered for an organization. I had always been involved in volunteer endeavors. Maybe this would be a good distraction and a way to meet people in a similar situation. It bloomed from there."

Patients thus struggle to cope with these technologies' ethical, social, and psychological challenges and respond in various ways. Not everyone is comfortable discussing their infertility. Prospective parents often try to "stay positive" and "be proactive" but also seek new supports—from psychotherapy to groups and organizations, whether formal or informal, in-person or online.

Yet challenges surface—including paying for mental health services and patients "moving on" over time. Would-be parents may be wary of psychotherapy or support and/or not receive information or encouragement about these possible aids. Enhanced guidelines, insurance, and patient, provider and broader public education can reduce these obstacles.

As we shall see, patients respond to the challenges posed by these technologies through not only psychological but broader, spiritual, and religious perspectives as well.

16 | "Meant to Be?"
CHOOSING SPIRITUALITY AND RELIGION

"What did I do in my life to deserve this?" Amanda, the Delaware high school administrator, asked. "Some days, I figured maybe God did not want me to have a baby. At moments, I was definitely ticked at God, and thought: 'Just let me get pregnant, it would be easier on both of us.' I prayed, begged, pleaded, and tried not praying—the whole spectrum. I like the tradition of religion, but don't necessarily believe in religion. I believe in God. But if there is one God, there would be one religion, and no matter what religion you are, it should [unify], not [divide] people.

"After six failed IUIs, I thought: maybe my body isn't meant to have a child. So, deciding whether to go through one more procedure was really difficult. I don't have a relationship with my mother, so I wondered if karma had come back to bite me. Logically, I knew that was not it. I have endometriosis, so *that* was a contributing factor. Mostly I kept it in perspective. But emotionally, what had brought this upon me?

"People have told me, 'If God wanted it to happen, it would happen. It's God's way. If you're meant to get pregnant, you'll get pregnant. If it was meant to happen, it would happen on its own, without any kind of assistance.' But if I got pregnant, they'd say that was great—*meant* to be. Some people just believe what's meant to happen is meant to happen—don't *mess* with it.

"I took it as 'a *sign*' when three eggs fertilized well. But the day of the transfer, the embryologist suggested that one of the three embryos wasn't strong. 'These *two* have a good shot.' I thought, this is what was meant to be!

"Yet by the time we got to IVF, I just felt that maybe it wasn't meant to happen. As a protective measure for myself, if I didn't think it would work, I wouldn't be disappointed when it didn't. I tried that strategy, but it wasn't successful."

In confronting these technologies and trying to make sense of these unpredictable reproductive journeys, Amanda and many others draw on long-standing established religions but also on more non-specific spiritual and metaphysical ideas, wrestling with questions about cosmological order and fairness. At times, they seek "omens," either good or bad. If they expect failure, they won't be as upset afterward. These questions of whether having a child is "meant to be" can in turn influence treatment decisions. Yet patients may be unsure about how to view, weigh, or question these notions.

All major religions contain myths about the creation of life—including of human beings—which indeed seems miraculous, an act of God. From Sarah in the Bible through 12-step programs, people have appealed to "higher powers." Countless religious practices, rituals, and beliefs seek to remedy infertility. Many religions also promulgate strong beliefs about when life begins that can shadow and impede infertility treatment.

Patients, family, and friends can disagree about even non-specific metaphysical notions of whether children are "meant to be."

With healthcare decisions, more broadly, religious and spiritual views affect various provider and patient decisions. When treatment is futile, Catholic and Jewish internists, for instance, are less likely than Protestant ones to withdraw patients' life support.[1] Forty-five percent of gynecological oncologists think their religious/spiritual beliefs affect their medical decisions, 87% consider patients' religious/spiritual beliefs when discussing end-of-life issues, 68% feel their spiritual/religious beliefs help them in death with feelings about death, and 62% feel that their spiritual religious beliefs comfort them as oncologists.[2] Among obstetrics and gynecology specialists, those who attended religious services twice a month or more or were Catholics or Evangelical Protestants or thought religion was important were more likely to oppose offering patients contraception.[3]

But do religious and spiritual beliefs ever impact infertility patients' decisions and, if so, how? Remarkably, these questions have not heretofore been examined.

Many prospective parents turn out to invoke religion and spirituality, to cope and/or make sense of their predicament. Patients confront not only religious but related spiritual and metaphysical issues, as well as religious objections to certain treatments, from clergy, family members, and friends. Religion can thus provide solace and social support but also impede care. Opponents of ART argue that children, if intended by God, would already be born. Patients generally resent such comments. Yet between and within religions, clergy may view these objections differently. Patients may consequently follow religious prohibitions, change, or proceed with treatment, at times secretly.

Religion in Coping

Patients range widely from rejecting to pondering to adopting various religious perspectives. Sizable numbers feel that the cause of their infertility is not at all metaphysical but simply biological, even if unknown. "Religion has nothing to do with whether or not I'm able to have a baby," Francine, the Mexico-bound legal assistant, argued. "I ended up with infertility. My sister ended up with cancer, probably due to something in our environment. There's nothing I could have done. I had endometriosis from age 13. I don't feel any self-blame. Things just happened for me when I was ready for them to happen. They happened when they happened for *a reason*, because I wasn't ready at another time to explore those options."

Yet Amanda—who had asked "Why me?"—and many other patients wrestle with religious questions. John, from Texas, tries, "to console my wife, telling her 'it will happen. You just have to have a little faith. Things happen for a purpose and are pre-written.' My pastor just brings it back to being in God's hands." Patients may invoke metaphysical beliefs in seeking solace and/or assessing why they are infertile, what treatments they should choose or avoid, and how to view ongoing failures. Religious organizations can also provide beneficial social support. Cathy's parish priest "used to visit once or twice a week. It was good to have somebody there."

Why Me?: Metaphysical Notions of Fate and Explanation

Many prospective parents, such as Amanda, see their infertility as unfair and seek some larger explanation. Patients often consider or adopt metaphysical perspectives, looking for something to blame, even if they realize that these beliefs are not wholly logical. Patients often ponder notions of fate, karma, or a larger sense of purpose in the universe—that having children is or is not somehow "meant to be," fated, and preordained or not. Still, who or what is predetermining these outcomes, and how, is generally unclear.

Many patients conclude that God did not will their infertility but that some larger metaphysical reason may be involved. "I don't believe God wanted this," said Roxanne, the Jewish Michigan marketer who underwent eight years of ART, through which she now has a son. "Because there is no reason for it. It's definitely unfair and stinks. Why does it happen to certain people? I don't know. I don't think anyone has that answer."

Beliefs in fate may be related to broad notions of cosmic justice, unlinked to religion per se. "I'm not a very religious person," Wendy, who was raised Catholic, said. "I just take life as it comes. *If it's meant to be, it's meant to be.* If it's going to happen, it will happen."

Roles of Religion in Choosing Procedures

Religious and metaphysical beliefs can affect whether patients pursue these interventions, and if so, which . Clergy, families, and friends may sanction certain procedures but oppose others and vary widely, even within any one faith. Patients may not have strong religious values but nonetheless see IVF as fulfilling what God wants. Patients may see IVF as "natural," as John said, "just doing what God intended."

Other patients confront stauncher moral or religious objections either to ARTs as a whole or to specific procedures, and consequently do not seek treatment. Wendy, Amanda, and other patients who are "not very religious" may nonetheless encounter other people's religious disapproval, which can require considerable pluck to counter. "People are ignorant about IVF. The religious ones feel you're messing with God," Wendy stated. "When I had my first IVF, my boss was excited, but was also a religious freak: 'Don't you feel that you're tampering with what God [wants] you to do?' I said, 'I'm not very religious. If there's a God, there's a God. This is *my body.* This is what *I* want. I will do what it takes to get what I want.' She just looked at me and said, 'But you're tampering with what *God* wants!' I said, 'If God wanted me to *not* have a child, believe me, He'd let me know.' She said, 'Don't you see this is His way?' I said, 'No. If He wanted me to not have a child, He would have made me sterile. He wouldn't have given me any ovaries.'" Patients who in fact consult ART clinics have generally already come to terms with these qualms for themselves.

"Religious views affect patients a lot," Brenda, a psychotherapist, observed. "But a lot of that we'll never see in the clinic. They take themselves out and don't set foot here."

Patients may struggle with their desires for a child versus their religious views, with varying outcomes. "The patients we see figure out a way to work around them," Steve, the Virginia reproductive endocrinology and infertility specialist (REI), found. "It's sad when people we know we could help don't pursue ARTs for religious reasons."

Many religious patients and their clergy feel that only certain relatively "low-tech" procedures are acceptable—even if less effective. "Extremely religious Jews have to talk to the rabbi about everything," Diane, a nurse, observed. "Their rabbis will say, 'No, you can't do this.' The couple will have a diagnosis, but the rabbi will only let them do *low-tech* interventions. The rabbi ties the doctor's hands. You watch the couple go through disappointment after disappointment. It's not our place to tell them they should believe or do something else."

Depending on the specific details involved, religious leaders may oppose a variety of procedures, from small to large. "When told they need IVF, some Catholics will say, 'I'm sorry, we can't do that,'" Diane explained. "Many Muslims don't use donor eggs. Rabbis may also say 'No' to a particular treatment, though the couple would benefit—donor eggs, for instance."

Due to religious concerns, patients may want to avoid creating extra, unused embryos. "Occasionally, patients want us to limit the number of eggs we fertilize because 'life begins with fertilization,'" Nicholas, the New Jersey REI, said. The number of embryos that a woman will need cannot be predicted.

Clergy can also disagree with providers concerning whether and how much to reduce the number of fetuses in a pregnancy. One rabbi, for instance, permitted a woman, pregnant with quadruplets, to reduce to three—instead of only one or two fetuses. Sam, a patient advocate, "felt horrible" about one case: "The patient transferred three embryos. I did not discourage her. She got pregnant *with four*; one was twins. The rabbi said he 'will *only* allow reduction to three.' But triplets are not a good outcome. She's now at 14 weeks. She's going to have a tough pregnancy, and need a lot of help." Embryo testing appeals to pro-life patients at risk of genetic disease. But while Orthodox rabbis often permit it, clergy in other religions may not. Some Orthodox Jewish patients, Sam thought, "will not undergo an abortion, even if the fetus has a severe mutation."

Disagreements Within Religions

Within any particular religion, leaders may clash about not only abortion but amnio and PGD since the moral considerations regarding various approaches can differ. A rabbi, for instance, may oppose IVF and PGD since they require collection of sperm through masturbation. "How does the husband produce and bring the sperm, and is he allowed to produce sperm that way?" Sam asked. "Most IVF centers allow men to bring it in a special condom that doesn't have spermicide. But some rabbis don't allow that kind of condom. One prominent rabbi will not allow it if the couple is fertile. He allows it with a condom if someone is *infertile*, whose only option to have kids is through IVF. But for PGD, the issue is *sick* kids, not just any kids. . . . Thank God, we figured out a different way to get sperm—by retrieving it from the woman a half-hour later. If there's a life-threatening illness, most rabbis will agree to PGD; but what if the couple already has four kids? Should they do PGD, or not have any more kids? One rabbi can tell you: CVS [chorionic villus sampling]. For the same couple, the other rabbi would tell you, 'Only PGD or do nothing—rely on God!' "

Which clergy a patient consults can thus constitute a key decision. "The issue is which rabbi they're going to choose," Sam added. "Once they decide the rabbi, they'll definitely listen to him."

Patients, providers, and others may also misunderstand a religion's perspectives, making incorrect assumptions. "There's a lot of misinformation on what you could do," Sam observed. "When I tell people about PGD or CVS, they say, 'You can do that?' No one would believe it's permitted or allowed in any circumstances. Most people assume that in Judaism you could not do CVS and an abortion for anything, even for Tay-Sachs. People become very judgmental: 'Abortions are killing the fetus.' But many rabbis will allow you to abort a Tay-Sachs child. . . . PGD is such a blessing: you can do it!"

Rabbis may thus differ on specific embryo tests, even for medical conditions, and on the relative undesirability of particular severe diseases. "Which is worse? Fragile X or Tay-Sachs?" Sam asked. "Most rabbis say the babies suffer and die—the diseases are very bad. But one rabbi believes: with Tay-Sachs, suffering ends fairly quickly. It's very sad. But with fragile X, the child and the family suffer for years and years. How do you measure suffering? There are different opinions. I don't think there will ever be a ruling where everyone agrees."

Religious Opposition from Others

Clergy may also block treatment because they are not well informed or do not understand all the medical issues involved. Consequently, patients can end up frustrated and feel forced to choose or reject approaches that may be less effective and/or endanger themselves or their future offspring. Sam described a couple who "went to a rabbi who was very uninformed and uneducated. They had a child die from spinal muscular atrophy"—a severely debilitating disease. The couple could use PGD to prevent the transmission of the mutation to future children. "But the rabbi said, 'Do nothing. I'm not allowing you to do anything.' The couple listened. They have a healthy child, but I don't think it was a smart gamble."

Ill-informed advice from clergy can prompt misconceptions and not serve the patient's best interest. "Some regular OBs and midwives say to Orthodox patients, "We want to do the Ashkenazi panel of genetic mutations that are common among Ashkenazi Jews," Sam explained. "The patient says, 'I don't know what that is. I'm calling my rabbi.' The rabbi says, 'Don't do it. How is it going to help you?'"

Patients "Working Around" Objections

Patients can follow or avoid such prohibitions and frequently wrestle with, but ultimately ignore, these religious qualms. As a physician, Nicholas "has treated Catholics and rabbis. Somehow, people do things privately that they don't talk about."

Patients have "somehow made peace with the religious or ethical conflicts for themselves, because they just want to have a baby," Suzanne said. They weigh their religion against their desires for offspring and choose the latter. "Some have religious conflicts," she continued, "and do not tell their families what they're doing, because they expect they'll be censured." Patients may simply not inform family members, friends, or others who might oppose it. Religion can thus impede even discussions about possible ART use.

Over time, patients may also vary their treatment, following, rejecting, or compromising their beliefs, weighing these against ongoing treatment failures and/or successes. Given the gambles on both sides, many patients vacillate over time.

Clinicians' Responses

Providers are frequently unsure how to interact with patients' religious leaders and beliefs, especially when clergy, not the patients, are making the decisions. "For physicians, it's challenging, because they don't know how to advise patients," Sam stated. "It becomes frustrating when the patient says, 'I'm going to the rabbi who will make a decision for me.'"

Providers may not fully grasp the particular religious issues. "Clinicians have to understand that the rabbi is going to make the decision, and there's nothing they can do about it," Sam said.

Discrepancies between clergy within a religion can further frustrate clinicians. "Doctors ask, 'How can it be that this rabbi allows this, and *that* rabbi does not?'" Sam explained. "That this rabbi says 'Don't do any testing. It's a dominant disease; it's going to kill your family. No one is going to marry your family. It's going to be the curse of your life.' And the other rabbi says, 'Do it: it's only 50%. If she's not a carrier, it's good.'"

Given variations among religious leaders, knowledgeable providers may in fact suggest that patients consult with different clergy. Some couples "will be savvy enough to go *rabbi shopping*, to get one to say 'Yes,'" Diane explained. "We have our list of rabbis and say, 'Why don't you consult Rabbi so-and-so?'"

Religious, spiritual, and metaphysical issues can therefore help patients cope with these technologies and/or affect whether patients pursue treatments and, if so, which. Prospective parents can clash with religious objections from not only clergy but family members, friends, or others and then have to determine how to proceed. Patients may forego care, "work around" objections and/or obtain treatment covertly. Providers may discuss these issues with patients, suggesting consultations with other, more sympathetic clergy. Such interactions may, according to some critics, lie beyond the clinicians' role, but nonetheless assist perspective parents. Providers should never argue with a patient about these issues, but certain patients may benefit from learning about these options and be grateful.

Written thousands of years ago, traditional religious texts do not explicitly address ARTs. Hence, clergy may interpret these documents as applying to ARTs in differing ways. Abortion debates also form a highly charged backdrop that can obscure relevant differences and distinctions among these medical procedures.

Clergy, who end up making these decisions for many patients, may not fully understand these technologies and the trade-offs involved. Religious leaders may oppose certain ARTs but accept others, in ways that patients and providers may not realize. Within one religion, clergy can differ dramatically. Heightened awareness of such possible alternative interpretations of religious doctrines can aid both clinicians and patients. Providers should consider enlightening certain patients about these possibilities. Professional organizations could help providers and patients address these challenges.

Clinicians should recognize how much religious uneasiness can hamper patients, partly due to misunderstandings. Mental health and other providers can also aid patients in confronting questions of "Why me?" and whether children are "meant to be." Similar religious and metaphysical questions confront patients with other medical problems as well but take on particular forms here, given the miracle of bringing new life into the world.

V | Choosing Our Future

17 | Choosing Education

"To be infertile myself, and understand what it was like made me a much better person and doctor," Karen, the California infertility doctor and patient, said. "I am much more emotionally supportive. During my training, I thought that all those women trying to get pregnant were crazy, neurotic. Now, I understand, and treat them much differently. I can now truly empathize—anticipate the pain they are going to face, and help them prepare."

More education and awareness about the challenges of infertility and its treatment can aid patients, their loved ones, physicians in several fields, and the general public. While some patients are savvy about these topics, many are not, and misunderstand key aspects of assisted reproductive technology and genetics. Both the form and content of such education need improvement.

What to Teach

Patients, families, and the public at large would gain from heightened knowledge in several areas, from basic physiology and causes of infertility to specific procedures. "Women and men are amazingly ignorant about how their bodies work," Diane said.

Many women have to learn about infertility on their own—gaining background. "I'm learning it on my own online as I go," Nancy said, "about follicle sizes, what's a good size, how many, and how big follicles start hurting smaller ones." Misunderstandings arise about several specific areas, especially statistics, biological clocks, and genetic testing.

Patients may not know about key aspects of hormonal and reproductive physiology and the pros and cons of treatments, such as increased complications from carrying twins. However, the details can be intricate. "I'm not sure people understand," Amanda said. "Like the FSH [follicle-stimulating hormone] test coming back quickly, but not being 100% accurate. Or how little PGD can really screen for. The concepts are tough to understand. You can tell people the statistics, but if they're gung-ho for a baby, they will probably ignore them. How do you get people to pay attention?" Patients may, understandably, be overly optimistic. Jennifer tries to "help couples be thoughtful. Not just, 'the technology's available, jump on board,' but what's involved."

Yet misunderstandings about statistics are common. "The concept of odds is difficult to explain to patients," Henry, the Midwest medical center infertility physician, found. "Most people hear of 2% or 3% lab error, and dismiss it, and don't take it to heart. They assume they're in the 97%. So just making that more real to them—that it *could* happen to them!"

Patients must comprehend large amounts of material, which can take considerable effort. "Education is needed *over and over again*, to help patients process this," Brenda said. "It's a lot of information." When she gives patients percentages, "it doesn't necessarily have meaning or emotional resonance for them."

Patients have trouble, too, extrapolating and applying to themselves population-wide estimates. They may feel, for instance, that they are healthier than other people, such that these numerical averages do not apply to them. "For any individual, the chance of getting pregnant is either 100% or 0%," Suzanne stated. "You're either going to get pregnant or not. It's not population statistics. It's hard for people to process the likelihood at their age—particularly since people feel healthy. A patient has always gotten her period, exercised, is incredibly healthy, doesn't smoke, or drink alcohol or caffeine. How could treatment *not* work? If I say, 'But, you're in your 40s! There are additional obstacles,' she says, 'Yes. But not for *me*. I'm very healthy.' I try to explain that there's an aging process: her eggs were all there when she was born, but over the years she has been exposed to different environment things, and our bodies age."

Torn between hope and harsher realities, older patients, in particular, wrestle with how optimistic to be, trying to apply to themselves overall success statistics but discovering impediments due to age. "Patients wouldn't be investing $15,000 in IVF cycles if they thought they had a 0% chance," Suzanne said. "They often predict 0% for themselves to protect themselves from being too hopeful. But if they don't get pregnant, they are still very disappointed."

Many patients want data, which do not yet exist, on specific diagnoses and clinics. Francine, the Miami legal secretary, wants "to see information about success rates by diagnosis. If I have endometriosis, how am I going to have genetically normal embryos? Will I have a 40% birth rate if I transfer one embryo? They just don't have the statistics, because they're not studying it." Information is often limited due to scant research funding and doctors resisting or undervaluing such data collection, frustrating prospective parents.

Patients may have difficulty understanding important aspects of genetics as well, such as recessive genes (i.e., being a carrier but not getting sick oneself) and genetic variants of unknown significance, for which the likelihood and extent of symptoms remain uncertain. "There are some complicated things that I don't understand myself—different types of translocations," Henry admitted. "Many more embryos than we'd expect don't develop well. Consequently, the outcomes are worse than what we'd anticipate."

He and other providers encounter difficulty conveying to patients the odds of multiple options. "What to do with expected normal versus abnormal embryos? What to do with carrier embryos that won't be affected by the disease, but risk passing on CF [cystic fibrosis] or autosomal recessive diseases—implanting them, but explaining that sometimes even just carrier status can manifest the disease?"

Additional education about specific procedures can be critical, allaying patients' fears. "I didn't realize IVF was not so terrible," Roxanne said. "The fear of it initially—the unknown—prevented me from doing it earlier. If women knew it wasn't so terrible, maybe more would do it quicker." Clearly, patients and providers also confront ethical quandaries for which simple answers may be elusive.

Challenges in Educating Patients

"Patients are never prepared for having no embryos to transplant," Steve said. "Human capacity to hope is tremendous. You can show a 45-year-old all the data in the world that she has less than a 1% chance of a 'take-home baby.' But she still wants to do 12 days of shots and egg retrieval for $10,000 to $15,000, and basically torture herself. It's *irrational* behavior—the power of the human heart."

"Plenty of education is available," he added. "But patients may not understand or avail themselves of it. It may not be in the necessary language. Many of my patients don't read the consent forms. But when things go wrong, they say, 'you never told me that.' We fall back on, 'Yes we did. It was right here.' It would be better if they took it a little bit more seriously."

Clinicians face several obstacles in educating patients, who feel desperate, eager for children. "Doctors tell patients the likelihood of success as best it can be estimated, but people don't always hear it," Suzanne said. Nonetheless, "Doctors can improve communication, or get feedback from patients on how much they understand." Other times, though, Suzanne thought, "patients are not really told some of the risks."

Providers vary in how well they educate patients. "Some clinics are better at it than others," Nicholas said. Physicians can be overly optimistic and not always communicate risks. They should give all patients evidence and reasons for or against a procedure. Roger saw needs to "mandate that the evidence be reviewed every year, and reported in layman's terms, just as HIPAA [Health Insurance Portability and Accountability Act] information gets posted in every hospital hallway, or you're violating a regulation."

REIs regularly introduce new procedures, for which evidence of effectiveness is still being gathered. Physicians must walk a very fine line and not "oversell" these interventions. "A lot of very promising technologies are currently undergoing evaluation or experimental protocols. There are reasons to be hopeful," Marvin, the Massachusetts REI, reported.

Enhanced public education about women's biological time clocks is also essential but tricky, potentially colliding with public health messages to prevent teenage pregnancy. Public health campaigns that emphasize the difficulty of getting pregnant might lead some teenagers to forgo contraception. "You don't want younger girls thinking 'There's not much chance of me getting pregnant,'" Helen said. "That goes right against what they learn in high school: '*It's easy to get pregnant. Be careful!*' The message could be: 'It's easy to get pregnant until you're 32.' But 16-year-olds may then say, 'I need to get pregnant now, just in case it doesn't happen later.' A very targeted, sophisticated ad campaign is needed."

Amanda, for instance, was surprised at the relatively low odds of getting pregnant. "My doctor told me that 'normal people' who decide to have a baby and have no medical problems have only a 15% chance of getting pregnant in any given month. That's stunning! I told my 16-year-old niece, 'That's only for *old* people like me!'"

Patient education campaigns about infertility are difficult, too, since reproduction and sex are sensitive and controversial topics. Education carefully needs to avoid encouraging women to focus on pursuing motherhood rather than careers. "If you try to educate people about their fertility," Brenda stated, "some people are going to say: 'Wait! Women's roles are beyond reproduction! Don't turn women into breeders.' The flip argument is, 'No, we're empowering women to make choices that are important to them.' But that gets lost."

Helen described one such controversial ad that "showed an hourglass-type egg timer saying, 'time is running out.' There was a tremendous backlash— all sorts of politics. God forbid we mention in high school health class that fertility drops with age," Brenda added. "That will never happen."

Types and Sources of Education

"The media loves to cover reproduction because it's so 'sexy' and a lot of money is involved," Roger said. "But they know almost nothing about normal biology. It's pathetic." Patients receive information from not only clinicians but television, newspapers, magazines, and websites. The quality, however, ranges. These varied sources provide messages about infertility and treatment that are commonly misleading or false. "Talk shows say, 'Smear natural hormones all over your body, but don't take hormone replacement therapy because that's poisonous,'" Roger noted. "They think everybody's a quack except for them. They have no degree, and are making a fortune."

Patient advocacy websites attempt to post relevant evidence-based information, but innumerable patients fail to access it. "One very consumer-based organization does a lot of education," Helen said, "and hopes people will become well informed and empowered." Organizations disseminate such information through several modalities. Ginger got "a blog every day about adoption, new medical technologies and environmental contaminants." Such websites can potentially give useful information. Yvonne, the Philadelphia psychologist, looks for "reassuring magazine articles. Cover stories about a famous actress going through fertility treatments are good, normalizing."

But patients may spend long hours reading confusing or incorrect messages. "The Internet is amazing, but a double-edged sword," Yvonne commented. "Some people on online chats give up every type of food, just to make their bodies as healthy as possible." Patients may admonish one another to constrain these online searches but in vain. "We all cull the Internet and share these little factoids," Yvonne continued. "Hoarding online information is a natural reaction, but it sends women spinning. You can drive yourself insane. At a certain point, you have to stop looking, get information from only one place, and trust that you don't have to look elsewhere or oversearch. Women say to one another, 'Don't go on the Internet tonight.' People ask, 'What does my FSH level mean?' You could find a website that says anything: it's great or it's not, or one woman had it and was fine."

Online pros, cons, and hype about treatment can be hard to disentangle. Website content can be partially or wholly inaccurate and/or contradictory. Amanda "looked at 3,000 websites. Some sites say, 'A cup of coffee a day while you're trying to conceive is OK, but caffeine may contribute to miscarriage.' Others say, 'You can have a cup or two a day' or 'Do not take even one sip of caffeine!' You start getting crazy! When I started, I went to one of the bulletin board websites and thought, 'These women are desperate and crazy!' Then five months later, I was typing, 'Is a runny nose a possible sign of pregnancy?' It's painful trying to interpret all your physical symptoms—are they pregnancy symptoms or premenstrual syndrome? The conflicting information is overwhelming. Somewhere you're going to find what you're looking for, and get 3,000 different opinions. We like information that confirms our beliefs."

Given these downsides of online searches, providers occasionally try to aid or caution patients. "You're going to look online," Amanda's physician told her. He urged her to be very careful, " 'because you could get into knowing too much about this or that. Just make sure it ends in .edu or .gov. Don't ask me about anything you find anyplace else.' " Amanda followed this advice. "Eventually I only went to one clinic's website because I trusted it. They had good information. Later, I shut myself off from the computer."

Yet, inaccurate online information may remain uncorrected. "People post questions and answers online that sometimes are helpful and other times comical," Amanda concluded. " 'Do itchy eyes mean you're pregnant?' That's ridiculous, but thirty people will say, 'My eyes were itchy, and I was pregnant with my son!' Either that, or you have allergies!"

Patients regularly struggle to interpret the validity of Internet information. Wendy had to figure out, "Is that website right? Let me find out from somebody else. You need to look around."

Most patients grasp these limitations, but many nonetheless feel nervous. "Cyberchondria is very pervasive—diagnosing all of your problems online," Steve said. "The Internet causes contagious anxiety. I try to drive patients to useful sites. We recommend certain links, but information overload is a problem for our whole society."

Websites range in the types and quality of not only their content but their organization and structure. Sally, the Maryland website manager, liked one site that "retains its history of discussion threads for a long time, so you can find something from a year ago. Another site is trying to save word space, so has already deleted threads a couple of weeks old. That's poor web management."

How to Improve Education

Patients would benefit from not just general overall information but targeted materials to help navigate these complex choices. Such information could cover multiple medical, psychological, emotional, and financial issues but require funds to design optimally. "The biggest need for us as a community is to help patients think through these very difficult decisions," Jennifer argued. "To develop educational materials, decision-making aids, brochures, and videos of patients who've been through the experience, with flow diagrams and decision trees, so that patients can really understand the process: 'These are your options. How would you handle that?' Hearing a story directly from another patient sticks with them. They remember. Tools in a couple of languages could help different types of people." Objective informational websites could also facilitate patient and broader public understandings.

Strong emotions and prior failures also sway patients' choices. Many decisions "are based on previous bad outcomes," Steve, the physician, said. "It's a wonderful technology, but the fear can blind them. Making a 20-page consent form understandable at the *5th grade level* is not easy. We dumb down 35 years of science into 45 minutes. I'm routinely amazed that patients have no questions. They should be asking, 'Will this improve my take-home baby rate?' " Improved informed consent is vital.

Developing optimal videotaped information can also help in certain ways. Patients might learn much by hearing from not just providers but other patients. Steve suggests using "real patients or actors representing patients—not doctors or nurses—in different scenarios, talking about their decisions and experiences: examples of patients whose selective reduction went well or poorly, or who lost a pregnancy, or thought twins were going to be OK, but financial stress led to divorce."

Breaking down the process into smaller, more manageable steps can also aid patients. Ginger, the mental health professional who underwent intrauterine inseminations for her first child, said that her practice manager, "told a young couple what an IVF cycle would look like: 'You'll come to the clinic four times for sonograms, then for the retrieval. Afterwards, the eggs will be in a petri dish with the sperm. Then you're going to have to wait three to five days, depending on how well these fertilized eggs do.' She walked them through it. They seemed comfortable, asking questions. I thought, 'They'll call when they're not sure of something.' We do a lot through email. Patients can get answers more quickly without waiting for a phone call."

Patients can enhance their comprehension by taking written materials home and/or bringing a partner or friend. "Patients are so anxious in the doctor's office that they have trouble processing a lot of information," Suzanne, the psychotherapist, reported. "They should leave with printed material they can read later. At the next appointment, doctors should get feedback: Any questions? Is the information too high or too low? Interactive, not passive education is most effective."

Broader Public Education

"The population at large needs to be educated how much infertility is a deep, emotional loss," Suzanne continued. "They take fertility for granted, and make hurtful jokes." Public awareness remains limited concerning the emotional tribulations of treatment, and the fact that these treatments and responses can require years. The very length of these journeys poses problems. "People in our society have a limited tolerance for grief," Suzanne added. "Patients are supposed to move on. That's hard."

Public education about male infertility is lacking, too. "The American Urologic Society has not done as much as the OB/GYN societies in giving out information on male fertility," Suzanne observed. "People are just not ready."

Enhanced public grasp of these issues would assist individuals *before* they become patients. "You can't do education simply by showing up at an IVF clinic one day when you're infertile and want a baby," Marvin explained. Nicholas saw needs to convey several specific topics, including "general infertility; how common it is, and what the current standard is for evaluation and treatment."

Genetics and embryo testing present challenges, too. "My job would be easier if the community accepted people with genetic diseases," Sam explained. "They don't understand how genetics works. They don't believe that if one person's a carrier and the spouse is not, that their kids will be free of the disease. It makes no difference that they hear it from ten doctors. They still think, 'There's something wrong in that family. There are crazy kids in the family.'" Public ignorance and limited education perpetuate stigma.

Reviewing a family's genealogy carefully can help. "The community doesn't want education," Sam continued. "Even if I sit down for an hour or two, some people will not understand about carrier status, because they just don't have the medical knowledge. If I sit down with a chart and show them, most people will get it." Unfortunately, communities often do not seek or receive adequate education about these issues.

Fear and stigma impede learning. "People say, 'Please don't put it in the record,'" Sam reported. "'We don't talk about these things.' It's bad to discuss. 'Better not to know.' But that creates stigma!"

Professional Education

Physicians in and outside of fertility also have knowledge gaps. Non-specialists often don't comprehend patient options well. "A lot of doctors don't want to look stupid in front of their patients, so don't bring infertility up," Jennifer explained. "If a patient asks, maybe they'll make a referral, but they're not going to bring it up. A lot of patients have never heard about it."

Education about a range of topics, from infertility generally to specific procedures, can help. Potential patients should be aware of all their choices and of physician hype, anticipate these stresses, and try to ameliorate them—from setting realistic expectations to establishing strong social supports. Ideally, patient and provider organizations should work both together and separately to instruct prospective parents and clinicians. Sadly, such endeavors remain underfunded but can be relatively inexpensive. Public health campaigns about reproduction can also help. Education is vital, though other changes are essential as well.

18 | Choosing Policies

"Because of 'Octomom,' California was going to pass a bill that doctors can only fertilize three eggs, and only transfer one embryo," Bill, the New England infertility doctor, said. "But just because there's one bad apple, you don't need more laws." Octomom, who gave birth to octuplets in 2009 while on public assistance, and "Octodoc" triggered discussions about tighter guidelines and regulations. The American Society for Reproductive Medicine (ASRM) changed several guidelines, but Bill and other providers generally oppose any regulation or lessened independence. Questions linger, however, over whether the ASRM's consequent modifications suffice.

Debates rage over whether to increase oversight of these technologies and, if so, how much and by whom. The US regulates ARTs far less than do most Western countries, most of which have more nationalized health insurance, linked to regulations. US states vary in whether they have any laws and, if so, what. Enhanced guidelines and regulations are needed.

Professional guidelines have begun addressing several areas, including oversight and screening of gamete donors, numbers of embryos to implant, age cutoffs for mothers, embryo selection for sex or other non-medical characteristics, and monitoring of providers. But professional organizations could potentially issue stronger, more forceful statements.

Moreover, new technologies continue to develop, including gene editing of embryos through Clustered Regularly Interspaced Short Palindromic Repeats (CRISPR) and other methods, and mitochondrial replacement therapy (producing so-called three-parent babies). At some point in the future, more children will surely be born through gene editing, exacerbating these concerns. Tensions surface over how to avoid either over- or underregulation and how much to strengthen free markets versus government or professional input.

Overall, national and professional policies need to be enhanced in several areas, including egg and sperm donation, egg donor agencies, numbers of embryos transferred into wombs, gestational surrogacy, oversight of providers, data collection, and insurance.

Need for New Guidelines or Regulations?

Past scandals have ignited calls for change. Additional guidelines or regulations, though staunchly opposed by some physicians, could nonetheless benefit countless patients and providers. Possible improvements vary based on who would develop such policies and what these would address. Clinicians and patients frequently express strong opinions—pro, con, or mixed—about potential guidelines or regulations regarding both specific procedures and more general practices, and about how flexible or enforceable such input should be.

Opposition to Monitoring Providers

"Octomom brought to the forefront some of the bad actors in our field," Peter, the West Coast physician who was sued for an inaccurate PGD result and who now screens embryos only for sex, commented. "That somebody would be so callous as to allow the octuplets to occur is a major event. Every day, we hold in our hands these time bombs for potential disasters. Another doctor said on TV that he will 'give people designer babies.' Just amazing. He is not a scientist, merely a practitioner out there, without any regard for ethics, gaming the system. . . . There's no penalty."

Doctors generally recognize problems in the field but argue that industry self-regulation, rather than government policy, was adequate. "Private clinics slip through the cracks," Peter added. "We need to tighten up those loopholes. But what we are doing now is best: professional guidelines from a professional body like ASRM."

He and others saw perils in relying on the political process: anti-abortion activists might push their own agenda. "The pro-life movement has hijacked these guidelines, rules, and reproductive rights arguments," he continued. "When these issues come up, these groups are always trying to limit abortion. Then, pro-choice groups fight back. So, any kind of reasonable discussion on how we should regulate fertility treatment falls through the cracks."

Physicians aver that current, individual state systems of medical licensure and board certification can best handle these problems. As Peter said about

Octodoc, "If he were board-certified in OB/GYN, then ACOG [American College of Obstetricians and Gynecologists] could expel him; but he's not board-certified. So, ACOG cannot throw him out." Licensing boards have to "take that into account at the state level. Sex with a patient gets all kinds of alarms, but misleading advertising, or merely being a slime ball, doesn't trigger anything."

Not all providers follow current guidelines and regulations. "Rogue" providers' scandals can taint the entire field. If gross violations occur, state medical boards can discipline physicians, revoking licenses. But professional organizations can potentially do more, by working more closely with state medical boards.

Improved oversight at several levels can also help but faces hurdles. Clinics within hospitals and academic medical centers are more likely to have formal ethics committees. Yet most IVF providers are unaffiliated with such institutions, and instead operate in less monitored private clinics.

Many clinicians insist that significant change is unnecessary and that both government regulations and additional professional guidelines have major disadvantages. "I don't really want someone telling me what to do," Jill added. "There are a reasonable number of guidelines. Physicians relish autonomy. With time, the guidelines may have to change; but right now, I think they're very appropriate."

Providers thought that their colleagues were mostly well intentioned and that a few outliers shouldn't trigger regulations. "Some doctors are a little bit more snake-oily," Brenda, the Vermont psychotherapist, confessed. "But most doctors are very conscientious and well meaning. You don't legislate on the few or the one Octodoc."

Significant problems, doctors felt, remain relatively rare. "You hear about these cases only once every five to six years," Roger, an REI, said. "Yet each year, we treat hundreds of thousands of cases. So, doctors are overwhelmingly doing the right thing. There's always going to be some crackpot. If you had a zillion regulators, Octodoc might still do his thing. Overall, we are doing a pretty good job. The government needs to be involved *only* in educating the public."

As in all specialties, these doctors felt, such scandals invariably occur. "Rogues and moneymakers will always be out there," Diane, a nurse, stressed. "The doctor could have been an eye, ear, nose, and throat specialist and done the same thing. I worry that the arguments are emotional, rather than about individual procreative liberty. I wouldn't want to legislate on the few patients who have kids in their 60s."

Providers also argue that the media exaggerate scandals and sometimes overlook positive news and that sensational cases should not drive legislation. "Over 10 million babies are born with us," Steve said. "*That* fact is always lost in the one bad doctor or crazy patient on TV. Not everybody has quadruplets or octuplets. We're not all using our own sperm. *We help people achieve their dreams!* We have guidelines for most everything we do. Sadly, the media take every opportunity to demonstrate the negatives."

Wariness of government regulation also reflects larger national attitudes, even if these may be problematic. "Self-regulation is how Americans work!" Anne, the northern California REI, said. "People are supposed to follow professional guidelines. This is how we have always handled professional guidelines in the US. But there is no enforcement. I'm not sure it really works well."

Additional regulations would also be hard to develop because technologies are rapidly advancing beyond policymakers' ken. Laws have inherent limitations: clinical practice contains complexities. "Government agencies don't fully understand all the issues and nuances of situations that make black-and-white laws hard," Henry pointed out.

Once enacted, policies can also become rigid, while technologies surge ahead. Policies "get very concretized," Suzanne, the psychotherapist, said. "Then, as science advances, it's very hard to make any changes."

Government restrictions could also potentially delay patients' treatment. "I don't want patients waiting a year for answers," Sam, the patient advocate, said. "That's usually what happens when the government does things. In the UK, if a couple wants PGD for fragile X, doctors have to submit forms and the family history to the government, which decides if it's appropriate or not."

More government involvement could have unintended consequences as well. "I don't want to see the government, with their teeth involved," Joe, a physician, said. "Their teeth will all be infected and screw up the whole system. There's no way to enforce guidelines. Medicine has always been practiced by individuals. Enforcement of guidelines and regulations is intrinsically limited. As in all medicine, individual physician discretion and flexibility are important."

Doctors argue, too, that fuller government regulations abroad have impeded clinical practice and effectiveness, compelling patients to travel elsewhere. "Worldwide, legislation has not helped the field," Thomas, an East Coast academic hospital-affiliated physician, asserted. "In many European countries, legislatively limiting numbers of embryos to transfer, without autonomous decision-making by physicians and patients, has hindered the clinical efficacy of ART. So, patients are coming to the US because they can get more than two embryos and donor eggs, which they can't in Europe. In countries where sperm donors are no longer anonymous, sperm donation has dropped off considerably." Yet, in fact, sperm donation in the United Kingdom, though initially decreasing after anonymity was banned, subsequently returned to its earlier levels.[1] Perceptions of disadvantages of tighter regulations in Europe may thus be biased. Moreover, multiple complex factors may shape whether a nation permits particular individual procedures.

Doctors oppose more guidelines, too, because of fears of being liable if failing to follow these standards. "If you start making guidelines stricter," Nicholas, a physician, contended, "it increases the liability of a practice. We need flexibility."

Professional organizations such as the ASRM may also fear litigation if they issue or enforce guidelines too much. "Everybody's afraid of getting sued," Steve declared. "ASRM, if it censored a clinic for an unethical activity, would try to give the practice time to remediate. But if that information was not kept confidential and that doctor goes out of business, he or she could sue. It could be held up in court for months, years—for nothing. For doing the right thing!" Diane, the nurse, described, for example, a doctor on a professional Listserv who "was bipolar and making inappropriate comments. We wanted ASRM to bar his comments. They would just have had to keep his email address from going through. Easy. It could have been done by hitting a button. *But they wouldn't do it!* They were afraid of being sued. They're timid, and not willing to be a regulatory agency."

Only a few clinicians mentioned that they may have conflicts of interest (COIs) concerning these issues. "Professional societies, not politicians, should set the guidelines, but I'm obviously biased," Marvin said. Nonetheless, even though these providers may not be wholly objective, their perspectives are important to note.

Advantages of Improved Policies, Guidelines, and Monitoring

Yet despite this opposition, new guidelines and policies, both generally and regarding specific areas, can have several major advantages. Moreover, increasingly, physicians' and patients' decisions about ARTs have broader social ramifications that need to be considered. The patients' and providers' insights here have crucial implications concerning *who* should regulate *what* and *how*.

Overall, federal and state governments, professional associations, insurance companies, and employers can enhance policies and guidelines regarding providers, egg agencies, and others.

Several clinicians felt that the ASRM, in particular, could play more of a role in ways from big to small, partly to avoid government regulations. These individuals thought the profession should behave better, especially given potential bad publicity and desires to maintain self-regulation.

Scandals can result from only a few rogue providers but cause major crises and harms. "A couple of idiots who do really stupid things undermine everybody," Marvin said: "95% of this profession have been responsible; 5% haven't but have damaged the other 95%." Still, he supports self-regulation. Ultimately, he thought, each physician has to monitor him- or herself. "We need to do our jobs well, which means being socially responsible, and being *seen* as socially responsible, not just *telling* people that we are."

Providers' adherence to current guidelines and standards is not currently monitored but should be. "The problem with guidelines and SART [Society for Assisted Reproductive Technologies] is *enforcement*," Steve said. "One of the most disturbing things is that programs way below the national average still take care of patients. They should be either cited and remediated or shut down. It threatens all of us when bad things happen. ASRM's and SART's first goal is to keep government regulation out of ART. To do that, we have to *regulate* ourselves. These organizations have to take this on."

Hence, several providers see more active roles by ASRM and SART as essential to forestall more government involvement. "If we don't at least pull in the reins sometimes," Edward, the Ohio REI, said, "We'll end up with outliers and become regulated." Several clinicians agreed that professional associations should therefore do more. "The argument has always been that we're self-regulating more than any other area of medicine," Diane said. "We have to live up to the PR. The ASRM Ethics Committee takes a leadership role, issuing guidelines, but the process is very slow."

Professional associations could potentially further incentivize, mandate, and/or enforce physician behavior but may hesitate, not wanting to play these roles. "There should be guidelines and punishment, but we don't have the stick," Peter said. "ASRM has guidelines, but no teeth to back them up," Joe agreed.

Currently, not all providers are even ASRM or SART members or need to be. "All ASRM can do is make nice statements," Diane explained. "After Octomom, ASRM contacted the California Medical Board, which looked like it was going to rip up this guy's license." The ASRM expelled Dr. Kamrava after he transferred 12 embryos into Nadya Suleman in 2008. In 2011, the Medical

Board of California revoked his license.[2] The fact that infertility specialists may not even be members of these organizations may perhaps excuse these societies but needs to be addressed, potentially through certification.

The lack of other means of addressing colleagues' deficiencies bolsters the need for ASRM to assume a larger role. Individual doctors themselves feel they cannot improve substandard colleagues' performance. "There's no way," Steve said, "that I can say something to a physician in a wholly separate practice."

Specific Areas for New Guidance

Policies could be improved in several particular areas—from enhancing insurance coverage to collecting more data. Patients commonly feel that insurance should include ART more. "There should definitely be more insurance," Nancy, a 44-year-old IVF clinic nurse, said. "It should just be a benefit." Several states have gradually increased insurance coverage requirements, though the amounts generally remain insufficient.

Belgium and several other European governments cover more IVF but, in return, stipulate that doctors meet several conditions, such as transferring only single embryos.[3] Expanding insurance coverage for IVF can thus both heighten access and remedy certain public health problems by reducing multiple births. "We have guidelines on how many embryos you put back," Joe said. "But lots of IVF centers don't follow these carefully, because they want pregnancies, which bring more patients. More doctors would follow the guidelines if that economic incentive were removed."

Insurance industry associations should encourage companies to consider more fully the longer-term benefits of covering certain procedures such as PGD to prevent transmission of dangerous mutations. Insurers should also cover more mental health services for infertility patients.

But more coverage may require more transparency and accountability. "If the government's paying for it, they can tell you what to do," Bill, the New England REI, said.

ASRM and ACOG can clarify guidelines in several areas such as age cutoffs and need for patients to understand more fully that certain interventions may remain unproven and/or have profound psychological effects. Marvin thought that doctors need to "inform patients that a lot of this is *experimental* and has social and emotional impact."

REIs should question themselves more and avoid simply performing procedures or tests to maximize profit. ASRM should change its statement that clinicians should not provide futile treatment "solely for" their own benefit. The phrase "solely for" should be eliminated. Rather, doctors should not be "motivated by" or provide futile treatment at all for their own benefit. Guidelines should further spur providers to avoid procedures that will probably fail to offer clear benefit; and should clarify how and for what medical problems clinicians should not treat patients over age 50.

Enhanced professional organizations' recommendations can also aid providers in refusing certain patient requests. "Guidelines can, in some ways, actually be nice for doctors: we then don't have to say 'no' to people," Peter stated. "But, 'I'd really like to help you, Mrs. X, but can't because of the guidelines.'"

The US Food and Drug Administration and others should more actively oversee the use of new, expensive, unproven interventions. Desperate to have children, many prospective parents willingly pay for costly novel tests and procedures that doctors may widely order but that then prove to have little, if any, benefit for most of these patients—such as broad use of ICSI when male factor is absent, given that these injections increase risks of birth defects. Many doctors routinely ordered preimplantation genetic screening (PGS), for example, to screen out any embryos with abnormal numbers of chromosomes, in order to prevent recurrent miscarriage, even though no evidence supported this use. The Cochrane Database of Systematic Reviews collects, synthesizes and performs meta-analyses of evidence about whether particular interventions are effective[4] and along with the National Academy of Science (NAS) and others, can help evaluate such new ARTs and inform guidelines. REIs are beginning to order whole-genome sequencing of embryos, though doing so can yield massive amounts of ambiguous and confusing information. All people—and embryos—possess genes associated with slight or mildly increased rates of several diseases and variants of uncertain significance, including dozens of new mutations that lack clear implications. How patients and providers will understand and respond to these enormous amounts of data is wholly unknown. The American College of Medical Genetics, ACOG, National Society of Genetic CDC, and other organizations should develop recommendations concerning which genetic tests physicians should order for not only an adult (about his or her own health) but embryos.

The ASRM, ACOG, and other associations should also develop much stronger guidelines and requirements that doctors seek to ensure that women selling eggs fully understand the potential health risks. Children born through egg, sperm, and embryo donation should all have the option, as adults, of contacting these biological parents.

Since most egg agencies fail to follow current guidelines,[5] these companies should be required to undergo certification and monitoring, report data on their practices, and follow American Medical Association (AMA) guidelines for advertising and presenting the risks, not only the benefits of a woman selling her eggs.

The NAS, AMA, and other organizations should more strongly encourage self-regulating medical associations such as the ASRM and the ACOG to maintain the highest possible ethical standards and recognize and manage their conflicts of interest as vigorously as possible—more so than at present.

Professional and patient organizations such as the AMA or the National Infertility Association (RESOLVE) should establish websites where patients or others can document clinics' violations of AMA or ASRM guidelines concerning online advertising or other practices. Such websites will need to be curated to ensure the appropriateness of content; but patients associated with these organizations can potentially help monitor such activity, and users can report any potential misuses of the site.

Many employers, including Google, Facebook, and Amazon, are paying for freezing and storage of employees' eggs. Yet these employees must fully comprehend the potential limitations of this procedure, including the uncertain long-term viability of frozen eggs, to ensure that these women understand that this technology may fail and/or pose additional challenges. Unfortunately, as revealed when one freezer's temperature recently fluctuated unchecked, losing thousands of eggs, these young women often do not fully grasp these inherent limitations.[6]

Altering Embryos' Genes

"Within our lifetimes," Nicholas predicted, "you will be able to put a little drop of blood into vending machines and learn what lethal mutations you carry." We have essentially reached that point. Technologies for testing and altering genes of embryos are advancing, making several key questions imperative—regarding, for instance, selection of genes for non-medical reasons. Currently, doctors regularly perform non-medical testing of embryos to choose sex. Despite ongoing academic debates of whether it should be done, countless clinics are in fact routinely doing it. Discussions should thus shift from *whether* the practice should be allowed to *when* and *how*. Clinics often decide whether to perform sex selection based on the relative strength of its providers' opinions. But within a clinic, physicians may simply overrule nurses and mental health providers (MHPs). Doctors may be biased, earning income through sex selection.

Many clinicians wrestle with how much to weigh patients' rights to choose children's traits against the child's best interests and rights to an open future[7] and broader social justice concerns. Balancing such competing ethical principles can pit clinicians' personal profit against larger ethical concerns and professional responsibilities. While many doctors readily proceed, others, especially female physicians, feel awkward and uncomfortable.

Given these fundamental tensions, full consensus may prove elusive. Though ASRM now recommends that clinics should write policies about sex selection, challenges persist—for instance, whether not only physicians but nurses and MHPs should be involved. Non-physicians who might resist sex selection may feel uncomfortable opposing physician-employers. Professional organizations should clarify *when* exactly social sex selection is acceptable or not—perhaps permitting it for a family that already has four but not merely one child of a particular sex and stipulating that non-physicians should participate in these decisions.

The current, relatively weak regulatory structure and "free market" approach to ART in the US and several other countries bodes poorly for potential abilities to regulate future transfers of edited embryos into wombs. Though scientists from several countries have called for a moratorium on transfer of gene-edited embryos into wombs, the world lacks any system of global guidance to implement or enforce it. Even the United Nations, with representatives from every country, has restricted powers. At some point, the transfer of gene-edited embryos into human wombs will probably be approved for use on an experimental or wider clinical basis—once its risks are better characterized, even though dangers will probably persist. Many doctors and would-be parents prioritize patient autonomy over the rights of the future child, and minimize potential longer-term medical, psychological and social risks of ARTs. *Given, in addition, the lucrative and competitive, profit-driven ART industry and its overall resistance to strong guidelines, numerous doctors might thus end up performing these procedures, with little regulatory or administrative hindrance.*

Yet given the broader social implications (with the wealthy being able to alter their diseases and traits in ways that the poor cannot), and the potential long-term risks to children and their descendants, a different model is needed than one involving only consent of the physician and parents. Rather, broader social consensus and attempts to respond to larger ethical concerns is critical.

Clinicians should also recognize their possible biases. The involvement of hospital ethics committees in these varied decisions may alter the outcomes. Outsiders who are not ART providers and do not benefit financially from a procedure may consider more fully the social costs. In academic institutions, external individuals on clinical and research ethics committees can enhance transparency and accountability and counter COIs. Stronger guidelines and policies from the ASRM, the ACOG, American College of Medical Genetics and Genomics, and National Society of Genetic Counselors are vital here.

Need for Better Data

More data and research are imperative. As clinical practices shift, governmental agencies and professional organizations need to update the categories of statistics they collect from physicians, to include additional and more thorough data; require that clinics submit this information; and respond to problems that emerge.

The CDC's National ART Surveillance System has reported both individual clinic and national aggregate statistics, regarding five age groups of women (from younger than 35 to over 44), separated by whether embryos are from donors or non-donors and, if the latter, whether embryos are fresh or frozen.[8] These statistics include numbers of IVF cycles and embryo transfers, average numbers of embryos used, percent of transferred embryos that implant and that are singletons, and proportions of cycles resulting in live singleton, twin and other births, and normal-term and - weight singletons. Yet the data on egg donors are not separated by key sociodemographic factors such as age. The CDC also publishes data three years after the fact.

Recently, the CDC has published as well a few papers on certain additional national aggregate data, not separated by individual clinics, on side effects, including infections, hemorrhages, ovarian hyperstimulation syndrome (OHSS), severe OHSS, adverse drug effects, anesthesia-related complications, hospitalizations, patient deaths within 12 weeks of stimulation, and maternal deaths prior to infant birth. From 2000 to 2011, approximately 15,000 patients (1.3%) had complications, as did about 1 in 300 egg donors, with 82 donors requiring hospitalization.[9] The CDC has presented, too, some data on gestational surrogates, revealing that these women are more likely than ART women carrying their own children to be older and from outside the US.[10]

SART reports additional data on births of triplets or higher multiple births, percent of preterm and very preterm births, and outcomes from donated embryos. The CDC, but not SART, seeks to verify the information that doctors self-report by auditing 5%–10% of clinics. Yet CDC has not reported the specific results of these audits and what errors were found.

Public data depend on transparency, compliance, and lack of bias; yet in 2010, 6.5% of clinics failed to report to CDC and 28.1% did not report complete data to SART.[11] CDC lists non-reporting clinics but does not penalize them in any way. The most egregious non-reporters (the 5% with the highest rates of non-reporting) failed to disclose as many as 66% of their cycles, with an average of 37%, and thereby artificially heightened their pregnancy success rate—and consequently increased their market share by 19.9%. Between 2005 and 2010, their non-reporting increased by 124%. The CDC lists only cycle outcomes but not the number of cycles initiated. In calculating success rates, CDC does not use cycles that evaluate new procedures or aim to obtain eggs for freezing.[12] But the outlying clinics appear to be excluding older women without transferable embryos, thereby increasing reported pregnancy rates but excluding these women retrospectively.

Reporting of pregnancy rates has fostered competition between clinics. Though SART and CDC websites state in one location that patients should *not* use these numbers to choose clinics, patients nonetheless do so, encouraged even at another point on the CDC website, which calls the government Clinic Success Rates Report, "one tool consumers can use to identify clinics".[13] Current reporting requirements can thus incentivize "gaming" the system.

Major gaps persist in types of information gathered on, for example, shorter- and longer-term risks and benefits of procedures such as embryo testing. "It's unknown how often downsides occur," Thomas, who works at an academic hospital, said. "I don't get a separate report on PGD, so don't know our success rate. The patients are very heterogeneous." Even at academic medical centers, clinics may not fully examine their full data.

Edward listed several specific types of information that neither CDC nor SART now report but should: "why exactly PGD is done, what the embryos looked like, how many were biopsied, what the biopsies showed, how they survived, whether they were frozen, how they looked between days three and five, what percent died, what the accuracy and subsequent delivery rates were. Europe is much better at collecting data after birth because the government pays for most care. In the US, patients go off and deliver. SART collects data on whether it was a live birth or neonatal death, and the weight, gender, and any obvious birth defects. But that's it!" The European Society of Human Reproduction and Embryology (ESHRE), but not the US, collects and reports additional information about the longer-term effects of ARTs on infants.

ESHRE publishes data that the US does not on, for example, intrauterine insemination rates, Apgar scores (assessing newborn functioning), and details of PGD and PGS use, including numbers, methods, and outcomes of embryo testing and rates of subsequent miscarriages, deliveries, and misdiagnoses.[14] ESHRE divides PGD data by the purpose of the testing: selection of sex for social purposes or medical reasons, selection of genes associated with particular diseases (autosomal dominant and recessive), and human leukocyte antigen (HLA) both alone and with monogenic disease (to produce so-called savior siblings). ESHRE also segregates data on PGS by purpose—including advanced maternal age, recurrent miscarriage, and IVF failure, despite evidence that these uses lack benefit.[15] These European data reveal that from 2006 to 2014 rates of premature deliveries almost doubled and that rates of maternal deaths are increasing, presumably due to twin pregnancies, underscoring crucial needs for such ongoing follow-up.[16,17] ESHRE also reports data that the US does not, but should, on longer follow-up post-birth and whether infants ever left the hospital.

The most recent European report includes 1,279 clinics in 39 countries, of which 14, including the UK and the Netherlands, have mandatory data reporting—an approach that obtains information much more effectively than do voluntary systems.

Physicians perform several procedures far more in the US than in Europe and with different thresholds, making US-specific data critical.

Periodically, CDC and SART have changed a few reporting requirements, but should alter more, given improving and ever-newer technologies. "Octodoc slipped through because of oversights in how our professional organizations were looking at programs," Calvin said. He described a situation from a few years ago but one that illustrates several of the professional and interpersonal tensions and complexities involved. "SART was looking at success rates and programs with high multiple pregnancy rates, but not at doctors who were transferring more embryos or had fewer than 50 patients, because that wasn't thought to be statistically significant. In the early '90s, we had to transfer three and four embryos to get a 30% pregnancy rate in a young couple, and we got a lot of triplets and quadruplets. People said, 'We'll just look at triplet rates, because if doctors are transferring more embryos, what's the difference?' 'Octodoc' slipped through because he wasn't getting multiple pregnancy rates—because he wasn't getting pregnancy rates! And we weren't looking at programs with fewer than 50 patients." SART has since changed its requirements.

"SART sent a letter to everybody transferring an average of more than three embryos to women under age 35," Calvin, continued, "saying 'You're way off the mean. What are you doing about it?' SART has a mechanism to audit programs below the tenth percentile and pays $1,000 of the audit cost. The goal is to try to bring all programs up—not kick people out. About

30 programs out of 400 were averaging three or more embryos for patients over 35. It's not as bad as seven or eight years earlier, when at least 10 or 12 programs were transferring more than four, on average, to women under 35." Though, SART and CDC have thus revised the categories of data they collect, as Calvin highlights, these organizations should continue to reassess regularly their reporting guidelines, since these technologies continue to improve and evolve. Still, additional reporting requirements may face opposition from some clinicians–though aiding patients and the profession as a whole.

The CDC and SART should gather and report additional data to understand more fully physicians' current use of various treatments and the short- and long-term medical benefits and risks—for example, the specific types and amounts of embryo testing; the longer-term medical follow-up of patients, egg donors and recipients, and children born through ARTs; and socioeconomic data, such as patients' race or ethnicity across categories, costs and use of insurance, what types of patients employ gestational surrogates (e.g., what percent of the couples are gay), and whether donors are paid or unpaid. The CDC has not reported data per clinic on side effects or the socioeconomic status of these or other patients. Such reporting would help in identifying and assisting underperforming doctors, aid future patients, and address concerns.

Governmental agencies should also mandate reporting. "In some states, being kicked out of SART is a problem," Calvin continued. "In one state, the major insurance companies require that doctors are SART members for patients to be covered. But in *most* states, if you kick doctors out, they simply don't have to pay dues. They don't have to be part of SART. If you don't report to the CDC, it really doesn't matter—doctors just get listed as not reporting. There's not much bite in all this." Additional states should hence require SART participation.

State health departments obligate all physicians to report data on various other conditions and treatments, to monitor the spread of these disorders—especially where additional individuals may be involved—including cancer[18] and infectious diseases, such as cholera, encephalitis, hepatitis, malaria, meningitis, mumps, tetanus, and toxic shock syndrome.[19,20]

Enhanced reporting can also aid fertility clinics, increasing ART use. New York and numerous other states bar gestational surrogacy, for instance, largely because of concerns that poorer women would do it and be exploited and harmed. Yet no data exist to check whether this claim is true or false. Such data are thus essential–on who gestational surrogates in fact are (their age, race, ethnicity and socioeconomic status), and how they perceive and experience this work and whether they view it, retrospectively, favorably or regretfully. Such data, if they reveal few concerns, would prompt other states to legalize this practice, enabling more prospective parents to have children.

Given recent evidence that IVF and ICSI increase rates of intellectual disability—which often appears only when offspring enter school, rather than during infancy—longitudinal follow-up studies need to integrate ART data with medical and educational information. Several physicians hope that professional organizations will also use data collected more fully to improve standards in the field. "If you can't get them up to standards, it would be nice to say, 'Having this doctor doing IVF is not a service to the public,'" Calvin added. "Now, the most we can do is say, 'If you don't correct these things—which we'll help you do—you're out.'"

These data can spur improvements in both individual clinics and the field—"to tighten the bell curve," as Steve said. "The outlier practices may be great, or *cooking their books*. The really bad IVF practices need either to be fixed or stop. It's unfair to patients. Some practices have pregnancy rates that are half the national average! Those doctors need to be remediated and/or have two years to fix it. They need to pay consultants, bring the right people in, and invest in their own shop! Some just don't care: they'd rather go for the paycheck than the investment."

Medical boards could potentially use these data, too. "SART could lay out a doctor's problems, and if these issues weren't resolved, the medical boards could get involved," Calvin suggested. "I couldn't perform heart surgery," Bill added, "but any doctor can stick a needle in an ovary. Government should mandate that doctors be certified by professional bodies, and follow standards. Otherwise, they shouldn't practice."

Ever-widening use of electronic medical records can facilitate additional data collection on practices and risks of particular medical procedures. Doctors often use new drugs in unproven ways that can have unanticipated adverse effects over time. "Careful long-term monitoring is needed," said Sally, the pregnant Maryland website manager. "There should be some accountability. A lot of these medicines are used off-label. What are the long-term effects? I'm not sure. Whose responsibility is this?" Arguably, individual providers, professional associations, and government all have responsibilities and should collaborate here.

SART and the CDC should also alter reporting requirements to help patients better comprehend the realistic odds of success. Clinics should report success rate per embryo transferred, rather than per IVF cycle—which can be higher if extra embryos are transferred—to help patients understand the distinction and their own chances.

Augmented statistics and understandings are needed, too, regarding the risks of selling one's eggs to strangers and of certain other treatments, to help patients make informed decisions. Women should have the right to sell their eggs if they wish but should be aware of the potential harms, including OHSS.

The CDC and SART should also collect and report data on egg freezing, to gauge its effectiveness, including how often it leads to healthy children, after how many years of freezing. Longitudinal follow-up data on children born through this procedure are also essential to detect any potential dangers.

The fact that clinics do not all submit accurate statistics needs to be rectified. "Some clinics may not be as honest as they should be," Henry reported. "Some clinics say that they care for difficult-to-treat patients under 'research protocols' and do not report these data." ESHRE also states that clinics underreport adverse events,[21,22,23] highlighting the need for data transparency and mandatory reporting. Though the CDC verifies 5%–10% of the data, SART does not do any such checking, but should.

Doctors who sell embryos should be required to track purchasers' locations to avoid genetically-related progeny from unwittingly intermarrying in the future. Secure mandated registries should include information on children born through sales of eggs, sperm, or embryo so that these individuals can, as adults, locate their biological parents and full or half-siblings or obtain information about them, if they wish. Potentially, DNA from these newborns could be included—*if* appropriately strong confidentiality and data protections are established—to allow these individuals, once adults, to enter their genomic information to see if they were created in these ways. Such protections would be essential and are possible. Public health departments regularly collect and store other types of encrypted sensitive health information about individuals. Additional details would obviously need to be determined, with input from various types of providers (e.g., pediatricians and mental health professionals) and patients. Yet millions of Americans already now pay to give their genomic information to companies such as Ancestry.com and 23andMe, which are increasingly allowing identification of most Americans of European descent.[24]

More medical and psychological information on egg, sperm, and embryo donors, recipients, and offspring initially and over time is also vital: "following kids into the future," as Henry continued, "to make sure that what we're doing is *safe!*" Questions remain about medical and psychological effects, over several years, of certain procedures, such as selling eggs. "Paid egg donation is fine," Sally said, "but I am concerned about the long-term consequences of the medications. Are these women putting themselves at risk? Are they really aware of it, or just signing consent forms and not really reading them? Is anybody monitoring safety?" Providers, patients, and the public should grasp more fully the potential emotional complexities as well.

More long-term psychological follow-up and research should include patients' views of their experiences and ongoing effects on families. "We don't have 50-year studies on the implications of ART on these children's health," Brenda observed. "We all think the children are going to be fine, but we don't know. What are donor-conceived persons' needs? How are we meeting or not meeting those needs? Are offspring going to have access to information on their donors and, if so, what? With PGD, nobody follows up with families ten years later: 'Are you glad you did it, or have you lived with questions? Was it burdensome?'"

Data on donors' and gestational surrogates' motivations are also vital, "to make sure the surrogate is doing this for the healthy, right reasons, and the relationship is going forward," Brenda elaborated. "These areas are messy."

The lack of data confounds many patients. "I like statistics," Sally said. "I always want to know my chances of x or y. But I found very little about what my chances were. Was my treatment failure typical or an anomaly? It's frustrating that the stats for PGD—what percentage of cases work—are so loose. CDC tracks IVF cycles, but not for PGD. Women post their stats on a certain website: how many eggs, how many fertilized, how many tested, how many were OK, how many got put back, what the medicines and results were. I've kept that chart for several years. All these data should be put into a database to come up with a strategy: *this* works 25% of the time. My doctor said, 'We could try a different approach that worked for another couple sort of in your situation—they had a genetic disorder and PGD and just one embryo to put back. Maybe it will work.' It's frustrating that there's not enough research. There's no malice in that. Some of these treatments are just still relatively new.

"I asked my doctor, 'Can I talk to someone that's been through this?' He put me in touch with a very nice woman, but it *didn't work for her!* I thought, 'That's not a good sign.' I don't even know if they had a successful case."

On these uncertain journeys, such information can help patients make decisions, gauge their odds, and feel reassured.

Opponents of such data collection argue that gathering such information requires resources, but these costs are hardly prohibitive, and the additional statistics could simply be added to current CDC reporting requirements. Cost concerns can also serve as an excuse for not obtaining input that might reveal limitations of the status quo. Brenda, employed by an IVF clinic, said, "Who's going to pay for those data? It's not fair to burden practices: 'You have to follow-up.' Some people say: 'Clinics are making all the money.' But it isn't that much. A study would need to be multi-site, not just based at one fertility center. So, it's big bucks." As Marvin added, "I'd love someone to get a grant to make data gathering a funded mandate without costing clinics."

Yet, many IVF clinics have in fact been earning huge profits and are increasingly owned by venture capital companies. Moreover, clinics could potentially help fund such data collection in return for advantages they do and would collectively accrue. SART could manage the data. "If someone's willing to pay for more policing, I'm sure SART could provide the service," Edward said. "Some doctors have suggested that SART increase fees to cover monitoring; doctors could advertise that they're SART members. But that's easier said than done. Doctors are already complaining about the current fees!" Many infertility providers may resist such efforts and costs but earn very high profits and would reap long-term gains.

"Historically, the US government has lacked interest in funding anything to do with IVF," Henry reported. "The European data are largely reassuring for most outcomes," Unfortunately, due to very different social, cultural, legal, and financial contexts, data from European countries may not always directly apply elsewhere. Hence, US federal and state governments should collect more data.

While mandatory data reporting systems that include follow-up require government support, rapidly increasing ART use justifies such resources. In addition, the CDC already has various data but says it lacks funds to analyze them. Such funds are thus vital. In the meantime, to look at the CDC's data, researchers must travel to Atlanta to view the information and perform any analyses on the premises and cannot take any of the raw data with them, hampering the publication of these results. Allowing such access to these data by legitimate, appropriately credentialed scholars costs nothing but would benefit countless patients and the field as a whole.

Improving Practice

Physicians can also enhance their individual practices and one-to-one interactions with patients and staff. Only after becoming infertility patients themselves did several infertility providers fully grasp patients' perceptions and needs. Clinics can improve in other ways, bolstering their organization and offering more frequent appointments and better patient information.

In general, physicians have difficulty coping with treatment failures. When patients have stillbirths, for instance, obstetricians feel guilty.[25] Physicians also frequently feel badly about medical errors, hindering disclosure of these mistakes to patients.[26] Yet doctors' reticence can, more than they realize, antagonize and aggravate patients. Physicians need to become more aware of these difficulties. Patients often fail to appreciate providers' pressures and stresses. Enhanced mutual recognition of these challenges can help.

Potentially, doctors can involve non-physician staff, including psychotherapists, more fully to help convey challenges to patients and manage expectations. Insurers could cover brief professionally-led support groups or focused psychotherapy sessions concerning these issues, which need not cost much.

Providers should also be more careful about denigrating colleagues' prior approaches with patients, who may then be left confused. Professional norms generally deter providers from criticizing each other to patients, but doctors occasionally do so regarding ARTs—partly because technologies are relatively new, and consensus on standards of care in particular situations may be limited. Professional organizations can establish more consensus and more forcefully encourage providers to eschew hype, convey ambiguities, and assist patients in setting appropriate expectations.

Understandably, providers tend to defend this industry staunchly. After all, who would want to be monitored or regulated if they could avoid it? The effectiveness of guidelines is not always clear, but many clinicians are nonetheless wary, even if the status quo has problems.

Clinicians often feel conflicted and unsure about how the profession should proceed, aware of problems among colleagues but leery of scrutiny and of government. These tensions pit libertarian attitudes (opposing government involvement) against needs to protect patients and maintain public trust. "I'm very torn," Valerie, the IVF clinic mental health provider and single-mother-by-choice, conceded. "I'm not in favor of government intervention or control or paternalism. But 'Octomom' and *Jon & Kate Plus 8* [a reality TV show about a couple raising a set of twins and sextuplets] make people think it's OK to have multiples—that the kids will turn out hunky-dory."

Strengthened guidelines and policies from the ASRM, ACOG, the CDC, and others, though adamantly fought by many doctors, could provide, too, major advantages, such as increasing insurance coverage for patients. State boards could easily require ASRM membership and that ASRM and CDC could mandate reporting.

In sum, professional organizations, in conjunction with SART, CDC, and state medical boards, can and should improve the status quo in various ways. The ASRM, the SART, and other organizations may be limited in their abilities to enforce standards more stringently but could use both carrots and sticks, heightening the benefits of organizational membership and emphasizing the advantages of enhancing self-regulation. These suggestions are not necessarily easy and will surely encounter opposition but are critical. Resistance from some physicians should not block all such efforts. The ongoing advance and spread of these technologies in both the US and elsewhere make these needs ever more pressing.

19 | Designing Our Future World

The men and women here all embarked on complex journeys, and, in writing this book, I have, too. In retrospect, I realize now how little I initially knew.

In the end, I carefully considered my friend Abby's request to donate sperm to father her child, learning about and weighing the social, psychological, and ethical complexities—including how key decisions would be made in both creating and raising the child. I'd be involved in these activities and feel responsible and wasn't quite ready to take on these lifelong commitments. Ultimately, I told Abby no. I have often wondered if I made the right decision. I will never know. But the choice helped me understand the predicaments that countless potential parents confront.

As the patients and clinicians in this book have described, these swiftly evolving technologies are creating millions of new people but also medical, ethical, legal, cultural, and existential dilemmas. These interventions are spreading not only in the West but in developing countries, posing quandaries about choices of eggs, sperm, embryos, wombs, genes, children, parents, and providers.

The individuals here also highlight the benefits of examining the narratives of patients and providers—to grasp not only the medical but these crucial other conundrums. Studies of the human impact of technologies tend to emphasize broader social implications, rather than personal, private psychological effects that can be harder to discern. But the individuals here highlight these internal psychic ramifications. Nuanced understanding of these individuals' experiences suggests how technological advances can shape human lives in powerful, unexpected ways.

Having a baby is "the stuff of dreams." The desire for children propels individual quests and scientific developments but also presents quandaries that, as patients, providers, and a society, we are currently poorly equipped to address. Few road signs or maps exist.

Given human curiosity, hubris, greed, and desires for perfect children, new technologies, including CRISPR, also loom. Private companies are rapidly marketing new genetic tests that claim to predict traits from suicidality to opiate addiction, commonly based on limited, unreplicated studies of small samples, and lacking FDA approval for specific clinical uses. Yet countless physicians and patients insufficiently grasp these shortcomings.

In the infertility industry, laissez-faire market capitalism with little regulation has created a "Wild West." Numerous providers and patients have little heeded the broader social risks and implications of their choices, and the rights of the unborn children, often thinking little beyond their *own* immediate personal needs and desires. Yet, tens of thousands of infertility patients' daily decisions are affecting our country and species and, with the advent of CRISPR, will increasingly do so. Federal laws may not best address these concerns, especially when conservatives have immense political power that could threaten to eliminate many of these technologies' potential benefits.

While ASRM recommends that doctors not perform procedures "solely" for their own profit, larger questions emerge of what role, if any, such financial gain should play. No standard now exists. Any use of procedures or tests should be based on evidence of their potential to significantly benefit the patient. Yet at times such evidence is lacking.

Eventually, hopefully when research better comprehends CRISPR's risks, some scientist will undoubtedly transfer other genetically-edited human embryos into wombs, following Dr. He. Yet PGD already permits prospective parents to prevent transmission of dangerous mutations that they might otherwise pass on to their children. It is hard to imagine what medical benefits genetically-editing embryos will provide to the resulting child that PGD cannot already offer. I suspect, however, that CRISPR will be used not to prevent parents from transmitting deleterious genes, but to enhance progeny with socially-desirable traits. The men and women in this book suggest that physicians' and patients' personal and professional characteristics, and views of the relevant ethical, financial, social, cultural, religious and spiritual issues will profoundly shape these decisions; and that many providers will readily use CRISPR once it is approved as an experimental or clinical procedure, in a variety of ways, even if risks, though better described than at present, remain.

If the past is precedent, professional guidelines may lack bite. While the Asilomar conference of 1975 slowed the use of recombinant DNA[1], enormous profit-seeking companies now play major roles in DNA technologies and ARTs. In the US and some—though not all—other countries, the combination of providers and prospective parents valuing partial autonomy over potential risks to the future child and social costs, and free market approaches, with countless physicians pursuing profit and resisting regulations, suggest that many doctors will readily employ CRISPR, on an "experimental basis," once they feel it is possible to do so.

We need now to enhance discussion, awareness and understandings of these issues among patients, their families, clinicians in various fields, professional organizations, policymakers, and the public at large.

Even without CRISPR, we need to better appreciate the benefits and potential limitations and challenges of ARTs, and seek to improve the status quo. Each year, ARTs produce thousands of babies in the United States but also countless miscarriages, losses, discomforts, frustrations, anxieties, disappointments, expenses, second mortgages, bankruptcies, marital conflicts, and divorces. Physicians and professional organizations need to temper advertising and hype. Many providers recognize their great responsibilities in creating new life but also see their work as highly lucrative and routine and resist enhanced monitoring. We need to combat the stigma and shame surrounding both infertility and treatment, but also reduce problematic uses of these procedures.

These rapidly advancing technologies remain expensive and not universally available, widening inequalities between rich and poor and posing problems regarding social justice—from access to care to eugenics. I have focused here on patients who have in fact used these services, and thus have the financial resources to do so. Unfortunately, since these services remain expensive and little covered by insurance, financial constraints perpetuate economic and health disparities. Given possible selection of eggs, sperm, and embryos, with or without certain desired traits and genes, these technologies are altering how thousands of affluent people, but not poorer individuals, look and live. Wealthy individuals are eliminating certain lethal mutations from their offspring while the poor cannot. Several types of breast and colon cancers, anemias, and other diseases that now equally affect both the wealthy and the poor are becoming disorders more of the poor. Resources and research to battle these conditions may then fall, while stigma rises. Future parents of children with certain disorders may be asked why they did not simply screen out these diseases. Patients may be asked why their progenitors did not do so.

Hopefully, popular pressure will prompt governments to mandate more IVF coverage; but limits will remain on who receives treatment and how much—whether coverage will include, for instance, all women in their late 30s, as well as single, gay and lesbian parents.

Stronger guidelines and government-supported collection of better data are vital. Egg donors, in particular, may not fully grasp the risks they face but need to do so.

In the current US political climate, the pro-life religious right might get involved and impair, rather than enhance the status quo, impeding wider access to services. But that possibility should not serve as an excuse to avoid all improvements. Several years ago, doctors had feared that the religious right might prohibit PGD or other procedures, but that did not occur—in part, I suspect, because the notion of being able to control and improve one's child's genes, if you one afford to do so, appeals to libertarian beliefs.

Still, given the overall limited supply of healthcare resources, complex moral as well as economic trade-offs emerge of what medical services should be covered and for whom. As various US states and other governments expand coverage, they can and should consider coupling it to clinical practice standards and outcomes—for instance, to limiting transfers of extra embryos. The costs of lives, born and unborn, and political and social responses to these larger dilemmas will doubtlessly continue to shape this calculus.

Ultimately, these technologies trigger deep questions, too, about identity and metaphysics—who we are as human beings. ARTs contain odd dualities—whether making babies is a commercial enterprise (like manufacturing any other product, from shoes to smart phones) or something different, special, even sacred. This duality underpins controversies about reproduction, contraception, and abortion. After all, an embryo is arguably both an object *and* a potential life, both material *and* special, mere clumps of cells *and* a possible human being. By no means do I think that embryos *are* living creatures, with the same rights as full living human beings. But, as individuals and as a society, we struggle with precisely *how much* moral status, if any, embryos possess.

Potential genetic determinism and miscomprehensions loom. Compartmentalization of genetics and reproduction fuels misunderstandings and prejudice. Given websites selling human eggs, with drop-down menus of hair color, eye color, height, ethnicity, religion, education, and other characteristics, we need to consider carefully what is being sold.

Genetics lures us with the promise of fortune telling. Every day, millions of people read their horoscopes and pay palm readers to foretell the future. Yet, people also want to avoid feeling helpless, cursed, doomed or genetically determined, and rebel against notions that their fates are somehow sealed. Desires to know the future fuel hype. Companies peddle genetic tests of embryos, fetuses, babies, and adults that promise the power to take control of our own and our children's fates. But scientists will probably require decades, if not longer, to find and understand genes that strongly affect most diseases for most human beings. In the meantime, the allure of knowledge of our fate attracts many people who fail to grasp the inherent limitations of much genetic testing. The suggestions here can improve these fields, benefitting innumerable people.

We stand now at a great crossroads, able as never before to mold our descendants' and species' genetic futures. We must prepare. The men and women here provide rare insights on how we, as individuals and as a society, do and should respond and ready ourselves for these ever-advancing interventions, creating future human beings. These rapidly spreading technologies are altering our species. More awareness, education, guidelines, and research regarding these realms and efforts to reduce rather than expand disparities between the rich and poor are crucial.

As we evolve ever further into these brave new worlds, the individuals here, now wrestling with these conundrums, can aid us all.

APPENDICES

APPENDIX A | Glossary of Terms and Abbreviations

ADA: Americans with Disabilities Act

Amniocentesis ("amnio"): Test of the amniotic fluid that surrounds the fetus for genes or infections.

Aneuploidy: An abnormal number of chromosomes. The leading genetic cause of congenital birth defects in humans.

Ankylosing spondylitis: A serious, untreatable form of arthritis where spinal vertebrae fuse together, causing immobility, pain, and discomfort.

ART: Assisted reproductive technology

ASRM: American Society for Reproductive Medicine

Autosomal Dominant: A pattern of inheritance in which a disease occurs because a mutation exists on one non-sex chromosome.

Autosomal Recessive: A pattern of inheritance in which a disease occurs because a mutation exists on two non-sex chromosomes. Two parents may each thus be carriers of the mutation and be without symptoms but are at risk of transmitting both of these mutations to a child, who could thus get sick.

Azoospermic: The medical condition of a man not having any measurable level of sperm in his semen.

Beta-thalassemia: Autosomal recessive inherited blood disorder. Carriers have a reduced risk of acquiring malaria, but two carriers can have children with the disease.

BRCA: Breast cancer gene

CCCT: Clomid Challenge Test

CDC: Centers for Disease Control and Prevention, in the United States

CF: Cystic fibrosis. A progressive, fatal disease that causes persistent lung infections autosomal recessive.

Chromosome: Elongated coils of tightly packaged DNA. Humans have 23 pairs of chromosomes—one set from each parent—totaling 46.

Clomid: Clomiphene. Medication used to treat infertility in women. It stimulates an increase in the number of hormones that support the growth and release of a mature egg.

CMV: Cytomegalovirus

COIs: Conflicts of interest

CRISPR (clustered regularly interspaced short palindromic repeats): Technology that allows permanent modification of genes.

CVS: Chorionic villus sampling

Diamond-Blackfan anemia: An autosomal dominant inherited anemia, consisting of abnormally low numbers of red blood cells. Other medical problems occur as well in about half of patients.

DNA: Deoxyribonucleic acid, which constitutes genes and consists of four chemical bases: adenine (A), guanine (G), cytosine (C), and thymine (T). Human DNA consists of about 3 billion bases arranged in unique sequences for each person. Humans all share over 99% of their DNA.

Down syndrome: (See Trisomy 21)

EHR: Electronic health record

Emanuel syndrome: A genetic disorder in which parts of two chromosomes become translocated in ways that can be either balanced (and benign) or unbalanced (and cause severe symptoms).

Embryo: An organism in the earliest stages of development.

Endocrinology: The study of hormones within the body, including those in reproduction.

Endometriosis: A condition in which tissues that usually cover the uterus grow in the fallopian tubes, ovaries, or elsewhere, resulting in severe cramping, outer abdominal scarring, and, possibly, infertility.

Estradiol: A hormone that decreases follicle-stimulating hormone and increases luteinizing hormone.

Fanconi anemia (FA): A rare genetic disease that impairs bone marrow, red blood cell production, and other tissue functions.

FDA: Food and Drug Administration, in the United States

Follicles: Sacks in ovaries that contain mature eggs that develop over time.

Fragile X syndrome: A genetic condition that causes intellectual disability, behavioral and learning challenges, and various physical characteristics.

FSH: Follicle-stimulating hormone

Gamete: An egg or sperm.

Gene: Sections of DNA that encode instructions to make proteins that form cells, tissues, organs, and organisms. Humans have 20,000–25,000 genes.

GP: General practitioner

HD: Huntington's disease. A fatal neurological and psychiatric disorder caused by an autosomal dominant mutation.

Hemophilia: A genetic blood clotting deficiency most commonly occurring in males.

HEFA: Human Fertilisation and Embryology Authority, in the United Kingdom

HIPAA: Health Insurance Portability and Accountability Act, in the United States

Hormone: Chemicals made by animals and plants to regulate functions of organs or tissues.

Hysterosalpingogram (HSG): An X-ray to examine the uterus, fallopian tubes, and surrounding areas.

ICSI: Intracytoplasmic sperm injection

IUI: Intrauterine insemination

IVF: In vitro fertilization

Klinefelter syndrome: Caused by an extra chromosome in males and leading to various bodily symptoms.

Laparoscopy: Operation performed in the abdomen or pelvis through small incisions with the aid of a camera. It can either be used to inspect and diagnose a condition or to perform surgery.

Leukemia: A disease that causes the bone marrow to create high numbers of abnormal white blood cells.

Li-Fraumeni syndrome: A rare autosomal dominant genetic disorder that predisposes to cancer.

Luteinizing hormone: A hormone produced by the pituitary gland that triggers the release of eggs from ovaries

MHP: Mental health provider

MMPI: Minnesota Multiphasic Personality Inventory. A test to assess and diagnose mental illness.

Mosaicism: When an embryo's cells are not all be genetically identical.

Myotonic dystrophy: An autosomal dominant genetic disorder that impairs muscles, leading to muscle contractions loss and weakness.

NHS: National Health Service, in the United Kingdom

NICU: Neonatal intensive care unit

OB/GYNs: Obstetrics and gynecology specialists

OHSS: Ovarian hyperstimulation syndrome. A medical condition affecting the ovaries of some women who take fertility medication to stimulate egg growth. Most cases are mild, but in some cases, the condition is severe and can lead to serious illness or death.

Oocytes: Immature eggs.

Ovary: The female reproductive organ that produces eggs.

Ovulation: The release of female eggs.

PCP: Primary care provider

Penetrance: The degree to which a mutation, if present, will in fact cause symptoms.

PGD: Preimplantation genetic diagnosis

PGD/HLA: Screening embryos to find those that will produce a child who will be human leukocyte antigen–compatible with another individual (usually a sibling) to whom the child can thus donate tissues or organs, with less risk of rejection.

PGS: Preimplantation genetic screening (now referred to as PGT-A), used to test whether a cell has aneuploidy (contains an abnormal number of chromosomes).

Preeclampsia (or Toxemia): A condition in which a pregnant woman develops high blood pressure, accompanied by high protein levels in the urine and swelling of the hands and face.

REIs: Reproductive endocrinology and infertility specialists. A subspecialty of OB/GYN concerned with infertility.

Reiter's syndrome or reactive arthritis: Named for a Nazi war criminal who conducted unethical experiments in concentration camps, prompting efforts to stop using this name for this condition.

RESOLVE: A US national infertility association.

Retinoblastoma: A cancer in the retina, located in the rear of the eye, which is inheritable in around half of all cases.

SART: Society for Assisted Reproductive Technology

SET: Single embryo transfer; transfer of only a single embryo into the womb.

Sickle cell disease: Hereditary disease caused by abnormality in red blood cell proteins that carry oxygen.

Spinal muscular atrophy: a rare, autosomal recessive neuromuscular disorder

Sperm sorting: Procedures to try to determine the sex of a future child by sorting the sperm by whether they carry the male or female chromosome.

STD: Sexually transmitted disease

Storage disease: When the body fails to store properly enzymes that digest proteins and other chemicals.

Tay-Sachs Disease: A rare, inherited, and fatal childhood disorder that progressively destroys nerve cells (neurons) in the brain and spinal cord.

Trisomy: The presence of an extra chromosome, leading to three rather than a pair of chromosomes.

Trisomy 13 (or "Patau syndrome"): Severe genetic disorder caused by the presence of an extra, third copy of chromosome 13 that kills infants before their first birthday. Risk increases with the mother's age.

Trisomy 18 (or "Edwards syndrome"): Severe genetic disorder caused by the presence of an extra, third copy of chromosome 18. Risk increases with the mother's age.

Trisomy 21 (or "Down syndrome"): Serious genetic disorder caused by the presence of an extra third copy of chromosome 21. Risk increases with the mother's age.

Urologists: Specialists in male infertility.

Uterus: The womb—female organ in the lower body of a woman or female mammal where offspring are conceived and develop before birth.

Von Hippel-Lindau syndrome: Or familial cerebello retinal angiomatosis—a rare autosomal dominant genetic disorder characterized by multiple cysts and tumors that can become cancerous.

APPENDIX B | Methods

Since no prior studies have been published examining how patients and providers viewed and made decisions about many key aspects of infertility treatment, I used qualitative interview methods, which can best elicit the full range and typologies of attitudes, interactions, and practices involved and can inform subsequent quantitative studies. From a theoretical standpoint, the anthropologist Clifford Geertz[1] has advocated studying aspects of people's lives, decisions, and social situations not by imposing theoretical structures but by trying to understand these individuals' own experiences, drawing on their own words and perspectives to obtain a "thick description." For the methods for the present study, I thus adapted elements from "grounded theory"[2] using techniques of "constant comparison," comparing data from different contexts for similarities and differences, to see whether they suggest hypotheses. This technique generates new analytic categories and questions and checks them for reasonableness. These methods have been used in several other studies on key aspects of health behavior, doctor–patient relationships, and communications in genetics and other areas. During the ongoing process of interviewing, I constantly considered how participants resembled or differed from one another and how social, cultural, and medical contexts and factors might contribute to these differences. Grounded theory also involves both deductive and inductive thinking, building inductively from the data to an understanding of themes and patterns within the data and deductively drawing on frameworks from prior research and theories.

Participants

As shown in Table AP.1, I formally interviewed 37 infertility providers and patients, recruiting them through listservs, emails, and word of mouth. Providers were also recruited through national American Society for Reproductive Medicine meetings (e.g., preimplantation genetic diagnosis and mental health provider [MHP] interest group meetings). I approached these meeting attendees to ascertain whether they might be interested in participating in an interview study and, if so, I emailed them

	MALE	FEMALE	TOTAL
Physicians	14	3	17
Physicians who are also patients	0	1	1
Type of practice			
University-affiliated	5	1	6
Private practice	9	2	11
Other ART providers (e.g., MHPs, nurses)	1	9	10
Other providers who are also patients	0	3	3
Patients	1	9	10
Total	16	21	37

ART, assisted reproductive technology; MHP, mental health provider.

information about it. Most agreed to participate and did so. I also used a mental health listserv received by approximately 60 members (not all of whom are active), of whom 15 responded, and I then interviewed the first eight respondents. I conducted additional interviews as background, for informational purposes, with eight physicians, nine MHPs, and 14 patients, who informed, but were not included in, the final formal data analysis. I conducted interviews for the formal data analyses with each group until "saturation" was reached (i.e., "the point at which no new information or themes are observed in the data"[3]). These interviewees were from across the United States. The clinicians described interactions with multiple patients and colleagues, and patients often described interactions with multiple providers and other patients.

Instruments

I drafted the semistructured interview questionnaire, drawing on prior literature and exploring patients' and providers' views, experiences, and decisions concerning infertility treatment (see Appendix C for sample questions).

Data Analysis

Transcriptions and initial analyses of interviews occurred during the period in which the interviews were being conducted, enhancing validity and helping to shape subsequent interviews. Once the full set of interviews was completed, a trained research assistant and I conducted subsequent analyses in two phases. In phase I, we independently examined a subset of interviews to assess factors that shaped participants' experiences, identifying categories of recurrent themes and issues that were subsequently given codes. We read each interview, systematically coding blocks of text to assign "core" codes or categories. While reading the interviews, we inserted a topic name (or code) beside each excerpt of the interview to indicate the themes being discussed. We then worked together to reconcile these independently developed coding schemes into a single scheme. Next, we prepared a coding manual, defining each code and examining areas of disagreement until reaching consensus. New themes that did not fit into the original coding framework were modified for the manual when appropriate.

In phase II of the analysis, we independently content-analyzed the data to identify the principal subcategories and ranges of variation within each of the core codes. We reconciled the sub-themes we identified into a single set of "secondary" codes and an elaborated set of core codes. These codes assess subcategories and other situational and social factors.

We then used these codes and subcodes in analyzing all of the interviews. To ensure coding reliability, two coders analyzed all interviews. Where necessary, multiple codes were used. We assessed similarities and differences among participants, examining categories that emerged, ranges of variation within categories, and variables that may be involved. We examined areas of disagreement through closer analysis until we reached that consensus. We checked regularly for consistency, and accuracy in ratings was checked regularly by comparing earlier and later coded excerpts.

APPENDIX C | Sample Questions

For Providers

Background Information

- How would you describe your practice and your patient mix (i.e., by types of procedures, ages, diseases, insurance coverage, ethnicity, and religious beliefs)?
- Approximately how many patients do you see per year?
- Do you have a private or hospital-based practice?

General

- How do you view current uses of IVF and PGD?
- What challenges do you face in your work as an ART provider?
- How do you address these challenges?

Specific Procedures

- *PGD*
  - Have you ever recommended PGD/PGS to patients? For what indications? How often? How did you make these decisions?
  - Have you ever decided not to support use of PGD/PGS? Why?
  - Do you have any concerns about patients using PGD/PGS? If so, what?
  - Have any changes occurred over time in your PGD/PGS use? If so, what?
  - What obstacles exist in patients' use of IVF/PGD? Have you tried addressing these? If so, how? With what effect?
  - What are the most difficult IVF and/or PGD decisions you have faced? How did you resolve these?
  - What technical barriers to IVF and PGD exist?
  - What are your PGD and IVF success rates?

- Have you had any adverse outcomes associated with IVF or PGD?
- Have you faced challenges or concerns regarding PGD? Non-disclosing PGD? PGD for HLA typing?
- *Other procedures*
  - Have you faced challenges concerning procuring oocytes from donors? If so, when? What happened?
  - Have you ever faced challenges concerning egg donor agencies? If so, when? What happened?
  - Have you faced challenges concerning numbers of embryos transferred? Pregnancy reduction? If so, when? What happened?
  - Have you ever faced challenges related to other procedures? If so, what happened?

Doctor–Patient Relationships

- Have patients ever been uncertain or confused about using IVF and/or PGD and, if so, how?
- Have you seen any problems in provider or patient expectations and misunderstandings about IVF and/or PGD? If so, what?
- Have you ever faced challenges regarding referrals of ART patients for treatment, patients changing doctors, or doctor shopping? If so, when? What happened?
- Have you ever faced challenges related to other aspects of doctor–patient relationships? If so, what?

Patients' Characteristics

- Have you faced other challenges or concerns regarding patients? If so, what? Regarding ages of patients? Nontraditional combinations of parents? Patients' future parenting abilities? If so, when and how? What has been difficult about these situations? What did you do? How did you make these decisions?
- Have you ever considered not treating a patient? Why? What did you do?
- How have your patients viewed these issues?

Factors

- *Financial*
  - Have cost issues arisen with patients' use of IVF/PGD? If so, how?
  - Have you perceived financial barriers to IVF/PGD use? If so, what? How frequently?
  - What have been your experiences with patients using insurance for PGD? When have insurance companies agreed or not agreed to pay for IVF and/or PGD? Have you tried to address these issues? If so, how?
- *Institutional*
  - Does the clinical setting in which you work affect your use of IVF or PGD? If so, how?
- *Facilitators*

- Have any approaches helped to overcome barriers to use that may exist?
- Do you perceive any facilitators to use of IVF/PGD? If so, what (e.g., informational, financial, attitudinal, institutional)?
- *Personal*
  - Have you or anyone you have known well ever considered or undergone IVF? When and why? What were their experiences like? Did that influence you at all? If so, how?
  - Do you have religious, spiritual, or other beliefs that affect your views or approaches to IVF or PGD in any way? If so, how?

Implications

- Do you see areas for improving treatment? If so, what?
- In what areas, if any, do you think additional professional or public education would be helpful?
- What additional guidance, if any, do you think would be helpful?
- Do you have other thoughts on these issues?

For Patients

Background Information

- Age, gender, ethnicity, religion, geographic location, type of insurance, professional/educational background, socioeconomic status
- How many children do you have?
- What illness(es), if any, run in your family? Have you had any children born with birth defects? With other serious illnesses? If so, what? Have genetic causes been found? Do you think the illness(es) could recur in a future pregnancy? What do you think is the risk?
- Have you considered doing anything to prevent transmission of these illnesses in the next generation? If so, what? Amniocentesis/chorionic villus sampling? PGD? Donor egg/sperm? If so, when? Why?

Decision-Making Process

- Have you considered using IVF/PGD? Have you ever undergone IVF or PGD? If so, when? Why?
- Have you heard of PGD? Where did you first hear about it? What were your sources of information about PGD, and how helpful were they? Did anyone provide information on PGD (e.g., foundations/patient organizations, healthcare providers, friends, family)?
- What made you decide to undergo/not undergo IVF/PGD? How was your experience with considering IVF/PGD? When did you consider it? Why then? Why not earlier or later?

- How did you choose an IVF clinic?
- Do you think PGD can affect the risk of birth defects? If so, how?
- Did you also pursue prenatal diagnosis for the genetic condition? Before or after PGD? Why? What factors affected your decisions?
- Did you screen your embryos? If so, for what (e.g., aneuploidy, gender selection, single genetic disorder)? Who paid? How much?
- What other reproductive options did you consider (e.g., prenatal diagnosis, amniocentesis, donor egg/sperm, adoption/not having additional children)? How did you weigh these alternatives?
- How many cycles have you gone through thus far? Do you know how well they worked? Do you know how many embryos were successfully fertilized? Do you know how many were still viable at the time for transfer? Do you know what happened to the affected embryo(s)?
- Have you agreed to long-term storage of the unused embryos? Has anyone discussed with you issues concerning multiple births?
- How satisfied were you with the IVF/PGD process? How simple or difficult did you find it?
- Did religious views affect your decisions about PGD at all? What are your political beliefs? Did they affect your views of PGD at all?
- Did cost of the procedure play a role? What were you told was the "take-home baby" rate? Chance of PGD error? Risks associated with IVF/PGD?

Roles of Others in Decision-Making

- Did anything help you in making your decisions or make these more difficult? If so, what?
- How did your partner feel about the reproductive options? Did you disagree at all? If so, how? How did you resolve these differences?
- Did you discuss your choices with anyone besides your partner? If so, with whom? What did they say? How did you feel about that?
- Did you consult with religious advisors, family members, web resources, other media, genetic counselors, obstetricians, pediatricians, general practitioners, or other physicians, nurses, or healthcare providers? What did each tell you about PGD?
- Did you feel that your providers were sensitive to the issues you faced? How so?
- Do you know other people who have undergone PGD?
- Did their experiences affect you at all? If so, how?
- Are there others who you considered discussing it with but decided not to? If so, who? Why?

Informational, Financial, Institutional, and Attitudinal Barriers

- *Informational*
  - Did you feel you knew enough to make a decision?
  - Were you satisfied with the information you received from websites, patient organizations, or providers about IVF/PGD? Would you have wanted more? If so, what?

- *Financial*
  - Was cost at all an issue? How much did or would IVF/PGD cost you out-of-pocket? What type of insurance did you have?
  - What was your insurance company's policy on IVF/PGD? Did they cover the cost of IVF? Of PGD? If so, how much of it?
  - What limitations, if any, were involved? Were there any grants or donations available to cover any portion of the PGD costs? Who do you think should pay for PGD?
- *Attitudinal*
  - Have your religious or spiritual beliefs affected your views or approaches to PGD? If so, how?
  - How do you feel about abortion? Do your views about abortion relate in any way to your views about IVF/PGD? If so, how?
  - Would you consider terminating a pregnancy because of a prenatal diagnosis at 12–13 weeks of pregnancy?
- *Institutional*
  - Has the institution where you get your care (i.e., clinic of providers) affected your decision about or your use of IVF/PGD? If so, how?
- *Facilitators*
  - Did anything help you in making your decision about IVF/PGD? If so, what?
  - What source of information was most helpful? Did input from particular individuals prove helpful (e.g., IVF provider, obstetrician, geneticist, genetic counselor, patient organization, friends, family members)? If so, who? What did they say?
  - How important to you were each of the factors that affected your decision about IVF/PGD? Do you think other policies or guidelines should be developed concerning IVF/PGD use? If so, what?
  - What tests using PGD do you think should be available? What information do you think other patients would like concerning use of PGD? How important, on a scale of 1–10, do you think each of these issues was in your decision about whether to undergo PGD (i.e., costs, availability of information, risks, benefits, input from others)?

ACKNOWLEDGMENTS

I am deeply indebted to the men and women who spoke to me for this project for their candor and openness—their willingness to share their views and experiences, their stories of triumphs and failures, dreams and doubts.

I would like to thank Daniel Marcus-Toll, Bela Fishbeyn, Brigitte Burquez, Sayantanee Das, Kristina Khanh-Thy Hosi, Jiseop Kim, Sarah Kiskadden-Bechtel, Rebecca Pol, Andrew Rock, Charlene R. Sathi, Alexa A. Woodward, and most especially Patricia Contino, for their assistance with this project.

In addition, I am enormously grateful to Renée Fox, Joel Conarroe, Rick Hamlin, Jamie Marks, Melanie Thernstrom, Royce Flippin, Julie Danaher, and Stewart Moyer for their suggestions and comments concerning the text.

I want to express my profound appreciation to Peter Ohlin, Madeleine Freeman, Poppy Hatrick and Erin Cox at Oxford University Press, Susan Golombock, Mark Sauer, Eli Adashi and of course, Charles Bieber.

Research for this book was supported by the National Center for Research Resources (grant UL1 RR024156), the Greenwall Foundation, and the John Simon Guggenheim Memorial Foundation.

Portions of this material have appeared, though in different form and focus, in *Fertility and Sterility, Human Fertility, Human Reproduction, Reproductive Biomedicine, Journal of Generic Counseling, PLOS ONE, American Journal of Bioethics: Empirical Research, BMC Medical Ethics,* and *BMC Women's Health.*

ABOUT THE AUTHOR

Robert L. Klitzman, MD, is a Professor of Psychiatry in the College of Physicians and Surgeons and the Mailman School of Public Health, and the Director of the Masters of Bioethics Program at Columbia University. His prior books include *Am I My Genes?: Confronting Fate and Family Secrets in the Age of Genetic Testing, The Ethics Police?: The Struggle to Make Human Research Safe, When Doctors Become Patients,* and *A Year-long Night: Tales of A Medical Internship.* He has received several awards for his work, including fellowships from the John Simon Guggenheim Foundation and the Russell Sage Foundation, and is a gubernatorial appointee to the Empire State Stem Cell Board and a Distinguished Fellow of the American Psychiatric Association (see Figure A.1).

FIGURE A.1 The author with Louise Brown, the first "test tube baby"

NOTES

Chapter 1

1. Cobb, T., "An amazing 10 years: the discovery of egg and sperm in the 17th century," *Reproduction in Domestic Animals*, 47, Suppl 4 (2012): 2–6.

2. Spallanzani, Lazzaro, "Dissertations Relative to the Natural History of Animals and Vegetables," 1789, digitized and available through Columbia University, Department of Psychology, the John G. Curtis Library, 2010, accessed July 31, 2018, https://archive.org/stream/dissertationsrelo2spal/dissertationsrelo2spal_djvu.txt.

3. Ibid.

4. Belmont Gasking, Elizabeth, "Lazzaro Spallanzani," *Encyclopedia Britannica* (2008), accessed July 31, 2018, https://www.britannica.com/biography/Lazzaro-Spallanzani.

5. Wikipedia, "Lazzaro Spallanzani," page last edited May 10, 2018, accessed July 31, 2018, https://en.wikipedia.org/wiki/Lazzaro_Spallanzani.

6. Hard, Addison D, "Artificial Impregnation," *Medical World* 27 (1909): 163–64.

7. Ibid., 163–164.

8. Gregoire, A. T., and R. C. Mayer, "The Impregnators," *Fertility and Sterility* 16 (1965): 130–34.

9. Trounson, A., et al., "Pregnancy Established in an Infertile Patient After Transfer of a Donated Embryo Fertilised In Vitro," *British Medical Journal (Clinical Research Edition)* 286, no. 6368 (1983): 835–38.

10. Livingston, Gretchen, "A Third of U.S. Adults Say They Have Used Fertility Treatments or Know Someone Who Has," Pew Research Center, July 17, 2018, http://www.pewresearch.org/fact-tank/2018/07/17/a-third-of-u-s-adults-say-they-have-used-fertility-treatments-or-know-someone-who-has/.

11. Fox, Maggie, "A Million Babies Have Been Born in the U.S. with Fertility Help," *NBC News*, April 18, 2017, https://www.nbcnews.com/health/health-news/million-babies-have-been-born-u-s-fertility-help-n752506.

12. De Geyter Ch., et al., "Data Collection Systems in ART Must Follow the Pace of Change in Clinical Practice," *Human Reproduction* 31 (2016): 2160–63.

13. Sunderam, S., et al., "Assisted Reproductive Technology Surveillance—United States 2015," *Centers for Disease Control and Prevention Surveillance Summaries* 67, no. 3 (2018): 1–28, https://www.cdc.gov/mmwr/volumes/67/ss/ss6703a1.htm?s_cid=ss6703a1_w

14. Chandra, Anjani, Casey E. Copen, and Elizabeth Hervey Stephen, "Infertility Service Use in the United States: Data from the National Survey of Family Growth, 1982–2010," *National Health Statistics Reports* 73, no. 73 (2014): 1–20.

15. Martin, Joyce A., et al., "Births: Final Data for 2007," *National Vital Statistics Reports* 58, no. 24 (2010): 1–85.

16. Klitzman, Robert, *Am I My Genes? Confronting Fate and Family Secrets in the Age of Genetic Testing* (New York: Oxford University Press, 2012).

17. Centers for Disease Control and Prevention, "Assisted Reproductive Technology (ART) Data," 2015, https://nccd.cdc.gov/drh_art/rdPage.aspx?rdReport=DRH_ART.ClinicInfo&ClinicId=9999&ShowNational=1.

18. US Food and Drug Administration, "What You Should Know – Reproductive Tissue Donation," last modified March 26, 2018, accessed March 20, 2014, https://www.fda.gov/biologicsbloodvaccines/safetyavailability/tissuesafety/ucm232876.htm.

19. Centers for Medicare and Medicaid Services, last modified January 27, 2014, accessed February 14, 2014, http://www.cms.gov/.

20. Society for Assisted Reproductive Technology, last modified 2014, accessed February 12, 2014, http://www.sart.org/.

21. American Society for Reproductive Medicine, last modified 2014, accessed January 31, 2014, https://www.asrm.org/.

22. Klitzman, *Am I My Genes?*

23. Martin et al., "Births."

24. Daar, Judith, "Federalizing Embryo Transfers: Taming the Wild West of Reproductive Medicine," *Columbia Journal of Gender & Law* 23 (2012): 257.

25. Jackson, Rebecca A., Kimberly A. Gibson, Yvonne W. Wu, and Mary S. Croughan. "Perinatal Outcomes in Singletons Following in vitro Fertilization: A Meta Analysis," *Obstetrics & Gynecology* 103, no. 3 (2004): 551–63.

26. Hansen M, Bower C, Milne E, de Klerk N, Kurinczuk JJ. "Assisted Reproductive Technologies and the Risk of Birth Defects: A Systematic Review," *Human Reproduction* 20, no. 2 (2005): 328–38.

27. Hansen, Michelle, Katherine R. Greenop, Jenny Bourke, Gareth Baynam, Roger J. Hart, and Helen Leonard. "Intellectual Disability in Children Conceived Using Assisted Reproductive Technology," *Pediatrics* 142, no. 6 (2018): 328–38.

28. Zarembo, A., "An Ethics Debate over Embryos on the Cheap," *Los Angeles Times*, November 19, 2012, http://articles.latimes.com/2012/nov/19/local/la-me-embryo-20121120.

29. Keehn, Jason, et al., "Recruiting Egg Donors Online: An Analysis of In Vitro Fertilization Clinic and Agency Websites' Adherence to American Society for Reproductive Medicine Guidelines," *Fertility and Sterility* 98, no. 4 (2012): 995–1000.

30. Keehn, Jason, et al., "How Agencies Market Egg Donation on the Internet: A Qualitative Study," *Journal of Law, Medicine & Ethics* 43, no. 3 (2015): 610–18.

31. Holwell, Eve, et al., "Egg Donation Brokers: An Analysis of Agency Versus In Vitro Fertilization Clinic Websites," *Journal of Reproductive Medicine* 59 (2014): 534–41.

32. Levine, Aaron D., "Self-Regulation, Compensation, and the Ethical Recruitment of Oocyte Donors," *Hastings Center Report* 40, no. 2 (2010): 25–36.

33. Ibid.

34. Ethics Committee of the American Society for Reproductive Medicine, "Financial Compensation of Oocyte Donors," *Fertility and Sterility* 88, no. 2 (2007): 305–9.

35. Practice Committee of the American Society for Reproductive Medicine and Practice Committee of the Society for Assisted Reproductive Technology, "Recommendations for Gamete and Embryo Donation: A Committee Opinion," *Fertility and Sterility* 99, no. 1 (2013): 47–62.

36. Practice Committee of the American Society for Reproductive Medicine and Practice Committee of the Society for Assisted Reproductive Technology, "Recommendations for Practices Utilizing Gestational Carriers: A Committee Opinion," *Fertility and Sterility* 103, no.1 (2017): e1–8.

37. Lindsay Kamakahi v American Society for Reproductive Medicine, case no. C 11-01781 SBA, N. D. Cal, March 29, 2013, accessed August 17, 2016, http://www.leagle.com/decision/In%20FDCO%2020130409725.

38. Klitzman, Robert L., and Mark V. Sauer, "Kamakahi vs. ASRM and the Future of Compensation for Human Eggs," *American Journal of Obstetrics and Gynecology* 213, no. 2 (2015): 186–87.

39. LaFreniere, M., "Egg Donors Get Price Cap Removed in Class Action Lawsuit Settlement," *TopClass Actions*, February 2, 2016, accessed August 17, 2016. http://topclassactions.com/lawsuit-settlements/lawsuit-news/327105-eggdonors-get-price-cap-removed-in-class-action-lawsuit-settlement/.

40. Zarembo, A., "An Ethics Debate over Embryos on the Cheap," *Los Angeles Times*, November 19, 2012.

41. Wu, Alex K., et al., "Time Costs of Fertility Care: The Hidden Hardship of Building a Family," *Fertility and Sterility* 99, no. 7 (2013): 2025–30.

42. Katz, Patricia, et al., "Costs of Infertility Treatment: Results from an 18-Month Prospective Cohort Study," *Fertility and Sterility* 95, no. 3 (2011): 915–21.

43. Chambers, Georgina M., G. David Adamson, and Marinus J. C. Eijkemans, "Acceptable Cost for the Patient and Society," *Fertility and Sterility* 100, no. 2 (2013): 319–27.

44. Chambers, Georgina M., et al., "The Economic Impact of Assisted Reproductive Technology: A Review of Selected Developed Countries," *Fertility and Sterility* 91, no. 6 (2009): 2281–94.

45. Präg, Patrick, and Melinda C. Mills, "Assisted Reproductive Technology in Europe. Usage and Regulation in the Context of Cross-Border Reproductive Care," *Families and Societies* 43 (2015): 1–23.

46. Arons, Jessica, "Future Choices": Assisted Reproductive Technologies and the Law," Center for American Progress, December 17, 2007, accessed February 18, 2014, https://cdn.americanprogress.org/wp-content/uploads/issues/2007/12/pdf/arons_art.pdf.

47. Servick, Kelly, "CRISPR—A Weapon of Mass Destruction?" *Science* February 11, 2016, http://www.sciencemag.org/news/2016/02/crispr-weapon-mass-destruction.

48. Liang, P., et al., "CRISPR/Cas9-Mediated Gene Editing in Human Tripronuclear Zygotes," *Protein & Cell* 6, no. 5 (2015): 363–72.

49. Kolata, G., "Chinese Scientists Edit Genes of Human Embryos, Raising Concerns," *New York Times*, Accessed April 23, 2015. https://www.nytimes.com/2015/04/24/health/chinese-scientists-edit-genes-of-human-embryos-raising-concerns.html.

50. Cyranoski, David, and Sara Reardon, "Chinese Scientists Genetically Modify Human Embryos," *Nature News*, April 22, 2015.

51. National Academies of Sciences, Engineering, and Medicine, "International Summit on Gene Editing," Accessed December 3, 2015. http://nationalacademies.org/gene-editing/Gene-Edit-Summit/.

52. Callaway, Ewen, "UK Scientists Gain License to Edit Genes in Human Embryos," *Nature*, Accessed February 1, 2016. https://www.nature.com/news/uk-scientists-gain-licence-to-edit-genes-in-human-embryos-1.19270.

53. Ledford, H., "CRISPR Fixes Disease Gene in Viable Human Embryos: Gene-Editing Experiment Pushes Scientific and Ethical Boundaries," *Nature*, August 2, 2017.

54. Zimmer, K. "CRISPR Scientists Slam Methods Used on Gene-Edited Babies," *The Scientist* 12/3/18. https://www.the-scientist.com/news-opinion/crisprscientists-slam-methods-used-on-gene-edited-babies--65167. Accessed 2/19/19.

55. Sara Reardon. "World Health Organization panel weighs in on CRISPR-babies debate," *Nature* 567 (2019): 444–45.

56. Cyranoski, D., "Russian Biologist Plans More CRISPR-edited Babies," Science, 10 June 2019, Accessed June 20, 2019. https://www.nature.com/articles/d41586-019-01770-x.

57. Mullin, Emily, "FDA Cracks Down on Pioneering Doctor Who Created a Three-Parent Baby," *MIT Technology Review*, August 7, 2019, https://www.technologyreview.com/s/608583/fda-cracks-down-on-pioneering-doctor-who-created-three-parent-baby/

58. Levine, A. D., "Self-Regulation, Compensation, and the Ethical Recruitment of Oocyte Donors," *Hastings Center Report* 40 (2010): 25–36.

59. Greely, Henry T., *The End of Sex and the Future of Human Reproduction* (Cambridge, MA: Harvard University Press, 2016).

60. National Academies of Sciences, Engineering and Medicine, *Human Genome Editing: Science, Ethics, and Governance* (Washington, DC: National Academies Press, 2017).

61. Lotz, Mianna. "Feinberg, Mills, and the child's right to an open future," *Journal of Social Philosophy* 37, no. 4 (2006): 537–51.

62. Peterson, B. D., et al., "The Longitudinal Impact of Partner Coping in Couples Following 5 Years of Unsuccessful Fertility Treatments," *Human Reproduction* 24, no. 7 (2009): 1656–64.

63. Geertz, Clifford, *The Interpretation of Cultures* (New York: Basic Books, 1973).

64. Klitzman, Robert, et al., "Views of Internists Towards Uses of PGD," *Reproductive Biomedicine Online* 26, no. 2 (2013): 142–47.

65. Klitzman, Robert, et al., "Views of Preimplantation Genetic Diagnosis Among Psychiatrists and Neurologists," *Journal of Reproductive Medicine* 59, no. 7–8 (2014): 385–92.

66. Appelbaum, Paul S., Charles Lidz, and Robert Klitzman, "Voluntariness of Consent to Research: A Conceptual Model," *Hastings Center Report* 39, no. 1 (2009): 30–39.

67. Klitzman, Robert, "The Myth of Community Differences as the Cause of Variations Among IRBs," *AJOB Primary Research* 2, no. 2 (2012): 24–33.

68. Howell, Eve, et al., "Egg Donation Brokers: An Analysis of Agency vs. IVF Clinic Websites," *Journal of Reproductive Medicine* 59, no. 11–12 (2014): 534–41.

69. Saravelos, Sotiris H., et al., "Pain During Embryo Transfer Is Independently Associated with Clinical Pregnancy in Fresh/Frozen Assisted Reproductive Technology Cycles," *Journal of Obstetrics and Gynaecology Research* 42, no. 8 (2016): 684–93.

Chapter 2

1. Ulrich, Miriam, and Ann Weatherall, "Motherhood and Infertility: Viewing Motherhood Through the Lens of Infertility," *Feminism & Psychology* 10, no. 3 (2000): 323–36.

2. Schmidt, Lone, et al., "Does Infertility Cause Marital Benefit? An Epidemiological Study of 2250 Women and Men in Fertility Treatment," *Patient Education and Counseling* 59, no. 3 (2005): 244–51.

3. Lotti, Francesco and Mario Maggio. "Ultrasound of the Male Gential Tract in Relation to Male Reproductive Health," *Human Reproduction Update* 21, no. 1 (2015): 56–83.

4. Levine, Hagai, Niels Jorgensen, Anderson Martino-Andrade, Jaime Mendiola, Dan Weksler-Derri, Irina Mindlis, Rachel Pinotti, Shanna H. Swan. "Temporal Trends in Sperm Count: A Systematic Review and Meta-Regression Analysis," *Human Reproduction Update* 23, no. 6 (2017): 646–59.

5. Drosdzol, Agnieszka, et al., "Quality of Life and Marital Sexual Satisfaction in Women with Polycystic Ovary Syndrome," *Folia Histochem Cytobiol* 45, no. S1 (2007): S93–97.

Chapter 3

1. Greil, Arthur L., et al., "Race-Ethnicity and Medical Services for Infertility: Stratified Reproduction in a Population-Based Sample of U.S. Women," *Journal of Health and Social Behavior* 52, no. 4 (2011): 493–509.

2. Centers for Disease Control and Prevention, accessed May 1, 2015, https://www.cdc.gov/.

3. Greil, Arthur L., and Julia McQuillan, "'Trying' Times," *Medical Anthropology Quarterly* 24, no. 2 (2010): 137–56.

4. Moreau, Caroline, et al., "When Do Involuntarily Infertile Couples Choose to Seek Medical Help?" *Fertility and Sterility* 93, no. 3 (2010): 737–44.

5. Bunting, Laura, and Jacky Boivin, "Decision-Making About Seeking Medical Advice in an Internet Sample of Women Trying to Get Pregnant," *Human Reproduction* 22, no. 6 (2007): 1662–68.

6. Iaconelli, A., et al., "Main Concerns Regarding In Vitro Fertilization Techniques: Results of a Website Survey," *Fertility and Sterility* 100, no. 3 (2013): S66.

7. Chandra, Anjani, and Elizabeth Hervey Stephen, "Infertility Service Use Among U.S. Women: 1995 and 2002," *Fertility and Sterility* 93, no. 3 (2010): 725–36.

8. Präg, P., and M. C. Mills, "Assisted Reproductive Technology in Europe. Usage and Regulation in the Context of Cross-Border Reproductive Care," *Families and Socities* 43 (2015): 1–23.

9. Greil et al., "Race-Ethnicity and Medical Services for Infertility."

10. Marcus, Diana, et al., "Infertility Counselling—An Internet-Based Survey," *Human Fertility* 10, no. 2 (2007): 111–16.

11. Lass, Amir, and Peter Brinsden, "How Do Patients Choose Private In Vitro Fertilization Treatment? A Customer Survey in a Tertiary Fertility Center in the United Kingdom," *Fertility and Sterility* 75, no. 5 (2001): 893–97.

12. Forrest, Christopher B., et al., "Primary Care Physician Specialty Referral Decision Making: Patient, Physician, and Healthcare System Determinants," *Medical Decision Making* 26, no. 1 (2006): 76–85.

13. Lawrence, Ryan E., et al., "Obstetrician-Gynecologists' Views on Contraception and Natural Family Planning: A National Survey," *American Journal of Obstetrics and Gynecology* 204, no. 2 (2011): 124.e1–e7.

14. Klitzman, Robert, et al., "Views of Internists Towards Uses of PGD," *Reproductive Biomedicine Online* 26, no. 2 (2013): 142–47.

15. Klitzman, Robert, "The Need for Vigilance in the Marketing of Genomic Tests in Psychiatry," *Journal of Nervous and Mental Disease* 203, no. 10 (2015): 809.

16. Abbate, Kristopher J., et al., "Views of Preimplantation Genetic Diagnosis (PGD) Among Psychiatrists and Neurologists," *Journal of Reproductive Medicine* 59 (2014): 385.

17. Kurz, Richard S., and Fredric D. Wolinsky, "Who Picks the Hospital: Practitioner or Patient?" *Journal of Healthcare Management* 30, no. 2 (1985): 95–106.

18. Morgan, Tanya J., Lori W. Turner, and Lucy A. Savitz, "Factors Influencing Obstetrical Care Selection," *American Journal of Health Studies* 15, no. 2 (1999): 100.

19. Bretherick, Karla L., et al., "Fertility and Aging: Do Reproductive-Aged Canadian Women Know What They Need to Know?" *Fertility and Sterility* 93, no. 7 (2010): 2162–68.

20. Peterson, Brennan D., et al., "Fertility Awareness and Parenting Attitudes Among American Male and Female Undergraduate University Students," *Human Reproduction* 27, no. 5 (2012): 1375–82.

21. Twenge, Jean, "How Long Can You Wait to Have a Baby?" *The Atlantic* (July/August 2013).

22. Trop, Jaclyn, "More Women Over 40 Are Having Babies," *Fortune*, January 14, 2016.

23. Maheshwari, Abha, et al., "Women's Awareness and Perceptions of Delay in Childbearing," *Fertility and Sterility* 90, no. 4 (2008): 1036–42.

24. Marcus et al., "Infertility counselling."

25. Grumbach, Kevin, et al., "Resolving the Gatekeeper Conundrum: What Patients Value in Primary Care and Referrals to Specialists," *Journal of the American Medical Association* 282, no. 3 (1999): 261–66.

26. St. Peter, Robert F., "Gatekeeping Arrangements Are in Widespread Use," *Data Bulletin (Center for Studying Health System Change)* 7 (1997): 1.

27. Hartzell, Tristan L., et al., "Does the Gatekeeper Model Work in Hand Surgery?" *Plastic and Reconstructive Surgery* 132, no. 3 (2013): 381e–86e.

28. Bjornsson, Steinar, et al., "Gatekeeping and Referrals to Cardiologists: General Practitioners' Views on Interactive Communications," *Scandinavian Journal of Primary Healthcare* 31, no. 2 (2013): 79–82.

29. Lurie, Nicole, et al., "Physician Self-Report of Comfort and Skill in Providing Preventive Care to Patients of the Opposite Sex," *Archives of Family Medicine* 7, no. 2 (1998): 134.

30. National Institutes of Health, "Fertility Treatment for Females," January 31, 2017, accessed September 4, 2018, https://www.nichd.nih.gov/health/topics/infertility/conditioninfo/treatments/treatments-women.

31. Mayo Clinic, "In Vitro Fertilization (IVF)," March 22, 2018, accessed September 4, 2018, https://www.mayoclinic.org/tests-procedures/in-vitro-fertilization/about/pac-20384716.

32. Chandra, Anjani, Casey E. Copen, and Elizabeth Hervey Stephen, "Infertility Service Use in the United States: Data from the National Survey of Family Growth, 1982–2010," *National Health Statistics Reports* 73, no. 73 (2014): 1–20.

33. Boulet SL, Mehta A, Kissin DM, Warner L, Kawwass JF, Jamieson DJ. "Trends in use of and reproductive outcomes associated with intracytoplasmic sperm injection," *JAMA* 313, no. 3 (2015): 255–263.

34. Hansen M, Greenop KR, Bourke J, Baynam G, Hart RJ, Leonard H. "Intellectual Disability in Children Conceived Using Assisted Reproductive Technology," *Pediatrics* 142, no. 6 (2018): e20181269.

35. Baruffi, Ricardo LR, Ana L. Mauri, Claudia G. Petersen, Andréia Nicoletti, Anagloria Pontes, João Batista A. Oliveira, and José G. Franco. "Single-embryo transfer reduces clinical pregnancy rates and live births in fresh IVF and Intracytoplasmic Sperm Injection (ICSI) cycles: a meta-analysis," *Reproductive Biology and Endocrinology* 7, no. 1 (2009): 36.

36. Delvigne, Annick, and Serge Rozenberg, "Epidemiology and Prevention of Ovarian Hyperstimulation Syndrome (OHSS): A Review," *Human Reproduction Update* 8, no. 6 (2002): 559–77.

Chapter 4

1. Delvigne, Annick, and Serge Rozenberg, "Epidemiology and Prevention of Ovarian Hyperstimulation Syndrome (OHSS): A Review," *Human Reproduction Update* 8, no. 6 (2002): 559–77.

2. Bitette, N., "Indian Woman, 72, Gives Birth to First Child with Help of IVF." *New York Daily News*, May 12, 2016.

3. Gleicher, Norbert, et al., "The 'Graying' of Infertility Services: An Impending Revolution Nobody Is Ready For," *Reproductive Biology and Endocrinology* 12, no. 1 (2014): 63.

4. Le Ray, C., et al., "Association Between Oocyte Donation and Maternal and Perinatal Outcomes in Women Aged 43 Years or Older," *Human Reproduction* 27, no. 3 (2012): 896–901.

5. Gleicher et al., "The 'Graying' of Infertility Services," 63.

6. American Society for Reproductive Medicine, "Definitions of Infertility and Recurrent Pregnancy Loss: A Committee Opinion," February 1, 2013, accessed February 1, 2017, https://www.asrm.org/globalassets/asrm/asrm-content/news-and-publications/practice-guidelines/for-non-members/definitions_of_infertility_and_recurrent_pregnancy_loss-noprint.pdf.

7. Ethics Committee of the American Society for Reproductive Medicine, "Fertility Treatment when the Prognosis Is Very Poor or Futile: A Committee Opinion," *Fertility and Sterility* 98, no. 1 (2012): e6–9.

8. Kawwass, Jennifer F., et al., "Trends and Outcomes for Donor Oocyte Cycles in the United States, 2000–2010," *Journal of the American Medical Association* 310, no. 22 (2013): 2426–34.

9. Ethics Committee of the American Society for Reproductive Medicine, "Oocyte or Embryo Donation to Women of Advanced Age: A Committee Opinion," *Fertility and Sterility* 100, no. 2 (2013): 337–40.

10. American Society for Reproductive Medicine, "Definitions of Infertility and Recurrent Pregnancy Loss.".

11. Ethics Committee of the American Society for Reproductive Medicine, "Fertility Treatment when the Prognosis Is Very Poor or Futile.".

12. Ibid.

13. Ibid.

14. Human Fertilisation and Embryology Authority, "Advanced Clinic Search," 2012, accessed February 23, 2016. https://www.hfea.gov.uk/choose-a-clinic/what-to-look-for-in-a-clinic/.

15. Fertility SA, "What Is the Age Limit for IVF?" last modified 2014, accessed February 23, 2016, https://fertilitysa.com.au/faq/age-limit-ivf/.

16. Ubelacker, Sheryl, "Quebec's High Cost of Funding IVF Without an Age Limit, a Cautionary Tale: Study," *Global News*, October 15, 2015, accessed February 23, 2016, http://globalnews.ca/news/2284460/quebecs-high-cost-of-funding-ivfh without-an-age-limit-a-cautionary-tale-study/.

17. Zweifel, Julianne E., "Donor Conception from the Viewpoint of the Child: Positives, Negatives, and Promoting the Welfare of the Child," *Fertility and Sterility* 104, no. 3 (2015): 513–19.

18. Ibid.

19. Sieh, Dominik Sebastian, et al., "Risk Factors for Problem Behavior in Adolescents of Parents with a Chronic Medical Condition," *European Child & Adolescent Psychiatry* 21, no. 8 (2012): 459–71.

20. Lackey, Nancy R., and Marie F. Gates, "Adults' Recollections of Their Experiences as Young Caregivers of Family Members with Chronic Physical Illnesses," *Journal of Advanced Nursing* 34, no. 3 (2001): 320–28.

21. MacDougall, K., Y. Beyene, and R. D. Nachtigall, "'Inconvenient Biology: Advantages and Disadvantages of First-Time Parenting After Age 40 Using In Vitro Fertilization," *Human Reproduction* 27, no. 4 (2012): 1058–65.

22. Niemelä, Jussi, "What Puts the 'Yuck' in the Yuck Factor?" *Bioethics* 25, no. 5 (2011): 267–79.

23. Kass, Leon R., "The Wisdom of Repugnance: Why We Should Ban the Cloning of Humans," *Valparaiso University Law Review* 32, no. 2 (1998): 679–705.

24. Niemelä, "What Puts the 'Yuck' in the Yuck Factor?".

25. Kahneman, Daniel, *Thinking, Fast and Slow* (New York: Farrar, Straus and Giroux, 2011).

26. Practice Committee of the American Society for Reproductive Medicine, "Definitions of Infertility and Recurrent Pregnancy Loss: A Committee Opinion," *Fertility and Sterility* 99, no. 1 (2013): 63.

27. Robertson, John A., "Procreative Liberty and Harm to Offspring in Assisted Reproduction," *American Journal of Law & Medicine* 30, no. 1 (2004): 7–40.

28. Lotz, Mianna, "Feinberg, Mills, and the Child's Right to an Open Future," *Journal of Social Philosophy* 37, no. 4 (2006): 537–51.

29. US Department of Labor, "Age Discrimination Act of 1975 (42 U.S.C. Sections 6101-6107)," accessed February 23, 2016, http://www.dol.gov/oasam/regs/statutes/age_act.htm.

30. Klitzman, Robert, et al., "Preimplantation Genetic Diagnosis on In Vitro Fertilization Clinic Websites: Presentations of Risks, Benefits and Other Information," *Fertility and Sterility* 92, no. 4 (2009): 1276–83.

31. Sherman, S., B. A. Pletcher, and D. A. Driscoll, "Fragile X Syndrome: Diagnostic and Carrier Testing," *Genetics in Medicine* 7, no. 8 (2005): 584.

32. Lindheim, Steven R., and Mark V. Sauer, "Expectations of Recipient Couples Awaiting an Anonymous Oocyte Donor Match," *Journal of Assisted Reproduction and Genetics* 15, no. 7 (1998): 444–46.

33. Centers for Disease Control, "Assisted Reproduction Technology: National Summary Report. National Center for Chronic Disease Prevention and Health Promotion," 2013, accessed August 17, 2016, https://www.cdc.gov/art/pdf/2013-report/art_2013_national_summary_report.pdf.

34. Sawyer, Neroli, et al., "A Survey of 1700 Women Who Formed Their Families Using Donor Spermatozoa," *Reproductive Biomedicine Online* 27, no. 4 (2013): 436–47.

35. Furnham, Adrian, Natalie Salem, and David Lester, "Selecting Egg and Sperm Donors: The Role of Age, Social Class, Ethnicity, Height and Personality," *Psychology* 5, no. 3 (2014): 53033.

36. American Society for Reproductive Medicine Website, 2016, accessed August 17, 2016, https://www.asrm.org/?vs=1.

37. American Society for Reproductive Medicine, "Guidelines for Oocyte Donation," *Fertility and Sterility* 86, no. S1 (2006): S48–49.

38. Klitzman, Robert, "Buying and Selling Human Eggs: Infertility Providers' Ethical and Other Concerns Regarding Egg Donor Agencies," BMC Medical Ethics 17, no. 1 (2016): 71.

39. Ethics Committee of the American Society for Reproductive Medicine, "Financial Compensation of Oocyte Donors," *Fertility and Sterility* 88, no. 2 (2007): 305–9.

40. Practice Committee of the Society for Assisted Reproductive Technology, "Repetitive Oocyte Donation: A Committee Opinion," *Fertility and Sterility* 102, no. 4 (2014): 964–66.

41. Brinton, Louise A., Vikrant V. Sahasrabuddhe, and Bert Scoccia, "Fertility Drugs and the Risk of Breast and Gynecologic Cancers," *Seminars in Reproductive Medicine* 30, no. 2 (2012): 131–45.

42. Ethics Committee of the American Society for Reproductive Medicine, "Financial Compensation of Oocyte Donors."

43. Levine, Aaron D., "The Oversight and Practice of Oocyte Donation in the United States, United Kingdom and Canada," *HEC Forum* 23, no. 1 (2011): 15–30.

44. Centers for Disease Control and Protection, "Infertility," last updated July 15, 2015, accessed August 17, 2016, http://www.cdc.gov/nchs/fastats/infertility.htm.

45. National Center for Chronic Disease Prevention and Health Promotion, Division of Public Health Assisted Reproductive Technology, "National Summary Report," 2015, accessed August 17, 2016, https://www.cdc.gov/art/pdf/2015-report/ART-2015-National-Summary-Report.pdf.

46. Levine, Aaron D., "Self-Regulation, Compensation, and the Ethical Recruitment of Oocyte Donors," *Hastings Center Report* 40, no. 2 (2010): 25–36.

47. Keehn, J., et al., "Recruiting Egg Donors Online: An Analysis of In Vitro Fertilization Clinic and Agency Websites' Adherence to American Society for Reproductive Medicine Guidelines," *Fertility and Sterility* 98, no. 4 (2012): 995–1000.

48. Levine, "Self-Regulation, Compensation, and the Ethical Recruitment."

49. Alberta, Hillary B., Roberta M. Berry, and Aaron D. Levine, "Risk Disclosure and the Recruitment of Oocyte Donors: Are Advertisers Telling the Full Story?" *Journal of Law, Medicine & Ethics* 42, no. 2 (2014): 232–43.

50. Keehn et al., "Recruiting Egg Donors Online."

51. Holwell, Eve, et al., "Egg Donation Brokers: An Analysis of Agency versus In Vitro Fertilization Clinic Websites," *Journal of Reproductive Medicine* 59 (2014): 534.

52. Ibid.

53. Almeling, Rene, "Selling Genes, Selling Gender: Egg Agencies, Sperm Banks, and the Medical Market in Genetic Material," *American Sociological Review* 72, no. 3 (2007): 319–40.

54. Keehn, Jason, et al., "How Agencies Market Egg Donation on the Internet: A Qualitative Study," *Journal of Law, Medicine & Ethics* 43, no. 3 (2015): 610–18.

55. Brody, Jane, "Do Egg Donors Face Long-Term Risks?" *New York Times*, July 10, 2017, https://www.nytimes.com/2017/07/10/well/live/are-there-long-term-risks-to-egg-donors.html

56. Appelbaum, Paul S., Charles Lidz, and Robert Klitzman, "Voluntariness of Consent to Research: A Conceptual Model," *Hastings Center Report* 39, no. 1 (2009): 30–39.

57. Klitzman, Robert, "The Myth of Community Differences as the Cause of Variations Among IRBs," *AJOB Primary Research* 2, no. 2 (2012): 24–33.

58. Howell, "Egg Donation Brokers."

59. Saravelos, Sotiris H., et al., "Pain During Embryo Transfer Is Independently Associated with Clinical Pregnancy in Fresh/Frozen Assisted Reproductive Technology Cycles," *Journal of Obstetrics and Gynaecology Research* 42, no. 8 (2016): 684–93.

60. Woodriff, Molly, Mark V. Sauer, and Robert Klitzman, "Advocating for Longitudinal Follow-Up of the Health and Welfare of Egg Donors," *Fertility and Sterility* 102, no. 3 (2014): 662–66.

61. US Department of Health and Human Services, "HIPPA for Professionals," updated June 17, 2017, accessed August 24, 2018, https://www.hhs.gov/hipaa/for-professionals/index.html.

62. European Society of Human Reproduction and Embryology, *Comparative Analysis of Medically Assisted Reproduction in the EU: Regulation and Technologies— Final Report* (Grimbergen, Belgium: European Society of Human Reproduction and Embryology, 2008), accessed August 17, 2016, http://ec.europa.eu/health/blood_ tissues_organs/docs/study_eshre_en.pdf.

63. Klitzman, Robert, et al., "Disclosure of Information to Potential Subjects on Research Recruitment Web Sites," *IRB: Ethics & Human Research* 30, no. 1 (2008): 15–20.

64. Klitzman, R., and M. V. Sauer, "Creating and Selling Embryos for 'Donation': Ethical Challenges," *American Journal of Obstetrics and Gynecology* 212, no. 2 (2015): 167–70.

65. Ibid.

66. Ibid.

67. Sydell, Laura, "Silicon Valley Companies Add New Benefit for Women: Egg-Freezing," *NPR*, October 17, 2014, http://www.npr.org/sections/alltechconsidered/2014/ 10/17/356765423/silicon-valley-companies-add-new-benefit-for-women-egg-freezing

68. Tran, Mark, "Apple and Facebook Offer to Freeze Eggs for Female Employees," *Guardian*, October 15, 2014, https://www.theguardian.com/technology/2014/oct/15/ apple-facebook-offer-freeze-eggs-female-employees

69. Inhorn, Marcia C., et al., "Medical Egg Freezing: How Cost and Lack of Insurance Cover Impact Women and Their Families," *Reproductive Biomedicine & Society Online* 5 (2018): 82–92.

70. Anderson-Bialis, Jake, and Deborah Anderson-Bialis, "The Costs of Egg Freezing," FertilityIQ, accessed October 31, 2018, https://www.fertilityiq.com/egg-freezing/the-costs-of-egg-freezing.

71. Goldman KN, Kramer Y, Hodes-Wertz B, Noyes N, McCaffrey C, Grifo JA. "Long term cryopreservation of human oocytes does not increase embryonic aneuploidy," *Fertility and Sterility* 103, no. 3 (2015): .

72. Hauser, Christine, "4,000 Eggs and Embryos Are Lost in Tank Failure, Ohio Fertility Clinic Says," *New York Times*, March 28, 2018, https://www.nytimes.com/ 2018/03/28/us/frozen-embryos-eggs.html.

73. Almeling, Rene, *Sex Cells: The Medical Market for Eggs and Sperm* (Oakland: University of California Press, 2011).

Chapter 5

1. Madrigal, Alexis, "The Surprising Birthplace of the First Sperm Bank," *Atlantic*, April 28. 2014, https://www.theatlantic.com/technology/archive/2014/04/how-the-first-sperm-bank-began/361288/.

2. Agence France-Presse, "IVF Mix-Up: Wrong Sperm May Have Fertilised Eggs of 26 Women," *Guardian*, December 27, 2016, accessed August 2, 2017, https:// www.theguardian.com/society/2016/dec/28/ivf-mix-up-wrong-sperm-may-have-fertilised-eggs-of-26-women.

3. Kirkey, Sharon, "Switched Embryos and Wrong Sperm: IVF Mix-Ups Lead to Babies Born with 'Unintended Parentage.'" *National Post*, July 31, 2016, accessed August 2, 2017, http://nationalpost.com/health/ivf-mix-ups-lead-to-babies-born-with-unintended-parentage/wcm/98ab4e41-b516-41bb-af44-146668aca56c

4. Mroz, Jacqueline, "One Sperm Donor, 150 Offspring," *New York Times*, September 5, 2011, http://www.nytimes.com/2011/09/06/health/06donor.html.

5. Practice Committee of the American Society for Reproductive Medicine and Practice Committee of the Society for Assisted Reproductive Technology, "2008 Guidelines for Gamete and Embryo Donation: A Practice Committee Report," *Fertility and Sterility* 90, no. S5 (2008): S30–44.

6. National Institute for Health and Care Excellence, "Fertility Problems: Assessment and Treatment," first published 2013, updated 2016, accessed May 19, 2017, https://www.nice.org.uk/guidance/cg156.

7. Grady, Denise, "Sperm Donor Seen as Source of Disease in 5 Children," *New York Times*, May 19, 2018, https://www.nytimes.com/2006/05/19/health/19donor.html.

8. Boxer, Lawrence A., et al., "Strong Evidence for Autosomal Dominant Inheritance of Severe Congenital Neutropenia Associated with ELA2 Mutations," *Journal of Pediatrics* 148, no. 5 (2006): 633–36.

9. Ethics Committee of the American Society for Reproductive Medicine, "Informing Offspring of Their Conception by Gamete or Embryo Donation: A Committee Opinion," *Fertility and Sterility* 100, no. 1 (2013): 45–49.

Chapter 6

1. Baruch, Susannah, David Kaufman, and Kathy L. Hudson, "Genetic Testing of Embryos: Practices and Perspectives of U.S. In Vitro Fertilization Clinics," *Fertility and Sterility* 89, no. 5 (2008): 1053–58.

2. Human Fertilisation and Embryology Authority, "PGD Conditions Licensed by the HFEA," 2013, accessed February 22, 2017, https://www.hfea.gov.uk/pgd-conditions/.

3. Rich, Thereasa A., et al., "Comparison of Attitudes Regarding Preimplantation Genetic Diagnosis Among Patients with Hereditary Cancer Syndromes," *Familial Cancer* 13, no. 2 (2014): 291–99.

4. Klitzman, Robert, "Reducing the Number of Fetuses in a Pregnancy: Providers' and Patients' Views of Challenges," *Human Reproduction* 31, no. 11 (2016): 2570–76.

5. Abbate, Kristopher J., et al., "Views of Preimplantation Genetic Diagnosis (PGD) Among Psychiatrists and Neurologists," *Journal of Reproductive Medicine* 59 (2014): 385.

6. Kalfoglou, Andrea L., Joan Scott, and Kathy Hudson, "PGD Patients' and Providers' Attitudes to the Use and Regulation of Preimplantation Genetic Diagnosis," *Reproductive Biomedicine Online* 11, no. 4 (2005): 486–96.

7. Rich et al. "Comparison of Attitudes."

8. Quinn, Gwendolyn, et al., "Attitudes of High-Risk Women Toward Preimplantation Genetic Diagnosis," *Fertility and Sterility* 91, no. 6 (2009): 2361–68.

9. Ormondroyd, Elizabeth, et al., "Attitudes to Reproductive Genetic Testing in Women Who Had a Positive BRCA Test Before Having Children: A Qualitative Analysis," *European Journal of Human Genetics* 20, no. 1 (2012): 4–10.

10. Derks-Smeets, I. A. P., et al., "Decision-Making on Preimplantation Genetic Diagnosis and Prenatal Diagnosis: A Challenge for Couples with Hereditary Breast and Ovarian Cancer," *Human Reproduction* 29, no. 5 (2014): 1103–12.

11. Menon, U., et al., "Views of BRCA Gene Mutation Carriers on Preimplantation Genetic Diagnosis as a Reproductive Option for Hereditary Breast and Ovarian Cancer," *Human Reproduction* 22, no. 6 (2007): 1573–77.

12. Lavery, S. A., et al., "Preimplantation Genetic Diagnosis: Patients' Experiences and Attitudes," *Human Reproduction* 17, no. 9 (2002): 2464–67.

13. Drazba, Kathryn T., Michele A. Kelley, and Patricia E. Hershberger, "A Qualitative Inquiry of the Financial Concerns of Couples Opting to Use Preimplantation Genetic Diagnosis to Prevent the Transmission of Known Genetic Disorders," *Journal of Genetic Counseling* 23, no. 2 (2014): 202–11.

14. Järvholm, Stina, Ann Thurin-Kjellberg, and Malin Broberg, "Experiences of Pre-Implantation Genetic Diagnosis (PGD) in Sweden: A Three-Year Follow-Up of Men and Women," *Journal of Genetic Counseling* 26, no. 5 (2017): 1008–16.

15. Haude, K., et al., "Factors Influencing the Decision-Making Process and Long-Term Interpersonal Outcomes for Parents Who Undergo Preimplantation Genetic Diagnosis for Fanconi Anemia: A Qualitative Investigation," *Journal of Genetic Counseling* 26, no. 3 (2017): 640–55.

16. Verlinsky, Y., and A. Kuliev, "Preimplantation Diagnosis of Common Aneuploidies in Infertile Couples of Advanced Maternal Age," *Human Reproduction* 11, no. 10 (1996): 2076–77.

17. Gleicher, N., and R. Orvieto, "Is the Hypothesis of Preimplantation Genetics Screening (PGS) Still Supportable? A Review," *Journal of Ovarian Research* 10, no. 1 (2017): 21.

18. Rich et al. "Comparison of Attitudes."

19. Quinn et al., "Attitudes of High-Risk Women."

20. Menon et al., "Views of BRCA Gene Mutation Carriers."

21. Ormondroyd et al., "Attitudes to Reproductive Genetic Testing."

22. Derks-Smeets et al., "Decision-Making on Preimplantation Genetic Diagnosis."

23. Meiser, Bettina, "Psychological Impact of Genetic Testing for Cancer Susceptibility: An Update of the Literature," *Psycho-Oncology* 14, no. 12 (2005): 1060–74.

24. Croyle, Robert T., and Caryn Lerman, "Risk Communication in Genetic Testing for Cancer Susceptibility," *JNCI Monographs* 1999, no. 25 (1999): 59–66.

25. Klitzman, Robert, et al., "Attitudes and Practices Among Internists Concerning Genetic Testing," *Journal of Genetic Counseling* 22, no. 1 (2013): 90–100.

26. Fisher, P. B., J. F. Smith, and P. P. Katz, "Financial Burdens of Fertility Care: How Insurance Coverage and Perception of Cost Impact a Couple's Decision Making," *Fertility and Sterility* 96, no. 3 (2011): S30.

27. European Society of Human Reproduction and Embryology, *Comparative Analysis of Medically Assisted Reproduction in the EU: Regulation and Technologies—Final Report* (Grimbergen, Belgium: European Society of Human Reproduction and Embryology, 2008), accessed August 17, 2016, http://ec.europa.eu/health/blood_tissues_organs/docs/study_eshre_en.pdf.

28. Präg, Patrick, and Melinda C. Mills, "Assisted Reproductive Technology in Europe: Usage and Regulation in the Context of Cross-Border Reproductive Care," in *Childlessness in Europe: Contexts, Causes, and Consequences,* ed. Michaela Kreyenfeld and Dirk Konietzka (Cham, Switzerland: Springer International Publishing, 2017), 289–309.

29. European Society of Human Reproduction and Embryology, *Comparative Analysis*.

30. IVF Australia, "IVF Treatment Costs," 2017, accessed March 7, 2017. http://www.ivf.com.au/ivf-fees/ivf-costs.

31. Wilton, L., et al., "The Causes of Misdiagnosis and Adverse Outcomes in PGD," *Human Reproduction* 24, no. 5 (2009): 1221–28.

32. Verlinsky, Yury, et al., "Over a Decade of Experience with Preimplantation Genetic Diagnosis: A Multicenter Report," *Fertility and Sterility* 82, no. 2 (2004): 292–94.

33. Wilton, L., et al., "The Causes of Misdiagnosis and Adverse Outcomes in PGD," *Human Reproduction* 24, no. 5 (2009): 1221–28.

34. Towner, D., et al., "Miscarriage Risk from Amniocentesis Performed for Abnormal Maternal Serum Screening," *American Journal of Obstetrics and Gynecology* 196, no. 6 (2007): 608-e1–5.

35. Balasubramanyam, Sathya, "Knowledge and Attitudes of Women Towards Multiple Embryo Transfer, Fetal Reduction and Multiple Pregnancy," *International Journal of Infertility and Fetal Medicine* 1, no. 1 (2010): 31–34.

36. Kahraman, Semra, et al., "Successful Haematopoietic Stem Cell Transplantation in 44 Children from Healthy Siblings Conceived After Preimplantation HLA Matching," *Reproductive Biomedicine Online* 29, no. 3 (2014): 340–51.

37. Kakourou, Georgia, et al., "Complex Preimplantation Genetic Diagnosis for Beta-Thalassaemia, Sideroblastic Anaemia, and Human Leukocyte Antigen (HLA)-Typing," *Systems Biology in Reproductive Medicine* 62, no. 1 (2016): 69–76.

38. Khosravi, Sharifeh, et al., "Novel Multiplex Fluorescent PCR-Based Method for HLA Typing and Preimplantational Genetic Diagnosis of β-Thalassemia," *Archives of Medical Research* 47, no. 4 (2016): 293–98.

39. Zierhut, Heather A., et al., "Elucidating Genetic Counseling Outcomes from the Perspective of Genetic Counselors," *Journal of Genetic Counseling* 25, no. 5 (2016): 993–1001.

40. Lipton, Jeffrey M., and Steven R. Ellis, "Diamond-Blackfan Anemia: Diagnosis, Treatment, and Molecular Pathogenesis," *Hematology/Oncology Clinics of North America* 23, no. 2 (2009): 261–82.

41. Tur-Kaspa, Ilan, and Roohi Jeelani, "Clinical Guidelines for IVF with PGD for HLA Matching," *Reproductive Biomedicine Online* 30, no. 2 (2015): 115–19.

42. Verlinsky, Yury, et al., "Preimplantation Diagnosis for Fanconi Anemia Combined with HLA Matching," *Journal of the American Medical Association* 285, no. 24 (2001): 3130–33.

43. Verlinsky, Yury, et al., "Preimplantation HLA Resting," *Journal of the American Medical Association* 291, no. 17 (2004): 2079–85.

44. Van de Velde, Hilde, et al., "The Experience of Two European Preimplantation Genetic Diagnosis Centres on Human Leukocyte Antigen Typing," *Human Reproduction* 24, no. 3 (2008): 732–40.

45. Baruch, Susannah, David Kaufman, and Kathy L. Hudson, "Genetic Testing of Embryos: Practices and Perspectives of U.S. In Vitro Fertilization Clinics," *Fertility and Sterility* 89, no. 5 (2008): 1053–58.

46. Ibid.

47. Van de Velde et al., "Experience of Two European Preimplantation Genetic Diagnosis Centres."

48. Zierhut et al., "Elucidating Genetic Counseling Outcomes."

49. Devolder, Katrien, "Preimplantation HLA Typing: Having Children to Save Our Loved Ones," *Journal of Medical Ethics* 31, no. 10 (2005): 582–86.

50. Wagner, John E., and Jeffrey P. Kahn, "Preimplantation Testing to Produce an HLA-Matched Donor Infant," *Journal of the American Medical Association* 292, no. 7 (2004): 803–4; author reply 804.

51. Human Fertilisation and Embryology Authority, accessed March 3, 2017, https://www.hfea.gov.uk/.

52. Robertson, John A., Jeffrey P. Kahn, and John E. Wagner, "Conception to Obtain Hematopoietic Stem Cells," *Hastings Center Report* 32, no. 3 (2002): 34–40.

53. Robertson, John A., "Extending Preimplantation Genetic Diagnosis: Medical and Non-Medical Uses," *Journal of Medical Ethics* 29, no. 4 (2003): 213–16.

54. Tur-Kaspa, Ilan, and Roohi Jeelani, "Clinical Guidelines for IVF with PGD for HLA Matching," *Reproductive Biomedicine Online* 30, no. 2 (2015): 115–19.

55. Zierhut, Heather, et al., "More than 10 Years After the First 'Savior Siblings': Parental Experiences Surrounding Preimplantation Genetic Diagnosis," *Journal of Genetic Counseling* 22 (2013): 594–602.

56. Lotz, Mianna. "Feinberg, Mills, and the child's right to an open future." *Journal of social philosophy* 37, no. 4 (2006): 537–551.

57. Tur-Kaspa and Jeelani, "Clinical guidelines for IVF.".

58. Klitzman, Robert, et al., "Disclosures of Huntington Disease Risk Within Families: Patterns of Decision-Making and Implications," *American Journal of Medical Genetics Part A* 143, no. 16 (2007): 1835–49.

59. Wahlin, Tarja-Brita Robins, "To Know or Not to Know: A Review of Behaviour and Suicidal Ideation in Preclinical Huntington's Disease," *Patient Education and Counseling* 65, no. 3 (2007): 279–87.

60. Braude, Peter R., et al., "Non-Disclosure Preimplantation Genetic Diagnosis for Huntington's Disease: Practical and Ethical Dilemmas," *Prenatal Diagnosis* 18, no. 13 (1998): 1422–26.

61. Stern, Harvey J., et al., "Non-Disclosing Preimplantation Genetic Diagnosis for Huntington Disease," *Prenatal Diagnosis* 22, no. 6 (2002): 503–7.

62. Erez, Ayelet, et al., "The Right to Ignore Genetic Status of Late Onset Genetic Disease in the Genomic Era; Prenatal Testing for Huntington Disease as a Paradigm," *American Journal of Medical Genetics Part A* 152, no. 7 (2010): 1774–80.

63. Braude et al., "Non-Disclosure Preimplantation."

64. Erez et al., "Right to Ignore Genetic Status."

65. Ibid.

66. Ibid.

67. Bird, Thomas D., "Outrageous Fortune: The Risk of Suicide in Genetic Testing for Huntington Disease," *American Journal of Human Genetics* 64, no. 5 (1999): 1289–92.

68. Wahlin, "To Know or Not to Know.".

69. Myers, Richard H., "Huntington's Disease Genetics," NeuroRx 1, no. 2 (2004): 255–62.

70. Almqvist, Elisabeth W., et al., "A Worldwide Assessment of the Frequency of Suicide, Suicide Attempts, or Psychiatric Hospitalization After Predictive Testing for Huntington Disease," *American Journal of Human Genetics* 64, no. 5 (1999): 1293–1304.

71. Quaid, Kimberly A., et al., "Factors Related to Genetic Testing in Adults at Risk for Huntington Disease: The Prospective Huntington at-Risk Observational Study (PHAROS)," *Clinical Genetics* 91, no. 6 (2017): 824–31.

72. Lyerly, Anne Drapkin, et al., "Fertility Patients' Views About Frozen Embryo Disposition: Results of a Multi-Institutional U.S. Survey," *Fertility and Sterility* 93, no. 2 (2010): 499–509.

73. Samorinha, Catarina, et al., "Factors Associated with the Donation and Non-donation of Embryos for Research: A Systematic Review," *Human Reproduction Update* 20, no. 5 (2014): 641–55.

74. Arons, Jessica, *Future Choices: Assisted Reproductive Technologies and the Law* (Washington, DC: Center for American Progress, 2007).

75. Wilton, L., et al., "The Causes of Misdiagnosis and Adverse Outcomes in PGD," *Human Reproduction* 24, no. 5 (2009): 1221–28.

76. Ross, Laine Friedman, et al.; American Academy of Pediatrics, "Technical Report: Ethical and Policy Issues in Genetic Testing and Screening of Children," *Genetics in Medicine* 15, no. 3 (2013): 234–45.

77. Grody, Wayne W., et al., "ACMG Position Statement on Prenatal/Preconception Expanded Carrier Screening," *Genetics in Medicine* 15, no. 6 (2013): 482–83.

78. ACMG Board of Directors, "Scope of Practice: A Statement of the American College of Medical Genetics and Genomics (ACMG)," *Genetics in Medicine* 17, no. 9 (2015): e3.

79. Schoen, Cathy, et al., "New 2011 Survey of Patients with Complex Care Needs in Eleven Countries Finds that Care Is Often Poorly Coordinated," *Health Affairs* 30, no. 12 (2011): 2437–48.

80. Rebouché, Rachel, and Karen H. Rothenberg, "Mixed Messages: The Intersection of Prenatal Genetic Testing and Abortion," *Howard Law Journal* 55 no. 3 (2012): 983.

81. Katz, P., et al. "Costs of Infertility Treatment: Results from an 18-Month Prospective Cohort Study," *Fertility and Sterility* 95, no. 3 (2011): 915–21.

Chapter 7

1. Madan, Kamlesh, and Martijn H. Breuning, "Impact of Prenatal Technologies on the Sex Ratio in India: An Overview," *Genetics in Medicine* 16, no. 6 (2013): 425–32.

2. Sen, Amartya, "Missing Women," *BMJ: British Medical Journal* 304, no. 6827 (1992): 587.

3. Geneva Centre for the Democratic Control of Armed Forces, *Women in an Insecure World. Violence Against Women: Facts, Figures and Analysis*, ed. Marie Vlachová and Lea Biason (Geneva: Geneva Centre for the Democratic Control of Armed Forces, 2005), accessed July 25, 2016, http://www.unicef.org/emerg/files/women_insecure_world.pdf.

4. Sandel, Michael J., *The Case Against Perfection* (Cambridge, MA: Harvard University Press, 2009).

5. Sanghavi, Darshak M., "Wanting Babies Like Themselves, Some Parents Choose Genetic Defects," *New York Times*, December 5, 2006, accessed May 26, 2017 http://www.nytimes.com/2006/12/05/health/05essa.html.

6. Baruch, Susannah, David Kaufman, and Kathy L. Hudson, "Genetic Testing of Embryos: Practices and Perspectives of U.S. In Vitro Fertilization Clinics," *Fertility and Sterility* 89, no. 5 (2008): 1053–58.

7. Middleton, Anna, Jenny Hewison, and Robert F. Mueller, "Attitudes of Deaf Adults Toward Genetic Testing for Hereditary Deafness," *American Journal of Human Genetics* 63, no. 4 (1998): 1175–80.

8. Mand, Cara, et al., "Genetic Selection for Deafness: The Views of Hearing Children of Deaf Adults," *Journal of Medical Ethics* 35, no. 12 (2009): 722–28.

9. Gooding, Holly C., et al., "Issues Surrounding Prenatal Genetic Testing for Achondroplasia," *Prenatal Diagnosis* 22, no. 10 (2002): 933–40.

10. Ibid.

11. Kevles, Daniel J., *In the Name of Eugenics: Genetics and the Uses of Human Heredity* (Cambridge, MA: Harvard University Press, 1985).

12. Committee for Ethics, American Congress of Obstetricians and Gynecologists, "ACOG Committee Opinion No. 360: Sex Selection," *Obstetrics & Gynecology* 109, no. 2 (2007): 475–78.

13. Geneva Centre for the Democratic Control of Armed Forces, "Women in an Insecure World.".

14. Ethics Committee of the American Society for Reproductive Medicine, "Use of Reproductive Technology for Sex Selection for Nonmedical Reasons," *Fertility and Sterility* 103, no. 6 (2015): 1418–22.

15. Ibid.

16. Karabinus, David S., et al., "The Effectiveness of Flow Cytometric Sorting of Human Sperm (MicroSort®) for Influencing a Child's Sex," *Reproductive Biology and Endocrinology* 24, no. 12 (2014): 106.

17. Ethics Committee of the American Society for Reproductive Medicine, "Use of Reproductive Technology for Sex Selection."

18. Wilkinson, Stephen, "'Designer Babies,' Instrumentalisation and the Child's Right to an Open Future," In *Philosophical Reflections on Medical Ethics*, ed. Nafsika Athanassoulis, pp. 44–69 (Basingstoke, UK: Palgrave Macmillan, 2005).

19. Seavilleklein, Victoria, and Susan Sherwin, "The Myth of the Gendered Chromosome: Sex Selection and the Social Interest," *Cambridge Quarterly of Healthcare Ethics* 16, no. 01 (2007): 7–19.

20. Puri, Sunita, and Robert D. Nachtigall, "The Ethics of Sex Selection: A Comparison of the Attitudes and Experiences of Primary Care Physicians and Physician Providers of Clinical Sex Selection Services," *Fertility and Sterility* 93, no. 7 (2010): 2107–14.

21. Kalfoglou, A. L., et al., "Ethical Arguments For and Against Sperm Sorting for Non-medical Sex Selection: A Review," *Reproductive Biomedicine Online* 26, no. 3 (2013): 231–39.

22. Dahl, Edgar, et al., "Preconception Sex Selection Demand and Preferences in the United States," *Fertility and Sterility* 85, no. 2 (2006): 468–73.

23. Jain, Tarun, et al., "Preimplantation Sex Selection Demand and Preferences in an Infertility Population," *Fertility and Sterility* 83, no. 3 (2005): 649–58.

24. Missmer, Stacey A., and Tarun Jain, "Preimplantation Sex Selection Demand and Preferences Among Infertility Patients in Midwestern United States," *Journal of Assisted Reproduction and Genetics* 24, no. 10 (2007): 451–57.

25. Kalfoglou et al., "Ethical Arguments For and Against Sperm Sorting."

26. Savulescu, Julian, and Edgar Dahl, "Sex Selection and Preimplantation Diagnosis: A Response to the Ethics Committee of the American Society of Reproductive Medicine," *Human Reproduction* 15, no. 9 (2000): 1879–80.

27. Dahl, Edgar, "The 10 Most Common Objections to Sex Selection and Why They Are Far from Being Conclusive: A Western Perspective," *Reproductive Biomedicine Online* 14 (2007): 158–61.

28. Robertson, John A., "Preconception Gender Selection," *American Journal of Bioethics* 1, no. 1 (2001): 2–9.

29. Puri and Nachtigall, "The Ethics of Sex Selection."

30. Colls, Pere, et al., "Preimplantation Genetic Diagnosis for Gender Selection in the U.S.A.," *Reproductive Biomedicine Online* 19 (2009): 16–22.

31. Gleicher, Norbert, and David H. Barad, "The Choice of Gender: Is Elective Gender Selection, Indeed, Sexist?" *Human Reproduction* 22, no. 11 (2007): 3038–41.

32. Colls et al., "Preimplantation Genetic Diagnosis for Gender Selection."

33. Gleicher and Barad, "The Choice of Gender."

34. Puri, Sunita, et al., "'There Is Such a Thing as Too Many Daughters, but not Too Many Sons': A Qualitative Study of Son Preference and Fetal Sex Selection Among Indian Immigrants in the United States," *Social Science & Medicine* 72, no. 7 (2011): 1169–76.

35. Missmer and Jain, "Preimplantation Sex Selection Demand and Preferences."

36. Sharp, Richard R., et al., "Moral Attitudes and Beliefs Among Couples Pursuing PGD for Sex Selection," *Reproductive Biomedicine Online* 21, no. 7 (2010): 838–47.

37. Ibid.

38. Baruch, Kaufman, and Hudson, "Genetic Testing of Embryos."

39. Klitzman, Robert, et al., "Preimplantation Genetic Diagnosis on In Vitro Fertilization Clinic Websites: Presentations of Risks, Benefits and Other Information," *Fertility and Sterility* 92, no. 4 (2009): 1276–83.

40. Klitzman, Robert, et al., "Views of Internists Towards Uses of PGD," *Reproductive Biomedicine Online* 26, no. 2 (2013): 142–47.

41. Abbate, Kristopher J., et al., "Views of Preimplantation Genetic Diagnosis (PGD) Among Psychiatrists and Neurologists," *Journal of Reproductive Medicine* 59 (2014): 385.

42. Puri and Nachtigall, "The Ethics of Sex Selection."

43. Kalfoglou et al., "Ethical Arguments For and Against Sperm Sorting."

44. Ehrich, Kathryn, et al., "Choosing Embryos: Ethical Complexity and Relational Autonomy in Staff Accounts of PGD," *Sociology of Health & Illness* 29, no. 7 (2007): 1091–1106.

45. Ethics Committee of the American Society for Reproductive Medicine, "Use of Reproductive Technology for Sex Selection."

46. Savulescu and Dahl, "Sex Selection and Preimplantation Diagnosis."

47. Dahl, "10 Most Common Objections to Sex Selection."

48. Sanghavi, "Wanting Babies Like Themselves.".

49. Lotz, Mianna. "Feinberg, Mills, and the child's right to an open future." *Journal of social philosophy* 37, no. 4 (2006): 537-551.

50. Greely, Henry T., *The End of Sex and the Future of Human Reproduction* (Cambridge, MA: Harvard University Press, 2016).

Chapter 8

1. Osterman, Michelle J. K., et al., "Annual Summary of Vital Statistics: 2012–2013," *Pediatrics* (2015): 1115–25.

2. Pison, Gilles, Christiaan Monden, and Jeroen Smits, "Twinning Rates in Developed Countries: Trends and Explanations," *Population and Development Review* 41, no. 4 (2015): 629–49.

3. Muir, Hazel, "World Faces 'Epidemic' of Twins." *New Scientist*, July 16, 2001.

4. Davidson, C. M., "Octomom and Multi-Fetal Pregnancies: Why Federal Legislation Should Require Insurers to Cover In Vitro Fertilization," *William & Mary Journal of Women and the Law* 17, no. 1 (2010): 135–86.

5. Sazonova, Antonina, et al., "Neonatal and Maternal Outcomes Comparing Women Undergoing Two In Vitro Fertilization (IVF) Singleton Pregnancies and Women Undergoing One IVF Twin Pregnancy," *Fertility and Sterility* 99, no. 3 (2013): 731–37.

6. Ibid.

7. Gleicher, Norbert, and David Barad, "Twin Pregnancy, Contrary to Consensus, Is a Desirable Outcome in Infertility," *Fertility and Sterility* 91, no. 6 (2009): 2426–31.

8. Baruffi, Ricardo L. R., et al., "Single-Embryo Transfer Reduces Clinical Pregnancy Rates and Live Births in Fresh IVF and Intracytoplasmic Sperm Injection (ICSI) Cycles: A Meta-Analysis," *Reproductive Biology and Endocrinology* 7, no. 1 (2009): 36.

9. McLernon, D. J., et al., "Clinical Effectiveness of Elective Single versus Double Embryo Transfer: Meta-Analysis of Individual Patient Data from Randomised Trials," *BMJ: British Medical Journal* 341 (2010): c6945.

10. La Sala, Giovanni Battista, et al., "Two Consecutive Singleton Pregnancies versus One Twins Pregnancy as Preferred Outcome of In Vitro Fertilization for Mothers and Infants: A Retrospective Case–Control Study," *Current Medical Research and Opinion* 32, no. 4 (2016): 687–92.

11. Lawlor, Debbie A., and Scott M. Nelson, "Effect of Age on Decisions About the Numbers of Embryos to Transfer in Assisted Conception: A Prospective Study," *Lancet* 379, no. 9815 (2012): 521–27.

12. Kjellberg, Ann Thurin, Per Carlsson, and Christina Bergh, "Randomized Single versus Double Embryo Transfer: Obstetric and Paediatric Outcome and a Cost-Effectiveness Analysis," *Human Reproduction* 21, no. 1 (2005): 210–16.

13. Collins, John, "Cost Efficiency of Reducing Multiple Births," *Reproductive Biomedicine Online* 15 (2007): 35–39.

14. Wølner-Hanssen, P., and H. Rydhstroem, "Cost-Effectiveness Analysis of In-Vitro Fertilization: Estimated Costs per Successful Pregnancy After Transfer of One or Two Embryos," *Human Reproduction* 13, no. 1 (1998): 88–94.

15. Kissin, Dmitry M., Sheree L. Boulet, and Eli Y. Adashi, "Yes, Elective Single-Embryo Transfer Should Be the Standard of Care," In *Biennial Review of Infertility*, ed. Douglas T. Carrell et al., Vol. 4, pp. 177–87 (Cham, Switzerland: Springer International Publishing, 2015).

16. Leese, Brenda, and Jane Denton, "Attitudes Towards Single Embryo Transfer, Twin and Higher Order Pregnancies in Patients Undergoing Infertility Treatment: A Review," *Human Fertility* 13, no. 1 (2010): 28–34.

17. Griffin, Daniel, et al., "Impact of an Educational Intervention and Insurance Coverage on Patients' Preferences to Transfer Multiple Embryos," *Reproductive Biomedicine Online* 25, no. 2 (2012): 204–8.

18. Hope, Nicole, and Luk Rombauts, "Can an Educational DVD Improve the Acceptability of Elective Single Embryo Transfer? A Randomized Controlled Study," *Fertility and Sterility* 94, no. 2 (2010): 489–95.

19. Human Fertilisation and Embryology Authority, February 22, 2017, https://www.hfea.gov.uk/.

20. Zammit, Andrew, *Australia's Mothers and Babies 2012* (Canberra: Australia Institute of Health and Welfare, 2014).

21. Canadian Fertility and Andrology Society, *Human Assisted Reproduction 2014 Live Birth Rates for Canada* (Montreal: Canadian Fertility and Andrology Society, 2014), https://cfas.ca/_Library/media_releases_2/2014_press_release.pdf.

22. Stillman, Robert J., Kevin S. Richter, and Howard W. Jones, Jr., "Refuting a Misguided Campaign Against the Goal of Single-Embryo Transfer and Singleton Birth in Assisted Reproduction," *Human Reproduction* 28, no. 10 (2013): 2599–2607.

23. Human Fertilisation and Embryology Authority, *Fertility Treatment in 2013: Trends and Figures*, 2013, accessed April 4, 2017, https://www.hfea.gov.uk/media/2081/hfea-fertility-trends-2013.pdf.

24. De Neubourg, Diane, et al., "Belgium Model of Coupling Reimbursement of ART Costs to Restriction in Number of Embryos Transferred," *BMJ: British Medical Journal* 348 (2014): g1559.

25. Antsaklis, A., et al., "Pregnancy Outcome After Multifetal Pregnancy Reduction," *Journal of Maternal-Fetal & Neonatal Medicine* 16, no. 1 (2004): 27–31.

26. Practice Committee of the American Society for Reproductive Medicine and Practice Committee of the Society for Assisted Reproductive Technology, "Criteria for Number of Embryos to Transfer: A Committee Opinion," *Fertility and Sterility* 99, no. 1 (2013): 44–46.

27. Ibid.

28. Kulkarni, Aniket D., et al., "Fertility Treatments and Multiple Births in the United States," *New England Journal of Medicine* 369, no. 23 (2013): 2218–25.

29. Canadian Fertility and Andrology Society, *Human Assisted Reproduction 2014 Live Birth Rates for Canada* (Montreal: Canadian Fertility and Andrology Society, 2014), https://cfas.ca/_Library/media_releases_2/2014_press_release.pdf.

30. Kupka, M. S., et al., "Assisted Reproductive Technology in Europe, 2011: Results Generated from European Registers by ESHRE," *Human Reproduction* 31, no. 2 (2016): 233–48.

31. Ibid.

32. Ibid.

33. Centers for Disease Control and Prevention and US Department of Health and Human Services, *Assisted Reproductive Technology Success Rates: National Summary and Fertility Clinic Reports* (Atlanta, GA: Centers for Disease Control and Prevention, 2006).

34. Society for Assisted Reproductive Technology, *Final CSR for 2016.* (Birmingham, AL: Society for Assisted Reproductive Technology, 2017), https://www.sartcorsonline.com/rptCSR_PublicMultYear.aspx?reportingYear=2016.

35. Leese, Brenda, and Jane Denton, "Attitudes Towards Single Embryo Transfer, Twin and Higher Order Pregnancies in Patients Undergoing Infertility Treatment: A Review," *Human Fertility* 13, no. 1 (2010): 28–34.

36. Balasubramanyam, Sathya, "Knowledge and Attitudes of Women Towards Multiple Embryo Transfer, Fetal Reduction and Multiple Pregnancy," *International Journal of Infertility and Fetal Medicine* 1 (2010): 31–34.

37. Jungheim, Emily S., et al., "Embryo Transfer Practices in the United States: A Survey of Clinics Registered with the Society for Assisted Reproductive Technology," *Fertility and Sterility* 94, no. 4 (2010): 1432–36.

38. Martin, J. Ryan, et al., "Insurance Coverage and In Vitro Fertilization Outcomes: A U.S. Perspective," *Fertility and Sterility* 95, no. 3 (2011): 964–69.

39. Højgaard, Astrid, et al., "Patient Attitudes Towards Twin Pregnancies and Single Embryo Transfer—A Questionnaire Study," *Human Reproduction* 22, no. 10 (2007): 2673–78.

40. Greenfeld, Dorothy A., and Emre Seli, "Gay Men Choosing Parenthood Through Assisted Reproduction: Medical and Psychosocial Considerations," *Fertility and Sterility* 95, no. 1 (2011): 225–29.

41. Kalra, S. K., et al., "Infertility Patients and Their Partners: Differences in the Desire for Twin Gestations," *Obstetrics & Gynecology* 102, no. 1 (2003): 152–55.

42. Johnston, Josephine, Michael K. Gusmano, and Pasquale Patrizio, "Reducing Rate of Fertility Multiples Requires Policy Changes," *JAMA Pediatrics* 169, no. 3 (2015): 287.

43. Kahneman, D., *Thinking Fast and Slow* (New York: Farrar, Straus and Giroux, 2011).

44. Haas, Jigal, et al., "Pregnancy Outcome of Early Multifetal Pregnancy Reduction: Triplets to Twins Versus Triplets to Singletons," *Reproductive Biomedicine Online* 29, no. 6 (2014): 717–21.

45. Ibid.

46. Hasson, Joseph, et al., "Reduction of Twin Pregnancy to Singleton: Does It Improve Pregnancy Outcome?" *Journal of Maternal–Fetal & Neonatal Medicine* 24, no. 11 (2011): 1362–66.

47. Antsaklis et al., "Pregnancy Outcome After Multifetal Pregnancy Reduction."

48. Evans, Mark I., et al., "Fetal Reduction from Twins to a Singleton: A Reasonable Consideration?" *Obstetrics & Gynecology* 104, no. 1 (2004): 102–9.

49. Kupka et al., "Assisted Reproductive Technology in Europe, 2011."

50. Munks, Eryn B., et al., "IVF patients' Attitudes Toward Multifetal Pregnancy Reduction," *Journal of Reproductive Medicine* 52, no. 7 (2007): 635–38.

51. Balasubramanyam, "Knowledge and Attitudes of Women Towards Multiple Embryo Transfer."

52. Højgaard et al., "Patient Attitudes Towards Twin Pregnancies."

53. Jungheim et al., "Embryo Transfer Practices in the United States."

54. Kalra et al., "Infertility Patients and Their Partners."

55. Sazonova et al., "Neonatal and Maternal Outcomes."

56. Jungheim et al., "Embryo Transfer Practices in the United States."

57. Peters, Ellen, P. Sol Hart, and Liana Fraenkel, "Informing Patients: The Influence Of Numeracy, Framing, and Format of Side Effect Information on Risk Perceptions," *Medical Decision Making* 31, no. 3 (2011): 432–36.

58. Lloyd, A. J., "The Extent of Patients' Understanding of the Risk of Treatments," *Quality in Healthcare* 10, no. S1 (2001): i14–18.

59. Tversky, Amos, and Daniel Kahneman, "Judgment Under Uncertainty: Heuristics and Biases," in *Utility, Probability, and Human Decision Making*, ed. Dirk Wendt and Charles Vlek (Dordrecht, The Netherlands: D. Reidel, 1975), pp. 141–62.

60. Practice Committee of the American Society for Reproductive Medicine and Practice Committee of the Society for Assisted Reproductive Technology, "Criteria for Number of Embryos to Transfer."

61. Williams, R. Stan, et al., "Public Reporting of Assisted Reproductive Technology Outcomes: Past, Present, and Future," *American Journal of Obstetrics and Gynecology* 212, no. 2 (2015): 157–62.

62. Gleicher, N., "Is It Time to Limit IVF Transfers to One Embryo?" *Contemporary Ob/Gyn* 49 (2004): 73–85.

Chapter 9

1. *New York Times* Opinion, "Justice for All in the Baby M Case," February 4, 1988, accessed May 23, 2017, http://www.nytimes.com/1988/02/04/opinion/justice-for-all-in-the-baby-m-case.html.

2. Saul, Stephanie, "New Jersey Judge Calls Surrogate Legal Mother of Twins," *New York Times*, accessed May 23, 2017, http://www.nytimes.com/2009/12/31/us/31surrogate.html?_r=2.

3. RESOLVE, "Surrogacy," accessed October 10, 2017, http://resolve.org/what-are-my-options/surrogacy/?_sm_au_=iVVtnsPJKKDtQt7M

4. Hinson, Diane S., and Maureen McBrien, "Surrogacy Across America," *Family Advocacy* 34 (2011): 32.

5. Hinson, Diane S., "State-by-State Surrogacy Law Actual Practices," *Family Advocacy* 34 (2011): 36.

6. Arons, Jessica, *Future Choices: Assisted Reproductive Technologies and the Law* (Washington, DC: Center for American Progress, 2007).

7. Centers for Disease Control and Prevention, "Pregnancy Complications," last updated June 17, 2017, accessed October 10, 2017, https://www.cdc.gov/reproductivehealth/maternalinfanthealth/pregcomplications.htm

8. Caplan, Arthur, "Paid Surrogacy Is Exploitative," *New York Times*, September 23, 2014, https://www.nytimes.com/roomfordebate/2014/09/22/hiring-a-woman-for-her-womb/paid-surrogacy-is-exploitative

9. Mohapatra, Seema, "Stateless Babies & Adoption Scams: A Bioethical Analysis of International Commercial Surrogacy," *Berkeley Journal of International Law* 30 (2012): 412.

Chapter 10

1. Placek, Paul J., "National Adoption Data Assembled by the National Council for Adoption," in *Adoption Factbook V: The Most Comprehensive Source for Adoption Statistics Nationwide*, ed. Elisa A. Rosman, Charles E. Johnson, and Nicole M. Callahan (Alexandria, VA: National Council for Adoption, 2011), 3–68.

2. Jones, Jo., "Adoption Experiences of Women and Men and Demand for Children to Adopt by Women 18–44 Years of Age in the United States, 2002," *Vital and Health Statistics. Series 23, Data from the National Survey of Family Growth* 27 (2008): 1–36.

3. Park, Nicholas K., and Patricia Wonch Hill, "Is Adoption an Option? The Role of Importance of Motherhood and Fertility Help-Seeking in Considering Adoption," *Journal of Family Issues* 35, no. 5 (2014): 601–26.

4. US Department of State, Bureau of Consular Affairs, "Intercountry Adoption Statistics," accessed March 27, 2017, https://travel.state.gov/content/adoptionsabroad/en/about-us/statistics.html.

5. US Department of State, Bureau of Consular Affairs, *Intercountry Adoption. FY 2015 Annual Report on Intercountry Adoption* (Washington, DC: US Department of State, 2015).

6. Wetzstein, Cheryl, "Study: Families Trending Toward Open Adoptions," *Washington Times*, March 21, 2012, accessed March 27, 2017, http://www.washingtontimes.com/news/2012/mar/21/study-families-trending-toward-open-adoptions/.

7. Chandra, Anjani, et al., "Adoption, Adoption Seeking, and Relinquishment for Adoption in the United States," *Advance Data* 306, no. 11 (1999): 1–15.

8. Fisher, Allen P., "Still 'Not Quite as Good as Having Your Own'? Toward a Aociology of Adoption," *Annual Review of Sociology* 29, no. 1 (2003): 335–61.

Chapter 11

1. Dancet, E. A. F., et al., "The Patients' Perspective on Fertility Care: A Systematic Review," *Human Reproduction Update* 16, no. 5 (2010): 467–87.

2. Aarts, J. W. M., et al., "Professionals' Perceptions of Their Patients' Experiences with Fertility Care," *Human Reproduction* 26, no. 5 (2011): 1119–27.

3. van Empel, Inge W. H., et al., "Weaknesses, Strengths and Needs in Fertility Care According to Patients," *Human Reproduction* 25, no. 1 (2009): 142–49.

4. Aarts et al., "Professionals' Perceptions of Their Patients' Experiences."

5. Gameiro, Sofia, et al., "Why We Should Talk About Compliance with Assisted Reproductive Technologies (ART): A Systematic Review and Meta-Analysis of ART Compliance Rates," *Human Reproduction Update* 19, no. 2 (2012): 124–35.

6. Lo, Andrea Y., et al., "Doctor-Shopping in Hong Kong: Implications for Quality of Care," *International Journal for Quality Healthcare* 6, no. 4 (1994): 371–81.

7. Gameiro et al., "Why We Should Talk About Compliance."

8. Marcus, Hani J., Diana M. Marcus, and Samuel F. Marcus, "How do Infertile Couples Choose Their IVF Centers? An Internet-Based Survey," *Fertility and Sterility* 83, no. 3 (2005): 779–81.

9. Aarts et al., "Professionals' Perceptions of Their Patients' Experiences."

10. Lo et al., "Doctor-Shopping in Hong Kong."

11. Kasteler, Josephine, Robert L. Kane, Donna M. Olsen, and Constance Thetford, "Issues Underlying Prevalence of 'Doctor-Shopping' Behavior," *Journal of Health and Social Behavior* 17, no. 4 (1976): 328–39.

12. Gemmiti, Marco, et al., "Pediatricians' Affective Communication Behavior Attenuates Parents' Stress Response During the Medical Interview," *Patient Education and Counseling* 100, no. 3 (2017): 480–86.

13. Martinez, Kathryn A., et al., "Does Physician Communication Style Impact Patient Report of Decision Quality for Breast Cancer Treatment?" *Patient Education and Counseling* 99, no. 12 (2016): 1947–54.

Chapter 12

1. Golombok, Susan, Ann Spencer, and Michael Rutter, "Children in Lesbian and Single-Parent Households: Psychosexual and Psychiatric Appraisal," *Journal of Child Psychology and Psychiatry* 24, no. 4 (1983): 551–72.

2. Golombok, Susan, Fiona Tasker, and Clare Murray, "Children Raised in Fatherless Families from Infancy: Family Relationships and the Socioemotional Development of Children of Lesbian and Single Heterosexual Mothers," *Journal of Child Psychology and Psychiatry* 38, no. 7 (1997): 783–91.

3. Ethics Committee of the American Society for Reproductive Medicine, "Using Family Members as Gamete Donors or Surrogates," *Fertility and Sterility* 98, no. 4 (2012): 797–803.

4. *Buck v. Bell*, 274 U.S. 200 (1927).

5. Burgdorf, Robert L., Jr., and Marcia Pearce Burgdorf, "The Wicked Witch Is Almost Dead: Buck v. Bell and the Sterilization of Handicapped Persons," *Temple Law Quarterly* 50 (1976): 995.

6. *Buck v. Bell*, 274 U.S. 200.

7. Ethics Committee of the American Society for Reproductive Medicine, "Child-Rearing Ability and the Provision of Fertility Services: A Committee Opinion," *Fertility and Sterility* 100, no. 1 (2013): 50–53.

8. Ibid.

9. Practice Committee of the American Society for Reproductive Medicine and Practice Committee of the Society for Assisted Reproductive Technology, "Recommendations for Gamete and Embryo Donation: A Committee Opinion," *Fertility and Sterility* 99, no. 1 (2013): 47–62.

10. Braverman, Andrea Mechanick, "Mental Health Counseling in Third-Party Reproduction in the United States: Evaluation, Psychoeducation, or Ethical Gatekeeping?" *Fertility and Sterility* 104, no. 3 (2015): 501–6.

11. Ibid.

12. de Lacey, Sheryl L., Karen Peterson, and John McMillan, "Child Interests in Assisted Reproductive Technology: How Is the Welfare Principle Applied in Practice?" *Human Reproduction* 30, no. 3 (2015): 616–24.

13. Ibid.

14. Gurmankin, Andrea D., Arthur L. Caplan, and Andrea M. Braverman, "Screening Practices and Beliefs of Assisted Reproductive Technology Programs," *Fertility and Sterility* 83, no. 1 (2005): 61–67.

15. Sauer, Mark V., "American Physicians Remain Slow to Embrace the Reproductive Needs of Human Immunodeficiency Virus–Infected Patients," *Fertility and Sterility* 85, no. 2 (2006): 295–97.

16. Ethics Committee of the American Society for Reproductive Medicine, "Human Immunodeficiency Virus and Infertility Treatment," *Fertility and Sterility* 94, no. 1 (2010): 11–15.

17. Gurmankin, Caplan, and Braverman, "Screening Practices and Beliefs."

18. Ajmani, Sumeet, "North Coast Women's Care: California's Still-Undefined Standard for Protecting Religious Freedom," *California Law Review* 97, no. 6 (2009): 1867–76.

19. Lawrence, Ryan E., et al., "Obstetrician-Gynecologists' Beliefs About Assisted Reproductive Technologies," *Obstetrics & Gynecology* 116, no. 1 (2010): 127–35.

20. Budd, Karen S., and Michelle J. Holdsworth, "Issues in Clinical Assessment of Minimal Parenting Competence," *Journal of Clinical Child Psychology* 25, no. 1 (1996): 2–14.

21. Choate, Peter W., and Sandra Engstrom, "The 'Good Enough' Parent: Implications for Child Protection," *Child Care in Practice* 20, no. 4 (2014): 368–82.

22. Eve, Philippa M., Mitchell K. Byrne, and Cinzia R. Gagliardi, "What Is Good Parenting? The Perspectives of Different Professionals," *Family Court Review* 52, no. 1 (2014): 114–27.

23. Rodriguez, Christina M., and Michael J. Richardson, "Stress and Anger as Contextual Factors and Preexisting Cognitive Schemas: Predicting Parental Child Maltreatment Risk," *Child Maltreatment* 12, no. 4 (2007): 325–37.

24. Herman, Stephen P., "Special Issues in Child Custody Evaluations," *Journal of the American Academy of Child & Adolescent Psychiatry* 29, no. 6 (1990): 969–74.

25. Wolfe, David A., and Jeff St. Pierre, "Child Abuse and Neglect," in *Handbook of Child Psychopathology*, ed. Thomas H. Ollendick and Michel Hersen (New York: Springer, 1989), 377–98.

26. Clark, Lee Anna, Grazyna Kochanska, and Rebecca Ready, "Mothers' Personality and Its Interaction with Child Temperament as Predictors of Parenting Behavior," *Journal of Personality and Social Psychology* 79, no. 2 (2000): 274–85.

27. Robertson, John A., "Procreative Liberty and Harm to Offspring in Assisted Reproduction," *American Journal of Law & Medicine* 30, no. 1 (2004): 7–40.

28. Coleman, Carl H., "Conceiving Harm: Disability Discrimination in Assisted Reproductive Technologies," *UCLA Law Review* 50 (2002): 17–68.

29. Woodhouse, Barbara Bennett, "Talking About Children's Rights in Judicial Custody and Visitation Decision-Making," *Family Law Quarterly* 36, no. 1 (2002): 105–33.

30. De Sutter, Paul, "Gender Reassignment and Assisted Reproduction: Present and Future Reproductive Options for Transsexual People," *Human Reproduction* 16, no. 4 (2001): 612–14.

31. Jadva, V., et al., "A Longitudinal Study of Recipients' Views and Experiences of Intra-Family Egg Donation," *Human Reproduction* 26, no. 10 (2011): 2777–82.

32. Ibid.

33. Golombok, Spencer, and Rutter, "Children in Lesbian and Single-Parent Households."

34. Golombok, Tasker, and Murray, "Children Raised in Fatherless Families."

35. Stern, Judy E., et al., "Access to Services at Assisted Reproductive Technology Clinics: A Survey of Policies and Practices," *American Journal of Obstetrics and Gynecology* 184, no. 4 (2001): 591–97.

36. Ibid.

37. Ibid.

38. Gurmankin, Caplan, and Braverman, "Screening Practices and Beliefs."

39. Lawrence et al., "Obstetrician-Gynecologists' Beliefs."

40. Ethics Committee of the American Society for Reproductive Medicine, "Using Family Members as Gamete Donors or Surrogates."

41. Ibid.

42. Practice Committee of the American Society for Reproductive Medicine and Practice Committee of the Society for Assisted Reproductive Technology, "Recommendations for Gamete and Embryo Donation."

43. Ethics Committee of the American Society for Reproductive Medicine, "Using Family Members as Gamete Donors or Surrogates."

44. De Wert, G., et al., "Intrafamilial Medically Assisted Reproduction," *Human Reproduction* 26, no. 3 (2011): 504–9.

45. Centers for Disease Control and Prevention, "Nation Marriage and Divorce Rate Trends," last modified March 17, 2017, https://www.cdc.gov/nchs/fastats/marriage-divorce.htm.

46. Ethics Committee of the American Society for Reproductive Medicine, "Using Family Members as Gamete Donors or Surrogates."

47. Ibid.

48. Ibid.

49. Kahneman, Daniel, *Thinking, Fast and Slow* (New York: Macmillan, 2011).

50. Robertson, John A., "Procreative Liberty and Harm to Offspring in Assisted Reproduction," *American Journal of Law & Medicine* 30, no. 1 (2004): 7–40.

51. King, Joseph H., Jr., "Reconciling the Exercise of Judgment and the Objective Standards of Care in Medical Malpractice," *Oklahoma Law Review* 52 (1999): 49.

52. Lotz, Mianna, "Feinberg, Mills, and the Child's Right to an Open Future," *Journal of Social Philosophy* 37, no. 4 (2006): 537–51.

53. Ethics Committee of the American Society for Reproductive Medicine, "Using Family Members as Gamete Donors or Surrogates."

54. Practice Committee of the American Society for Reproductive Medicine and Practice Committee of the Society for Assisted Reproductive Technology, "Recommendations for Gamete and Embryo Donation."

55. Practice Committee of the American Society for Reproductive Medicine, and Practice Committee of the Society for Assisted Reproductive Technology, "Recommendations for Practices Utilizing Gestational Carriers: A Committee Opinion," *Fertility and Sterility* 107, no. 2 (2017): e3–10.

56. Hirschl, Ran, "'Negative' Rights vs. 'Positive' Entitlements: A Comparative Study of Judicial Interpretations of Rights in an Emerging Neo-Liberal Economic Order," *Human Rights Quarterly* 22, no. 4 (2000): 1060–98.

57. Fox, Renee C., "Medical Uncertainty Revisited," in *Handbook of Social Studies in Health and Medicine* (London: Sage Publications, 2000): 409–25.

Chapter 13

1. Fisher, P. B., J. F. Smith, and P. P. Katz, "Financial Burdens of Fertility Care: How Insurance Coverage and Perception of Cost Impact a Couple's Decision Making," *Fertility and Sterility* 96, no. 3 (2011): S30.

2. Chambers, Georgina M., G. David Adamson, and Marinus J. C. Eijkemans, "Acceptable Cost for the Patient and Society," *Fertility and Sterility* 100, no. 2 (2013): 319–27.

3. Marcus, Diana, et al., "Infertility Treatment: When Is It Time to Give Up? An Internet-Based Survey," *Human Fertility* 14, no. 1 (2011): 29–34.

4. Iaconelli, A., et al., "Main Concerns Regarding In Vitro Fertilization Techniques: Results of a Website Survey," *Fertility and Sterility* 100, no. 3 (2013): S66.

5. Chandra, Anjani, and Elizabeth Hervey Stephen, "Infertility Service Use Among U.S. Women: 1995 and 2002," *Fertility and Sterility* 93, no. 3 (2010): 725–36.

6. Wu, Alex K., et al., "Time Costs of Fertility Care: The Hidden Hardship of Building a Family," *Fertility and Sterility* 99, no. 7 (2013): 2025–30.

7. Smith, James F., et al., "Socioeconomic Disparities in the Use and Success of Fertility Treatments: Analysis of Data from a Prospective Cohort in the United States," *Fertility and Sterility* 96, no. 1 (2011): 95–101.

8. Kulkarni, Aniket D., et al., "Fertility Treatments and Multiple Births in the United States," *New England Journal of Medicine* 369, no. 23 (2013): 2218–25.

9. Johnston, Josephine, Michael K. Gusmano, and Pasquale Patrizio, "Reducing Rate of Fertility Multiples Requires Policy Changes," *JAMA Pediatrics* 169, no. 3 (2015): 287.

10. RESOLVE, "Insurance Coverage in Your State," 2017, accessed October 19, 2018, https://resolve.org/what-are-my-options/insurance-coverage/.

11. American Society for Reproductive Medicine, "State Infertility Insurance Laws," 2016, accessed October 19, 2016, https://www.reproductivefacts.org/resources/state-infertility-insurance-laws/?_ga=2.185504169.1834587782.155308696 8-800756311.1553086968.

12. Omurtag, Kenan, and G. David Adamson, "The Affordable Care Act's Impact on Fertility Care," *Fertility and Sterility* 99, no. 3 (2013): 652–55.

13. Devine, Kate, Robert J. Stillman, and Alan DeCherney, "The Affordable Care Act: Early Implications for Fertility Medicine," *Fertility and Sterility* 101, no. 5 (2014): 1224.

14. Ethics Committee of the American Society for Reproductive Medicine, "Disparities in Access to Effective Treatment for Infertility in the United States: An Ethics Committee Opinion," *Fertility and Sterility* 104, no. 5 (2015): 1104–10.

15. Ethics Committee of the American Society for Reproductive Medicine, "Financial 'Risk-Sharing' or Refund Programs in Assisted Reproduction: An Ethics Committee Opinion," *Fertility and Sterility* 106, no. 5 (2016): e8–11.

16. Cummins, James M., and Anne M. Jequier, "Treating Male Infertility Needs More Clinical Andrology, Not Less," *Human Reproduction* 9, no. 7 (1994): 1214–19.

17. Speier, Amy R., "Brokers, Consumers and the Internet: How North American Consumers Navigate Their Infertility Journeys," *Reproductive Biomedicine Online* 23, no. 5 (2011): 592–99.

18. Jungheim, Emily S., et al., "Embryo Transfer Practices in the United States: A Survey of Clinics Registered with the Society for Assisted Reproductive Technology," *Fertility and Sterility* 94, no. 4 (2010): 1432–36.

19. Owens Douglas, K., et al., "High-Value, Cost-Conscious Healthcare: Concepts for Clinicians to Evaluate the Benefits, Harms, and Costs of Medical Interventions," *Annals of Internal Medicine* 154, no. 3 (2011): 174–80.

20. Knox, Richard, "Poll: What It's Like to Be Sick in America," *All Things Considered*, National Public Radio, May 21, 2012, accessed May 25, 2017, http://www.npr.org/sections/health-shots/2012/05/21/153019327/poll-what-its-like-to-be-sick-in-america 2012.

21. Cafarella Lallemand, Nicole, "Health Policy Brief: Reducing Waste in Healthcare," *Health Affairs* 35, no. 10 (2012), accessed May 25, 2017, http://www.healthaffairs.org/healthpolicybriefs/brief.php?brief_id=82.

22. Halpern, Jodi, and Robert M. Arnold, "Affective Forecasting: An Unrecognized Challenge in Making Serious Health Decisions," *Journal of General Internal Medicine* 23, no. 10 (2008): 1708–12.

23. Gleicher, N., and R. Orvieto, "Is the Hypothesis of Preimplantation Genetics Screening (PGS) Still Supportable? A Review," *Journal of Ovarian Research* 10, no. 1 (2017): 21.

24. Klitzman, R., "How Infertility Patients and Providers View and Confront Religious and Spiritual Issues," *Journal of Religious Health* 57, no. 1 (2018): 223–39.

25. Institute of Medicine of the National Academies, *Best Care at Lower Cost: The Path to Continuously Learning Healthcare in America* (Washington, DC: National Academies Press, 2013).

26. James, J., "Health Policy Brief: Pay-for-Performance," *Health Affairs* October 11, 2012, accessed September 4, 2018, http://www.healthaffairs.org/healthpolicybriefs/brief.php?brief_id=78.

Chapter 14

1. Greil, A. L., K. Slauson-Blevins, and J. McQuillan, "The Experience of Infertility: A Review of Recent Literature," *Sociology of Health & Illness* 32, no. 1 (2010): 140–62.

2. Peterson, Brendan D., Lisa Gold, and Tal Feingold, "The Experience and Influence of Infertility: Considerations for Couple Counselors," *Family Journal* 15, no. 3 (2007): 251–57.

3. Domar, A. D., "Impact of Psychological Factors on Dropout Rates in Insured Infertility Patients," *Fertility and Sterility* 81, no. 2 (2004): 271–73.

Chapter 15

1. Kahneman, D., *Thinking Fast and Slow* (New York: Farrar, Straus and Giroux, 2011).

2. Van den Broeck, U., et al., "Counselling in Infertility: Individual, Couple and Group Interventions," *Patient Education and Counseling* 81, no. 3 (2010): 422–28.

3. Peterson, Brennan D., and Georg H. Eifert, "Using Acceptance and Commitment Therapy to Treat Infertility Stress," *Cognitive and Behavioral Practice* 18, no. 4 (2011): 577–87.

4. Van den Broeck et al., "Counselling in infertility."

5. Mohr, David C., et al., "Perceived Barriers to Psychological Treatments and Their Relationship to Depression," *Journal of Clinical Psychology* 66, no. 4 (2010): 394–409.

6. Clement, Sarah, et al., "What Is the Impact of Mental Health–Related Stigma on Help-Seeking? A Systematic Review of Quantitative and Qualitative Studies," *Psychological Medicine* 45, no. 1 (2015): 11–27.

7. Bergart, A. M., "The Experience of Women in Unsuccessful Infertility Treatment: What Do Patients Need When Medical Intervention Fails?" *Social Work in Health Care* 30, no. 4 (2000): 45–69.

8. Ibid.

Chapter 16

1. Christakis, N. A., and D. A. Asch, "Physician Characteristics Associated with Decisions to Withdraw Life Support," *American Journal of Public Health* 85, no. 3 (1995): 367–72.

2. Ramondetta, L., et al., "Religious and Spiritual Beliefs of Gynecologic Oncologists May Influence Medical Decision Making," *International Journal of Gynecological Cancer* 21, no. 3 (2011): 573–81.

3. Lawrence, R. E., et al., "Obstetrician-Gynecologists' Views on Contraception and Natural Family Planning: A National Survey," *American Journal of Obstetrics and Gynecology* 204, no. 2 (2011): 124.e1–7.

Chapter 18

1. Human Fertilisation and Embryology Authority, *Egg and Sperm Donation in the UK: 2012–2013* (London: Human Fertilisation and Embryology Authority, 2013), 10–13.

2. Duke, Alan, "Nadya Suleman's Doctor Loses California Medical License," CNN, June 2, 2011, accessed October 25, 2018, www.cnn.com/2011/US/06/01/california. octuplets.doctor.revoked/index.html.

3. De Neubourg, D., et al., "Belgium Model of Coupling Reimbursement of ART Costs to Restriction in Number of Embryos Transferred," *BMJ: British Medical Journal* 348 (2014): g1559.

4. Gleicher, N., and R. Orvieto, "Is the Hypothesis of Preimplantation Genetics Screening (PGS) Still Supportable? A Review," *Journal of Ovarian Research* 10, no. 1 (2017): 21.

5. Cochrane Library, "About Cochrane Reviews," accessed October 31, 2018, https://www.cochranelibrary.com/about/about-cochrane-reviews.

6. Keehn, Jason, et al., "Recruiting Egg Donors Online: An Analysis of In Vitro Fertilization Clinic and Agency Websites' Adherence to American Society for Reproductive Medicine Guidelines," *Fertility and Sterility* 98, no. 4 (2012): 995–1000.

7. Hauser, Christine, "4,000 Eggs and Embryos Are Lost in Tank Failure, Ohio Fertility Clinic Says," *New York Times*, March 28, 2018. https://www.nytimes.com/2018/03/28/us/frozen-embryos-eggs.html.

8. Lotz, Mianna. "Feinberg, Mills, and the child's right to an open future," *Journal of social philosophy* 37, no. 4 (2006): 537–551.

9. National Center for Chronic Disease Prevention and Health Promotion, Division of Public Health Assisted Reproductive Technology, "National Summary Report," 2016, accessed March 20, 2019, https://www.cdc.gov/art/pdf/2016-report/ART-2016-National-Summary-Report.pdf.

10. Kawwass, J. F., et al., "Safety of Assisted Reproductive Technology in the United States, 2000–2011," *Journal of the American Medical Association* 313, no. 1 (2015): 88–90.

11. Perkins, K. M., et al., "Trends and Outcomes of Gestational Surrogacy in the United States," *Fertility and Sterility* 106, no. 2 (2016): 435–42.

12. Kushnir, Vitaly A., et al., "The Status of Public Reporting of Clinical Outcomes in Assisted Reproductive Technology," *Fertility and Sterility* 100, no. 3 (2013): 736–41.

13. Centers for Disease Control and Prevention, American Society for Reproductive Medicine, and Society for Assisted Reproductive Technology, *2015 Assisted Reproductive Technology Fertility Clinic Success Rates Report* (Atlanta, GA: US Department of Health and Human Services, 2017).

14. Ibid.

15. De Rycke, M., et al., "ESHRE PGD Consortium Data Collection XIV–XV: Cycles from January 2011 to December 2012 with Pregnancy Follow-up to October 2013," *Human Reproduction* 32 (2017): 1974–94.

16. Gleicher, N., and R. Orvieto, "Is the Hypothesis of Preimplantation Genetics Screening (PGS) Still Supportable? A Review," *Journal of Ovarian Research* 10 (2017): 21.

17. Braat, D. D., et al., "Maternal Death Related to IVF in the Netherlands 1984–2008," *Human Reproduction* 25 (2010): 1782–86.

18. De Geyter Ch., et al., "Data Collection Systems in ART Must Follow the Pace of Change in Clinical Practice," *Human Reproduction* 31 (2016): 2160–2163.

19. Centers for Disease Control and Prevention, "National Program of Cancer Registries (NPCR)," accessed March 20, 2014, www.cdc.gov/cancer/npcr/about.htm.

20. New York State Department of Health, "Communicable Disease Reporting," February 2015, accessed October 31, 2018, https://www.health.ny.gov/professionals/diseases/reporting/communicable/.

21. Braat et al., "Maternal Death Related to IVF."

22. De Geyter et al., "Data Collection Systems in ART."

23. De Geyter, Ch., et al.; European IVF-Monitoring Consortium for the European Society of Human Reproduction and Embryology, "ART in Europe, 2014: Results Generated from European Registries by ESHRE: The European IVF-Monitoring

Consortium (EIM) for the European Society of Human Reproduction and Embryology (ESHRE)," *Human Reproduction* 33 (2018): 1586–1601, https://doi.org/10.1093/humrep/dey242

24. Erlich, Y., et al., "Identity Inference of Genomic Data Using Long-Range Familial Searches," *Science* 362, no. 6415 (2018): 690–94.

25. Gold, Katherine J., Angela L. Kuznia, and Rodney A. Hayward, "How Physicians Cope with Stillbirth or Neonatal Death: A National Survey of Obstetricians," *Obstetrics & Gynecology* 112, no. 1 (2008): 29–34.

26. Kaldjian, Lauris C., Elizabeth W. Jones, and Gary E. Rosenthal, "Facilitating and Impeding Factors for Physicians' Error Disclosure: A Structured Literature Review," *Joint Commission Journal on Quality and Patient Safety* 32, no. 4 (2006): 188–98.

Chapter 19

1. P Berg, D. Baltimore, S. Brenner, R. O. Roblin, and M. F. Singer, "Summary Statement of the Asilomar Conference on Recombinant DNA Molecules," *Proceedings of the National Academy of Sciences of the United States of America* 72, no. 6 (1975): 1981–1984.

Appendix B

1. Geertz, Clifford. *Interpretation of Cultures: Selected Essays* (New York, Basic Books, 1973).

2. Strauss, Anselm, and Juliet Corbin, *Basics of Qualitative Research: Techniques and Procedures for Developing Grounded Theory* (Newbury Park, CA: Sage Publications, 1990).

3. Guest, Greg, Arwen Bunce, and Laura Johnson, "How Many Interviews Are Enough? An Experiment with Data Saturation and Variability," *Field Methods* 18, no. 1 (2006): 59–82.

INDEX

For the benefit of digital users, indexed terms that span two pages (e.g., 52–53) may, on occasion, appear on only one of those pages.

abortion
  costs of, 87
  due to fetus's sex, 111
  of fetus with severe mutation,
    75–76, 81
    communication to others
      about, 221
  insurance companies' attitudes
    about, 87, 88
  partial (fetal reduction), 133
    and challenges for providers, 135
    decision-making about, 136
    and loss of entire pregnancy, 134
    patients' agreement with, 134
    patients' refusal of, 134, 136–37
    patients' wariness of, 134
    religious views on, 236
  past, patient's regrets about, 212
  patients' refusal of, 92, 134, 136–37
  religious views on, 236, 237
achondroplasia, 109
ACOG. *See* American College of
  Obstetricians and Gynecologists
acupuncture, 160, 229
ADA. *See* Americans with Disabilities
  Act (ADA)

adoption
  after successful infertility treatment,
    148, 149
  as alternative to assisted reproductive
    technology, 146, 147, 148, 151
  closed, 146, 147–48
  costs of, 146, 147–48, 150
  financial stress caused by, 201
  foreign (from other countries),
    147–48, 150
  of foster children, 147, 150–51
  of infants, 147
  obstacles to, 146, 150
  open, 146–48
  patients' rejection of, 149
  patients' wariness of, 148–50
  physicians as facilitators of, 151–52
  rates, 147
  screening of potential parents
    for, 186
  spousal disagreements about, 151
adult-onset disease, preimplantation
  genetic diagnosis for, 76
affective forecasting, 202
Affordable Care Act, 196–97
Against Perfection (Sandel), 109

age. *See also* biological clock
of adopted children, 147–48
of donor egg recipients, 42
of egg donors, 56–57
and fertility decline, 37
public education about, 244, 247
of first-time mothers, 5–6
and freezing of eggs, 68
and IVF with woman's own eggs, 42
lying about, 49
and number of embryos
transferred, 124–25
parental, and child's rights and well-
being, 47, 182
and pregnancy complications, 42
age discrimination, 52–53, 54
Age Discrimination Act of 1975, 54
age limits
combined age of parents and,
46–37, 47
for fathers, 45
for infertility treatment, 42–45, 52–53,
54, 182–83
Almeling, Rene, 57–58
alternative therapies, 160, 229
AMA. *See* American Medical
Association (AMA)
American College of Genetics and
Genomics, policies concerning
genetic testing, 105–6
American College of Obstetricians and
Gynecologists, 253–54
and ethical issues, 260
guidelines, enhancement of, need for,
258, 259, 261–62, 270
policy on sex selection, 110
American Medical Association (AMA)
guidelines, for health websites, 57
and regulation of infertility
treatment, 260
American Society for Reproductive
Medicine
on disclosure of treatment
costs, 196–97
and ethical issues, 260
fear of litigation, 256

guidelines, 10–11
advances in (future directions
for), 54
on age cutoffs for infertility
treatment, 42–43
on clinicians' financial benefit, 54
on compensation of, recruitment
of, and communicating risks/
benefits to egg donors, 12, 56–57
enhancement of, need for, 258–59,
261–62, 270
on explicit evidence-based policies,
42–43, 54
for number of embryos to transfer,
125, 126, 138, 252
on parental suitability, 191
on refusing to offer
treatment, 53, 54
in sperm donation, 71
on withholding services, 171
on insurance coverage, 196–97
on number of embryos transferred,
by age group, 124–25
and profit-taking by providers, 207
recommendations
on disclosure to donor-conceived
offspring, 74
on intrafamilial gamete donation,
180, 190, 191
and regulation of egg donor
agencies, 64
requirements for screening of egg
donors, 59
role of, providers' views on, 257–58
on sex selection, 110–12
on sperm sorting, 110
Americans with Disabilities Act
(ADA), 52–53
AMH. *See* anti-Müllerian hormone
Am I My Genes? (Klitzman), 6
amniocentesis, 81, 91
definition of, 279
patients' refusal of, 91
aneuploidy, definition of, 279
ankylosing spondylitis, 78–79, 90
definition of, 279

anonymity
  of gamete donors, 65
  of sperm donors, 11, 65, 255–56
anti-Müllerian hormone, definition
  of, 279
ART. *See* assisted reproductive
  technology
artwork, portraying fertility and
  pregnancy, 6
Ashkenazi Jews, genetic diagnosis
  for, 238
ASRM. *See* American Society for
  Reproductive Medicine
assisted reproductive technology
  advances in, 5
  choices about, 19–20
  decision-making about, 46
  dilemmas and challenges posed
    by, 272–76
  evaluation of, enhancement of, need
    for, 259
  historical perspective on, 5–6
  insurance coverage of, 12–13,
    158, 196–97
  lawsuits involving, 16
  number of babies born using, 5
  providers' lack of knowledge
    about, 33–34
  regulation of, 252
  religious views on, 235
  scandals involving, 16
  use rates for, 38–39
attitudinal barriers, experienced
  by patients, sample interview
  questions about, 291
autism, and sex selection, 116, 118
autonomy
  and non-disclosing preimplantation
    genetic diagnosis, 101
  and number of embryos transferred,
    131, 136–37, 138
  patients', 46–47
azoospermia, definition of, 279

Baby M, 141, 143
Baby Manji, 144–45

bedside manner, 161
  of office, 155
Benitez v. North Coast, 172
beta-thalassemia
  definition of, 279
  and savior siblings, 93
biological child, pursuit of,
  psychological costs of, 50
biological clock, 26, 31, 35
The Boys from Brazil, 16–17
Brave New World, 16–17
BRCA. *See* breast cancer gene *(BRCA)*
breast cancer gene *(BRCA)*, 83
  definition of, 279
  preimplantation genetic diagnosis for,
    76–77, 78
brother(s), as sperm donors, 72
Brown, Louise, 5, 296*f*
Buck v. Bell, 171

cancer patients, freezing of eggs and
  sperm for, 67, 68
carrier status, education about, need for,
  245, 250–51
CDC. *See* Centers for Disease Control
  and Prevention (CDC)
Centers for Disease Control and
  Prevention (CDC)
  data analysis by, lack of funds
    for, 269
  database of IVF statistics, 10
  data collection by, 262–63
  data on unconventional parenting,
    need for, 191
  and improved practice, 270–71
  National ART Surveillance
    System, 262
  and regulation of egg donor
    agencies, 64
  reporting policies of, 132, 138
    improvement of, need for, 264–65,
      266, 267, 270
CF. *See* cystic fibrosis
child(ren)
  biological, pursuit of, psychological
    costs of, 50

child(ren) *(cont.)*
    born through egg donation,
        disclosing status to, 66
    born through sperm donation,
        disclosing status to, 70–71, 74
    with donor-conceived status
        ethical right to know about, 74
        long-term follow-up of, need
            for, 268
        registry of, need for, 267
    rights and well-being of, 47, 53,
        121, 182
child abuse, forecasting, 172
child molestation, 175–76
chorionic villus sampling, 81, 91
    patients' refusal of, 91
    religious views on, 236–37
chromosomal abnormalities,
    preimplantation genetic diagnosis
        for, 76
chromosomal translocations
    balanced, 82–83
    unbalanced, 82–83
chromosome, definition of, 279
chromosome testing, 85
clergy. *See* religion
Clinical Laboratory Improvement Act, 10
clinician(s). *See* provider(s)
Clomid, 39, 279
clomiphene, 279
CMV. *See* cytomegalovirus (CMV)
COIs. *See* conflicts of interest
Coleman, Carl, 172
communication
    about infertility, men and, 27
    doctor–patient, 156, 161
        about infertility, 38
        about sex and reproduction, 38
        dynamic tensions in, 163, 168–69
    of statistics, by providers to
        patients, 164–65
conflicts of interest, management
    of, 260
constant comparison, 283
continuity of care, 162
coping strategies, 218, 230, 232–33

religion and, 234
for religious objections, 238
costs
    of abortion, 87
    of adoption, 146, 147–48, 150
    of amniocentesis, 87
    of donated eggs, 208
    of egg freezing, 67–68
    of fertility drugs, 206
    of gestational surrogacy, 141–42
    of infertility treatment, 33
        disclosure of, 196–97
        excess costs in, 197
    of in vitro fertilization (IVF),
        9–10, 12–13
    of sperm, 72–73
    of tubal reversal, 167
CRISPR (cluster of regularly interspaced
    short palindromic repeats), 13–14,
    15, 76, 252
    definition of, 280
cultural products, portraying fertility
    and pregnancy, 6
CVS. *See* chorionic villus sampling
cyberchondria, 248
cystic fibrosis, 48, 75–76, 83, 87, 88,
    91–93, 118
    definition of, 279
    preimplantation genetic diagnosis
        for, 76, 79
cytomegalovirus (CMV), 72–73

Darwin, Charles, 15
data collection, on assisted reproductive
    technology
    funding of, 268–69
    improvement of, need for, 262
    and inaccurate statistics from
        providers, 267
    long-term (longitudinal), need
        for, 267–68
    usefulness of, 265–66
deafness, preimplantation selection for,
    109, 118
designer babies, 16, 17, 20
desired traits, non-medical, 120

DET. *See* double embryo transfer

Diamond-Blackfan anemia
   definition of, 280
   and savior siblings, 93

Dilation and curettage (D & C),
      definition of, 280

disability(ies)
   and parenting, 171, 172
   preimplantation genetic diagnosis
      for, 76
   preimplantation selection
      for, 109, 118
      decision-making about, difficulties
         of, 119

disgust, and medical
      decision-making, 49

DNA, definition of, 280

doctor(s). *See* provider(s)

doctor–patient communication, 156, 161
   about infertility, 38
   about sex and reproduction, 38
   dynamic tensions in, 163, 168–69

doctor–patient relationships, sample
      interview questions about, for
      providers, 287

doctor shopping, 156, 168

donor registry(ies), 65–66

double embryo transfer, 124

Doud-Suleman, Nadya Denise, 16

Down syndrome. *See* trisomy 21 (Down
      syndrome)

dwarfism, 109
   preimplantation selection for, 118
      decision-making about, difficulties
         of, 119

education. *See also* patient education;
      public education
   about infertility and treatment, need
      for, 243
   of egg donors, 62
   of family and friends, 220
   improvement of, 249
   of providers, about infertility and
      treatment, 37, 251
   sources of, 247

types of, 247

Edwards syndrome. *See* trisomy 18
      (Edwards syndrome)

egg donation, 41
   donor recruitment for, 54
   by friends or family members, 55
   number of, limitations on, 60
   overuse of, 205, 208
   providers' approaches to, 50
   quandaries with, 42
   recipients' decision-making about, 47
   regulation of, 259
   by relative, 179–80
   religious views on, 236
   risks of, information on, need
      for, 267–68
   sister-to-sister, 179
   by strangers, 56

egg donor(s), 12
   agencies' characterizations
      of, 57–58
   age of, 56–57
   anonymity of, 65
   characteristics of, 61–62
   compensation of, 12, 56–57, 61–62, 63
   complications in, CDC data on, 262
   education of, 62
   informed consent from, 64
   motivations of, data collection on,
      need for, 268
   personality tests for, 60–61
   potential harms to, 61
   screening of, 59
   selection of, and eugenics, 15,
      42, 57–58
   traits of
      compensation based on, 57
      and which eggs to buy, 64
   veracity of, determination of, 60–61

egg donor agencies, 41, 57–58
   characterizations of egg donors, 57–58
   information about donors
      provided by, 63
   quality of, variations in, 58, 63
   regulation of, 63, 64, 260
   screening of egg donors, 59–60, 61

egg recipients
  age of, 42
  decision-making about donor eggs, 47
  misgivings and fears of, 49–50
eggs (human)
  buying/selling of, 11–12, 41, 42, 56
    use rates for, 38–39
  donated
    costs of, 208
    decision-making about using, 47, 159
    disclosing use of, 66
    obtaining, 54
  freezing, 67, 260
    information on, need for, 267
  freezing of
    age and, 68
    costs of, 67–68
    ethics committees and, 69
  for preimplantation genetic
      diagnosis, 80
  purchase of, 64
egg seller(s), 12
egg supplier(s), 12
Emanuel syndrome, 23, 84–85, 88
  definition of, 280
embryo(s), 7–8
  buying/selling of, 12
    rates of, 38–39
  definition of, 280
  discarding/destruction of, 101
  donation of
    to other prospective parents, 101
    for research, 101
  gene editing in, 13–14
  leftover, 101, 107
  manufacture of, 12
  moral status of, 107, 275
  non-medical selection of, 260 (see also
      sex selection)
  number of, religious views on, 236
  "on demand" creation of, 12
  storage of, 101
embryo screening, 75, 77, 78. See also
    embryo testing
  public education on, improvement
    of, 250

  religious views on, 236
embryo testing. See also embryo
    screening
  confirmation of
    need for, 91
    patients' refusal of, 91
  cost limitations and, 86
  and disclosure dilemmas, 98
  factors affecting, 83
  insurance coverage of, providers'
      advocacy for, 89
  patients' forgoing of, 90
  patients' out-of-pocket payment
      for, 90
  patients' understanding of, 84
  provider characteristics and, 83
  religious concerns and, 85
  without revealing results, 97
    provider and patient
      responses to, 99
embryo transfer
  numbers (how many to transfer),
      123–26 (see also double embryo
      transfer; single embryo transfer)
  decision-making about, 131, 137–38
  financial considerations and, 127–
      28, 133
  guidelines for, 132, 137–38
  patients' perspectives and, 126
  physicians' perspectives and, 128
  physicians versus committees
      and, 131
  policies for, 132
  quality assurance and, 131
  sham, 100, 101
endocrinology, 5
  definition of, 280
endometriosis, definition of, 280
entrepreneurism, among providers, 204
  factors affecting, 206
  possible solutions to, 207
Erez, Ayelet, 97
estradiol, definition of, 280
ethical issues, 11, 182, 189, 261
ethics committees, 52, 55, 95,
      178, 261–62

for assisted reproductive
technology, 108–9
and egg freezing for future use, 69
and leftover embryos, 105
and positive selection, 120
and sex selection, 116–17
eugenics, 15, 42, 57–58, 65, 109, 110, 113
European Society of Human
Reproduction and Embryology
data collection by, 263–64
guidelines, on intrafamilial gamete
donation, 180
exclusion testing, 97–98
eye color, selection for, 120

Facebook. *See* online support
facilitators, sample interview
questions about
for patients, 291
for providers, 288
failure, infertility as, 24–25
familial cerebello retinal angiomatosis.
See von Hippel-Lindau syndrome
family balancing
defining, challenges in, 114
providers' attitudes toward, 114
sex selection for, 108, 110, 113
family pressures, to have children, 24
Fanconi anemia (FA), 91, 96
definition of, 280
and savior siblings, 93
fate, metaphysical explanations
of, 234
father(s), age limits for, 45
FDA. *See* Food and Drug
Administration (FDA)
fertility drugs. *See also* medications;
specific drug
adverse effects and side effects
of, 209–10
black markets for, 201
and cancer, 56–57
costs of, 206
monitoring and data collection
on, 266
unused, resale of, 201

fertilization, artificial, historical
perspective on, 4–5
fetal reduction. *See* abortion, partial
(fetal reduction)
fetal testing. *See also* amniocentesis;
chorionic villus sampling
patients' refusal of, 91
financial factors, 197–98
ethical and moral challenges raised
by, 274
free and discounted treatments
and, 200
and parenting suitability, 173, 175
in providers' decision-making, 184
and referral to infertility
specialist, 36
sample interview questions about
for patients, 291
for providers, 288
stress caused by, 195–96, 201, 207–8
follicles, 8
definition of, 280
follicle-stimulating hormone (FSH),
levels of, and infertility treatment,
44–45, 130
Food and Drug Administration (FDA)
oversight by, enhancement of, need
for, 259
and regulation of egg donor
agencies, 64
and regulation of infertility
treatment, 10
requirements for screening of egg
donors, 59
foster care, adoption of children from,
147, 150–51
foster parenting, 147
fragile X syndrome, 56, 80
definition of, 280
religious views on, 237
freezing
of eggs (*see* eggs (human),
freezing of)
of eggs and sperm, for cancer
patients, 67, 68
friend(s), as egg donors, 56

friendships, stress on, with infertility treatment, 213, 217, 218
FSH. *See* follicle-stimulating hormone (FSH)
future directions (advances), sample interview questions about, for providers, 289

Galton, Francis, 15
gamete(s). *See also* eggs (human); sperm
  definition of, 280
  donated, 41
    disclosing use of, 66
    use of, 82
  frozen, disposition of, 102
gamete donation
  attitudes toward, male and female differences in, 71
  intergenerational, 181–82, 190
  intrafamilial, 179–80, 190
  regulation of, 259
gamete donors, anonymity of, 65
Gattaca, 16–17
Gaucher's disease, 79–80
gay couples/gay parents, 173, 175–76
  data on, need for, 191
  preference for twins, 125
  society's attitudes about, 188
  use of surrogates, 140, 143
Geertz, Clifford, 18, 283
gender differences, and stresses of infertility treatment, 212
gene(s), 6
  adding/removing, 13 (*see also* CRISPR (cluster of regularly interspaced short palindromic repeats))
  definition of, 280
  interactions of, 16
gene editing, 13–14, 15, 261
general practitioner(s), and referral to infertility specialist, 33
genetic counselors, lack of knowledge about preimplantation genetic diagnosis, 36
genetic determinism, 275–76

genetic disease, transmission by sperm donor, 74
genetic engineering, elective, 120
genetics
  dilemmas and challenges posed by, 275–76
  education about, need for, 245
  providers' understanding of, 84
  public education on, improvement of, 250–51
gestational surrogacy, 140, 141
  CDC data on, 262
  children born via
    abandoned in orphanages, 144–45
    outside U.S., and return to U.S., 145
  costs of, 141–42
  data collection on, usefulness of, 265–66
  in foreign countries, 144
  legal considerations with, 142, 143
  motivations for, data collection on, need for, 268
  paid, legal considerations with, 141–42
  and quadruplet pregnancy, 136
GP. *See* general practitioner(s)
Greely, Hank, 15, 122
grounded theory, 283
guidelines. *See also* American Society for Reproductive Medicine
  advantages of, 256
  enforcement of, 257
  for infertility treatment, 252, 253

hair color, selection for, 120
The Handmaid's Tale, 16–17
Hard, Addison Davis, 5
HD. *See* Huntington's disease
healthcare costs, 197
HEFA. *See* Human Fertilisation and Embryology Authority
hemophilia, 82
  definition of, 280
HIV-infected patients, 172, 179
HLA. *See* human leukocyte antigen(s) (HLA)
Holmes, Oliver Wendell, 171
Holocaust, 110

hormone(s), definition of, 280
HSG. *See* hysterosalpingogram
Human Fertilisation and Embryology
Authority
and age limits for infertility
treatment, 43
and multiple births, 124
human leukocyte antigen(s) (HLA). *See
also* PGD-HLA
and savior siblings, 93
Huntington's disease, 6, 97–98
definition of, 280
and depression, 100–1
and disclosure dilemmas, 98
exclusion testing for, 97–98
non-disclosing preimplantation
genetic diagnosis of, 97, 99
preimplantation genetic diagnosis
for, 76, 78
and suicide, 100–1
hysterosalpingogram, definition of, 280

ICSI. *See* intracytoplasmic sperm
injection (ICSI)
implantation rates, reporting of, 129–30
incest, 180, 183
infertility
emotional toll of, public education on,
improvement of, 250
as grief process, 22
prevalence of, 19
stress caused by, 22–23, 25, 28
for men, 26
infertility services, use rates for, 5
infertility treatment. *See also* treatment
failure
choice of procedures for, religion
and, 235
choices about, 19–20
cost of, 33
decision-making about, 46
dilemmas and challenges posed by,
10, 19–20, 272–76
disclosure of, to outsiders, 220
excess costs in, 197
free and discounted, 200

futile, ASRM on, 42–43
insurance coverage of, 12–13, 33
and overuse of specific procedures, 204
patient-related obstacles and, 36
procedures for, selection of, 38
public policy and, 10–11
racial/ethnic differences and, 33
refusing to offer
ASRM guidelines on, 53
ethical/moral considerations in,
182–83, 187, 189
regulation of, 10–11
respites in, 50
stopping
financial factors and, 196
providers' approaches to, 50, 51
who should make the decision
about, 51
traveling to other states/countries
for, 166
uncertainties of, 196
with very poor prognosis, ASRM
on, 42–43
withholding, policies on, 171, 187
informational barriers, experienced
by patients, sample interview
questions about, 290
informed consent
improvement of, 249
with intrafamilial arrangements, 181
with unconventional combinations, 181
institutional factors, sample interview
questions about
for patients, 291
for providers, 288
insurance coverage
for assisted reproductive technology,
12–13, 158, 196–97
controversies about, 203
of embryo testing, providers' advocacy
for, 89
enhancement of, need for, 258
increased, political factors in, 203
of infertility treatment, 12–13, 33
of IVF, 12–13, 86, 87, 126, 196–97
enhancement of, need for, 258

insurance coverage (*cont.*)
limitations of, 198–200
of mental health services,
enhancement of, need for, 258
of non-disclosing preimplantation
genetic diagnosis, 101
and non-infertility reimbursable
indications, 201
of preimplantation genetic diagnosis,
86, 106
enhancement of, need for, 258
providers' advocacy for, 89
for psychotherapy, 224
and SART membership, 265
uncertainties and stresses related to,
198–200, 207–8
intelligence
genes and other factors affecting, 16
and parenting suitability, 174
Internet. *See* online information; online
support
interviews (by Klitzman), 18
data analysis, 284
instrument for, 284
methods for, 283
participants in, 18, 283, 284*t*
sample questions
for patients, 289
for providers, 287
intracytoplasmic sperm injection
(ICSI), 8, 39
overuse of, 204–5, 208
intrauterine insemination (IUI), 7, 8
use rates for, 38–39
in vitro fertilization (IVF), 8, 31, 32
age cutoffs for, 43–45
complications and challenges
with, 39–40
costs of, 9–10, 12–13
with donor eggs, age cutoffs
for, 43–44
education about, need for, 245
failure, 81–82
financial stress caused by, 201
free and discounted treatments
and, 200

as future of reproduction, 15
historical perspective on, 4, 5
hormone shots for, 38
insurance coverage of, 12–13, 86,
87, 126
enhancement of, need for, 258
patient's decision-making about,
sample interview questions
about, 289
providers, selection of, 34
religious views on, 235, 236
standards of practice for, need
for, 266
success rates for, 9
with patients' own gametes, 41
providers' lack of knowledge
about, 35
use rates for, 38–39
with woman's own eggs
age cutoffs for, 43–44
birth rates for, 42
maternal age and, 42
Ishiguro, Kazuo, 16–17
IUI. *See* intrauterine insemination (IUI)
IVF. *See* in vitro fertilization (IVF)

Kahneman, Daniel, 49, 64
Kamakahi, Lindsey, 12, 58, 208
Kamrava, Michael, 16, 257–58
The Kids Are All Right (film), 66
Klinefelter syndrome, 80
definition of, 281

laparoscopy
definition of, 281
overuse of, 204, 206, 207
lawsuits, 164
legal regulation
advantages of, 256
of infertility treatment, 252, 253
opposition to, 253
lesbians/lesbian parents, 172, 173, 175, 176
data on, need for, 191
providers' attitudes about, 179–80
and referral to fertility specialist, 33
society's attitudes about, 188

leukemia
    definition of, 281
    and savior siblings, 93
Li-Fraumeni syndrome
    definition of, 281
    preimplantation genetic diagnosis
        for, 77
lymphocyte immunization
        therapy, 166–67

male infertility
    prevalence of, 26
    psychological effects of, 71
    public education on, improvement
        of, 250
marital instability, 171
marital tensions
    infertility and, 27
    stress and, 213
mate selection, 113
media, misleading information in, 37
medical problems, and parenting
        suitability, 173–74
medications. *See also* fertility drugs
    adverse effects and side effects
        of, 209–10
    in infertility treatment, 39
    use rates for, 38–39
men
    and communication about
        infertility, 27
    stresses of infertility for, 26
mental disorders, and parenting
        suitability, 173–74
mental health provider(s)
    and evaluation of parenting
        suitability, 171–72, 185
    and family-balancing decisions, 117
    functions of, 222
    as gatekeepers, 61, 171, 185, 187, 190–91
    interviews of, 283–84
    role in infertility treatment, 221, 225
mental health services. *See also*
        psychotherapy
    insurance coverage of, enhancement
        of, need for, 258

metaphysical perspectives, on
        infertility, 234
MHP. *See* mental health provider(s)
Minnesota Multiphasic Personality
        Inventory (MMPI), 60–61
    definition of, 281
miscarriage, 25
    amniocentesis and, 91
    fetal testing and, 92
    partial abortion (fetal reduction)
        and, 134
mitochondrial replacement therapy,
        14, 252
MMPI. *See* Minnesota Multiphasic
        Personality Inventory (MMPI)
money-back guarantees, 200
mosaicism, 91
    definition of, 281
multiple births
    due to IVF, 123, 124, 125
    patients' perspectives and, 126
    risks and problems with
        media reports on, 127
        physicians' perspectives on, 128
muscular dystrophy, 83
mutation(s), 19
    misdiagnosis of, and lawsuits, 83–84
    transmission of
        prevention by donor egg use, 48
        prevention by donor sperm use, 48
    variable expression of, 79
    variable penetrance of, 78–79
myotonic dystrophy, 80, 90
    definition of, 281

National Society of Genetic Counselors,
        guidelines, enhancement of,
        need for, 261–62
neurotic patients, and parenting
        suitability, 174
Never Let Me Go (Ishiguro), 16–17
non-maleficence, and number of
        embryos transferred, 137
Nuffield Council on Bioethics, 14
nurse(s), engagement in patients' care,
        161, 162–63

OB/GYNs, 5
   definition of, 281
   as gatekeepers, 34
   and referral to infertility specialist,
      33, 34, 36
Octodoc, 253–54, 264. *See also* Octomom
Octomom, 16, 123, 127, 252, 253, 257–58
OHSS. *See* ovarian hyperstimulation
   syndrome
online information, 247–48
online support, 217, 226–28
oocytes, 8
   definition of, 281
out-of-pocket payment, for
   preimplantation genetic
   diagnosis, 90
ovarian hyperstimulation syndrome, 39,
      42, 56–57, 61–62
   CDC data on, 262
   definition of, 281
ovary(ies), definition of, 281
ovulation, definition of, 281

Pancoast, William, 4–5
parenting ability/suitability
   concerns about, 171, 189–90
   evaluation of, mental health providers
      and, 171–72, 185
   provider decision-making about, 183,
      184, 187
   providers' questions about, 173
   questions about, 171
parenting competence, forecasting, 172
parents
   good enough, 170, 191
   screening
      for adoption, 186
      for infertility treatment, 186–87
   suboptimal, right to procreate vs.
      right to receive ART, 190
   traditional vs. nontraditional
      arrangements, 170
   unconventional arrangements, 170,
      179, 188
      types of, 180
   would-be, characteristics of, 19

Parker, Sarah Jessica, 141
Patau syndrome. *See* trisomy 13 (Patau
   syndrome)
paternalism, provider, 177–78
patient(s)
   attitudinal barriers experienced by,
      sample interview questions
      about, 291
   background information, sample
      interview questions about, 289
   characteristics, sample interview
      questions about, for
      providers, 287
   decision-making process
      others' roles in, sample interview
         questions about, 290
      sample interview questions
         about, 289
   financial barriers experienced by,
      sample interview questions
      about, 291
   informational barriers experienced
      by, sample interview questions
      about, 290
   institutional barriers experienced
      by, sample interview questions
      about, 291
   regrets about past decisions, 211–12
patient advocacy organizations, 230
   educational outreach by, 247
patient-centered care, 156
patient education
   challenges in, 245
   improvement of, 249
   need for, 243
PCP. *See* primary care provider(s)
peer pressures, to have children, 24
penetrance
   definition of, 281
   variable, 78–79
Pergonal (menotropins), 39
personal factors, sample interview
   questions about, for providers, 289
PGD. *See* preimplantation genetic
   diagnosis
PGD-HLA, 93

PGS. *See* prenatal genetic screening
physician(s). *See* provider(s)
polycystic kidney disease, 85
positive selection, 109, 118
  decision-making about, difficulties
    of, 119
poverty, and parenting suitability,
    173, 175
practice, quality improvement for, 269
preeclampsia, definition of, 281
pregnancy, postmenopausal, 42–43
pregnancy complications, with
    twins, 123–24
pregnancy rate
  preimplantation genetic diagnosis
    and, 83
  providers' reporting of, 129–30, 263
    improvement of, need for, 266
preimplantation genetic diagnosis, 9,
    48, 75–76. *See also* PGD-HLA
  applications of, 78
  cost limitations and, 86
  data collection on, need for, 268
  and disclosure dilemmas, 98
  factors affecting, 83
  insurance coverage of, 86, 106
    enhancement of, need for, 258
    providers' advocacy for, 89
  non-disclosing, 97
    informed consent for, 101
    insurance coverage of, 101
    provider and patient
      responses to, 99
  patients' abandonment of, 81–82
  patient's decision-making about,
    sample interview questions
    about, 289
  patients' forgoing of, 90
  patients' out-of-pocket payment
    for, 90
  patients' understanding of, 84
  patient support for, 76–77, 78
  provider characteristics and, 83
  providers' lack of knowledge about,
    33–34, 36
  provider support for, 76–77, 78

quandaries posed by, 77, 78, 80–81
  reasons for, 76
  religious views on, 236, 237–38
  sample interview questions about, for
    providers, 287
  for two diseases, 80
  use rates for, 38–39
  varying attitudes toward, 81–82
prenatal diagnosis, non-disclosing, 98
prenatal genetic screening, 77, 259
  overuse of, 205, 208
  sample interview questions about, for
    providers, 287
primary care provider(s)
  as gatekeepers, 38
  and referral to infertility specialist, 33
professional education, 251
profit-making
  factors affecting, 206
  and overuse of specific procedures,
    204, 208
  possible solutions to, 207
provider(s)
  and age cutoffs for infertility
    treatment, 43–44, 45, 52–53
  approaches to stopping infertility
    treatment, 50, 51
  changing, medical factors and, 159
  and decision-making about infertility
    treatment, 46
  decision-making about treatment
    requests, 183, 184, 187
  education of, about infertility and
    treatment, 37
  engagement in patients' care, 161
  entrepreneurism among, 204
    factors affecting, 206
    possible solutions to, 207
  as gatekeepers, 185, 186, 190–91
  lack of knowledge about infertility
    treatment, 33–35
  male, and referrals to fertility
    specialists, 33
  monitoring of
    advantages of, 256
    opposition to, 253

provider(s) (*cont.*)
  as patient, lessons learned by, 155
  and patient education, 245
  and patients' religious issues, 238
  patients' selection of, 156
  reimbursement of (*see*
    reimbursement, to providers)
  religious, and referrals to fertility
    specialists, 33
  stress among, caused by treatment
    failures, 211, 269
  understanding of genetics, 84
psychotherapists. *See* mental health
  provider(s)
psychotherapy, 221, 270
  ambivalence about, 223
  functions of, 222
public education
  about infertility and treatment, need
    for, 37, 243
  challenges in, 246
  improvement of, 250
public policy. *See also* legal regulation
  and infertility treatment, 10–11

qualitative research, 283
quality assurance, and embryo
  transfer, 131
quality improvement, 269

rabbi(s). *See also* religion
  views on infertility treatment,
    236, 237–38
  variations in, 236, 239
rabbi shopping, 239
rare disease(s), online support
  with, 228
reactive arthritis, definition of, 281
recall bias, 219
referral(s), 32
  impediments to, 35
  obstacles to, 37–38
  patient-related obstacles and, 36
  providers who make, 34
registry(ies)
  of adult offspring of sperm
    donors, 73–74

  of children with donor-conceived
    status, need for, 267
  donor, 65–66
  sperm donor, 73, 74
reimbursement, to providers. *See*
  also financial factors; insurance
  coverage; profit-taking
  based on outcomes, 208
  transparency in, 208
REIs, 5, 34–35, 36
  definition of, 281
Reiter's syndrome, 81–82, 89
  definition of, 281
relationship instability, and parenting
  suitability, 175
religion, 232–34
  and choice of procedures, 235
  clinicians' responses to, 238
  in coping, 234
  and use of donor eggs, 50
  and "working around" objections, 238
religious right
  and access to services, 275
  and leftover embryos, 104–5
renting wombs, 11–12, 141
  rates of, 38–39
reproductive tourism, 11–12
RESOLVE, definition of, 282
retinoblastoma, 79
  definition of, 282
rights, positive vs. negative, 190
Robertson, John, 53, 172

Sandel, Michael, 109
SART. *See* Society for Assisted
  Reproductive Technology
savior siblings, 77, 93
science fiction, 16–17
second opinions, 157–58
secrecy, effects on prevention, 221
self-selection, patients', 185
semistructured interview questionnaire,
  284. *See also* interviews
SET. *See* single embryo transfer
sex preferences
  ethnic groups and, 111
  in United States, 111

sex selection, 76, 83–84, 260–61
  for family balancing, 108, 110, 113
  and leftover embryos, 103
  social
    decision-making process about,
      challenges in, 116
    and male vs. female preferences, 111
    providers' attitudes toward, 111–12
    public opinion on, 111
    reasons for and against, 112
    regulation of, opposition to, 118
    requests for, 112
shame, about infertility, 24–25
sickle-cell disease, 80, 83, 96
  definition of, 282
  and savior siblings, 93
single embryo transfer, 124, 125
  financial considerations and,
    127–28, 133
  patients' opposition to, 125
  patients' perspectives and, 126
  physicians' perspectives and, 128
single parents
  providers' attitudes about, 179–80,
    184, 187–88
  support system for, 184, 185–86
single-parents-by-choice, 175
  and co-parent, 184
  data on, need for, 191
  providers' attitudes about, 179–80,
    184, 188
  society's attitudes about, 188
single patient(s), 173
  refusal to treat, 172
single women, and referral to fertility
  specialist, 33
sister(s), as egg donors, 54
SMA. See spinal muscular atrophy
sniff test, 117
social factors, and parenting
  suitability, 175
social justice, and number of embryos
  transferred, 137
social pressures, to have
  children, 24
social stress, with infertility
  treatment, 213, 217

Society for Assisted Reproductive
  Technology, 10, 69
  data collection by, 262–63
  data on unconventional parenting,
    need for, 191
  and improved practice, 270–71
  mandated participation in, need
    for, 265
  and monitoring of providers, 269
  and regulation of egg donor
    agencies, 64
  regulatory power of, 132–33
  reporting policies of, improvement of,
    need for, 264–65, 266, 267
  role of, providers' views on, 257–58
Spallanzani, Lazzaro, 4
sperm
  buying/selling of, use rates for, 38–39
  collection through masturbation,
    religious views on, 236–37
  costs of, 72–73
sperm donation, 41, 48
  anonymous, 11, 65, 255–56
  guidelines for, 71
  infertile fathers' concerns about, 71
  regulation of, 10, 259
  by relative, 179–80
sperm donor(s)
  adult offspring of, registries of, 73–74
  anonymity of, 11, 65, 255–56
  anorectal, 73
  assessment of, 73
  brother as, 183
  brothers as, 72
  genetic disease transmission by, 74
  motivations of, data collection on,
    need for, 268
  screening of, 73
  selection of, 70, 72
  self-reports, trustworthiness of, 73
sperm donor registry, 73, 74
sperm sorting, 110, 115–16
  definition of, 282
sperm washing, for HIV-infected
  men, 179
spinal muscular atrophy, 80, 237–38
spirituality, 232–34

spousal support. *See also* marital tensions
  in infertility treatment, 212, 213
state medical board(s)
  discipline of providers, 254
  and improved practice, 270–71
  and profit-making by providers, 207
  and standards or certification for ART providers, 266
statistics
  communication of, by providers to patients, 164–65
  misunderstandings about, 244
  reported by providers, misleading aspects of, 130
stem cell research, leftover embryos and, 104
Stern, William and Elizabeth, 141
steroids, and sterility, 27
stigma
  of genetic disease, 250–51
  of infertility, 25–26, 220
storage disease, 91
  definition of, 282
stress(es). *See also* financial factors
  alternative therapies for, 229
  of infertility treatment, 209–10
    effects on marriages, 213
    gender differences and, 212
    social effects of, 213
  on providers, 269
  of treatment failure, 211
substance abuse, 175–76, 185
support
  alternative therapies as, 229
  from family and friends
    inadequate, responding to, 220
    with infertility treatment, 213, 217, 218
  new forms of, 221
  psychotherapy as, 221
  for single parents, 184, 185–86
support group(s), 226, 270
  need for, 224
surgery, in infertility treatment, rates of, 38–39

surrogacy. *See also* gestational surrogacy
  bans on, 141
  gay couples and, 140, 143
  intrafamilial, 179, 180
  opponents of, 144
  paid, state laws and, 142, 144
  reasons for using, 141
  risks to surrogate in, 142
  traditional, 140, 141
    legal considerations with, 142
surrogate(s)
  counseling for, 143–44
  friends and relatives as, 142–43
  motives of, 144
  and prospective parents
    legal disputes between, 143 (*see also* Baby M)
    tensions between, 143
  unpaid, 143

Tay-Sachs disease
  definition of, 282
  preimplantation genetic diagnosis for, 76
  religious views on, 237
thalassemia major, 95
thick description, 18, 283
three-parent baby, 14, 252
Tiresias, 121
toxemia, definition of, 281
transgender parents, 177
treatment failure
  social stress of, 214
  stresses of, 211
triplet pregnancy
  and fetal reduction, 133, 136
  high rates of, 137
trisomy, definition of, 282
trisomy 13 (Patau syndrome), definition of, 282
trisomy 18 (Edwards syndrome), definition of, 282
trisomy 21 (Down syndrome), 85, 91–92
  definition of, 282
trisomy 22, 36, 88, 92, 221
tubal reversal, costs of, 167

twins/twin pregnancy
    complications with, 123–24
    epidemic of, due to IVF, 123, 125, 137
    and fetal reduction, 133, 136
    patients' preferences for, 125, 137
        financial considerations in, 127–28
    risks and problems with, 137
        physicians' perspectives on, 128, 137

unmarried patient(s)
    providers' attitudes about, 179–80
    refusal to treat, 172
unmarried women, and referral to
    fertility specialist, 33
urologists, definition of, 282
uterus, definition of, 282

Vergara, Sophia, 16
von Hippel-Lindau syndrome
    definition of, 282

preimplantation genetic diagnosis
    for, 77

Walker-Warburg syndrome, 78
website(s). See also online information;
    online support
    for documentation of clinics'
        violations of guidelines, 260
Whitehead, Mary Beth, 141
whole-genome sequencing, 76, 259
womb. See uterus
work strains, infertility treatment and, 215
wrongful life lawsuit, 83–84

X-linked disease, preimplantation
    genetic diagnosis for, 76

Yamada, Ikufumi and Yuki, 144–45
yuck response, and medical decision-
    making, 49